4G WIRELESS VIDEO COMMUNICATIONS

Wiley Series on Wireless Communications and Mobile Computing

Series Editors: Dr Xuemin (Sherman) Shen, *University of Waterloo, Canada*
Dr Yi Pan, *Georgia State University, USA*

The "Wiley Series on Wireless Communications and Mobile Computing" is a series of comprehensive, practical and timely books on wireless communication and network systems. The series focuses on topics ranging from wireless communication and coding theory to wireless applications and pervasive computing. The books provide engineers and other technical professionals, researchers, educators, and advanced students in these fields with invaluable insight into the latest developments and cutting-edge research.

Other titles in the series:

Misic and Misic: *Wireless Personal Area Networks: Performance, Interconnection, and Security with IEEE 802.15.4*, January 2008, 978-0-470-51847-2

Takagi and Walke: *Spectrum Requirement Planning in Wireless Communications: Model and Methodology for IMT-Advanced*, April 2008, 978-0-470-98647-9

Pérez-Fontán and Espiñeira: *Modeling the Wireless Propagation Channel: A simulation approach with MATLAB®*, August 2008, 978-0-470-72785-0

Ippolito: *Satellite Communications Systems Engineering: Atmospheric Effects, Satellite Link Design and System Performance*, August 2008, 978-0-470-72527-6

Lin and Sou: *Charging for Mobile All-IP Telecommunications*, September 2008, 978-0-470-77565-3

Myung and Goodman: *Single Carrier FDMA: A New Air Interface for Long Term Evolution*, October 2008, 978-0-470-72449-1

Hart, Tao and Zhou: *Mobile Multi-hop WiMAX: From Protocol to Performance*, July 2009, 978-0-470-99399-6

Cai, Shen and Mark: *Multimedia Services in Wireless Internet: Modeling and Analysis*, August 2009, 978-0-470-77065-8

Stojmenovic: *Wireless Sensor and Actuator Networks: Algorithms and Protocols for Scalable Coordination and Data Communication*, September 2009, 978-0-470-17082-3

Qian, Muller and Chen: *Security in Wireless Networks and Systems*, January 2010, 978-0-470-51212-8

4G WIRELESS VIDEO COMMUNICATIONS

Haohong Wang

Marvell Semiconductors, USA

Lisimachos P. Kondi

University of Ioannina, Greece

Ajay Luthra

Motorola, USA

Song Ci

University of Nebraska-Lincoln, USA

A John Wiley and Sons, Ltd., Publication

Library of Congress Cataloging-in-Publication Data:

4G wireless video communications / Haohong Wang ... [et al.].
    p. cm.
  Includes bibliographical references and index.
  ISBN 978-0-470-77307-9 (cloth)
  1. Multimedia communications. 2. Wireless communication systems. 3. Video
telephone.   I. Wang, Haohong, 1973-
  TK5105.15.A23 2009
  621.384 – dc22
                                                    2008052216

A catalogue record for this book is available from the British Library.

ISBN   978-0-470-77307-9 (H/B)

Typeset in 10/12pt Times by Laserwords Private Limited, Chennai, India.
Printed and bound in Great Britain by CPI Antony Rowe, Chippenham, Wiltshire.

Contents

Foreword   xiii

Preface   xv

About the Authors   xxi

About the Series Editors   xxv

1   Introduction   **1**

1.1   Why 4G?   1
1.2   4G Status and Key Technologies   3
       1.2.1   3GPP LTE   3
       1.2.2   Mobile WiMAX   4
1.3   Video Over Wireless   5
       1.3.1   Video Compression Basics   5
       1.3.2   Video Coding Standards   9
       1.3.3   Error Resilience   10
       1.3.4   Network Integration   12
       1.3.5   Cross-Layer Design for Wireless Video Delivery   14
1.4   Challenges and Opportunities for 4G Wireless Video   15
References   17

2   Wireless Communications and Networking   **19**

2.1   Characteristics and Modeling of Wireless Channels   19
       2.1.1   Degradation in Radio Propagation   19
       2.1.2   Rayleigh Fading Channel   20
2.2   Adaptive Modulation and Coding   23
       2.2.1   Basics of Modulation Schemes   23
       2.2.2   System Model of AMC   25
       2.2.3   Channel Quality Estimation and Prediction   26
       2.2.4   Modulation and Coding Parameter Adaptation   28

	2.2.5	Estimation Error and Delay in AMC	30
	2.2.6	Selection of Adaptation Interval	30
2.3	Orthogonal Frequency Division Multiplexing		31
	2.3.1	Background	31
	2.3.2	System Model and Implementation	31
	2.3.3	Pros and Cons	33
2.4	Multiple-Input Multiple-Output Systems		34
	2.4.1	MIMO System Model	34
	2.4.2	MIMO Capacity Gain: Multiplexing	35
	2.4.3	MIMO Diversity Gain: Beamforming	35
	2.4.4	Diversity-Multiplexing Trade-offs	35
	2.4.5	Space-Time Coding	36
2.5	Cross-Layer Design of AMC and HARQ		37
	2.5.1	Background	38
	2.5.2	System Modeling	39
	2.5.3	Cross-Layer Design	41
	2.5.4	Performance Analysis	44
	2.5.5	Performance	45
2.6	Wireless Networking		47
	2.6.1	Layering Network Architectures	48
	2.6.2	Network Service Models	50
	2.6.3	Multiplexing Methods	51
	2.6.4	Connection Management in IP-Based Data Networks	53
	2.6.5	QoS Handoff	54
2.7	Summary		55
References			56
3	**Video Coding and Communications**		**59**
3.1	Digital Video Compression – Why and How Much?		59
3.2	Basics		60
	3.2.1	Video Formats	60
		3.2.1.1 Scanning	60
		3.2.1.2 Color	61
		3.2.1.3 Luminance, Luma, Chrominance, Chroma	64
3.3	Information Theory		64
	3.3.1	Entropy and Mutual Information	65
	3.3.2	Encoding of an Information Source	66
	3.3.3	Variable Length Coding	68
	3.3.4	Quantization	71
3.4	Encoder Architectures		73
	3.4.1	DPCM	73
	3.4.2	Hybrid Transform-DPCM Architecture	77
	3.4.3	A Typical Hybrid Transform DPCM-based Video Codec	79
	3.4.4	Motion Compensation	82
	3.4.5	DCT and Quantization	83
	3.4.6	Procedures Performed at the Decoder	84

3.5   Wavelet-Based Video Compression   86
        3.5.1   Motion-Compensated Temporal Wavelet Transform Using Lifting   90
References   94

4   4G Wireless Communications and Networking   97

4.1   IMT-Advanced and 4G   97
4.2   LTE   99
        4.2.1   Introduction   101
        4.2.2   Protocol Architecture   102
                4.2.2.1   E-UTRAN Overview Architecture   102
                4.2.2.2   User Plane and Control Plane   102
                4.2.2.3   LTE Physical Layer   106
        4.2.3   LTE Layer 2   107
        4.2.4   The Evolution of Architecture   110
        4.2.5   LTE Standardization   110
4.3   WIMAX-IEEE 802.16m   112
        4.3.1   Network Architecture   113
        4.3.2   System Reference Model   114
        4.3.3   Protocol Structure   114
                4.3.3.1   MAC Layer   114
                4.3.3.2   PHY Layer   120
        4.3.4   Other Functions Supported by IEEE 802.16m for Further Study   125
4.4   3GPP2 UMB   125
        4.4.1   Architecture Reference Model   126
        4.4.2   Layering Architecture and Protocols   127
Acknowledgements   133
References   133

5   Advanced Video Coding (AVC)/H.264 Standard   135

5.1   Digital Video Compression Standards   135
5.2   AVC/H.264 Coding Algorithm   138
        5.2.1   Temporal Prediction   139
                5.2.1.1   Motion Estimation   140
                5.2.1.2   P and B MBs   142
                5.2.1.3   Multiple References   143
                5.2.1.4   Motion Estimation Accuracy   143
                5.2.1.5   Weighted Prediction   144
                5.2.1.6   Frame and Field MV   144
                5.2.1.7   MV Compression   145
        5.2.2   Spatial Prediction   147
        5.2.3   The Transform   148
                5.2.3.1 4×4 Integer DCT and Inverse Integer DCT Transform   149
                5.2.3.2 8×8 Transform   150
                5.2.3.3   Hadamard Transform for DC   151

5.2.4   Quantization and Scaling   151
5.2.5   Scanning   151
5.2.6   Variable Length Lossless Codecs   152
        5.2.6.1   Exp-Golomb Code   153
        5.2.6.2   CAVLC (Context Adaptive VLC)   154
        5.2.6.3   CABAC   154
5.2.7   Deblocking Filter   155
5.2.8   Hierarchy in the Coded Video   156
        5.2.8.1   Basic Picture Types (I, P, B, B_R)   157
        5.2.8.2   SP and SI Pictures   157
5.2.9   Buffers   158
5.2.10 Encapsulation/Packetization   159
5.2.11 Profiles   160
        5.2.11.1   Baseline Profile   160
        5.2.11.2   Extended Profile   162
        5.2.11.3   Main Profile   162
        5.2.11.4   High Profile   162
        5.2.11.5   High10 Profile   163
        5.2.11.6   High 4:2:2 Profile   163
        5.2.11.7   High 4:4:4 Predictive Profile   163
        5.2.11.8   Intra Only Profiles   163
5.2.12 Levels   163
        5.2.12.1   Maximum Bit Rates, Picture Sizes and Frame Rates   164
        5.2.12.2   Maximum CPB, DPB and Reference Frames   164
5.2.13 Parameter Sets   167
        5.2.13.1   Sequence Parameter Sets (SPS)   167
        5.2.13.2   Picture Parameter Sets (PPS)   167
5.2.14 Supplemental Enhancement Information (SEI)   167
5.2.15 Subjective Tests   168
References   168

6   **Content Analysis for Communications**   **171**

6.1   Introduction   171
6.2   Content Analysis   173
        6.2.1   Low-Level Feature Extraction   174
                6.2.1.1   Edge   174
                6.2.1.2   Shape   176
                6.2.1.3   Color   177
                6.2.1.4   Texture   177
                6.2.1.5   Motion   178
                6.2.1.6   Mathematical Morphology   178
        6.2.2   Image Segmentation   179
                6.2.2.1   Threshold and Boundary Based Segmentation   181
                6.2.2.2   Clustering Based Segmentation   181
                6.2.2.3   Region Based Approach   181
                6.2.2.4   Adaptive Perceptual Color-Texture Segmentation   182

	6.2.3	Video Object Segmentation	185
		6.2.3.1 COST211 Analysis Model	187
		6.2.3.2 Spatial-Temporal Segmentation	187
		6.2.3.3 Moving Object Tracking	188
		6.2.3.4 Head-and-Shoulder Object Segmentation	190
	6.2.4	Video Structure Understanding	200
		6.2.4.1 Video Abstraction	201
		6.2.4.2 Video Summary Extraction	203
	6.2.5	Analysis Methods in Compressed Domain	208
6.3	Content-Based Video Representation		209
6.4	Content-Based Video Coding and Communications		212
	6.4.1	Object-Based Video Coding	212
	6.4.2	Error Resilience for Object-Based Video	215
6.5	Content Description and Management		217
	6.5.1	MPEG-7	217
	6.5.2	MPEG-21	219
References			219

7	**Video Error Resilience and Error Concealment**	**223**
7.1	Introduction	223
7.2	Error Resilience	224
	7.2.1 Resynchronization Markers	224
	7.2.2 Reversible Variable Length Coding (RVLC)	225
	7.2.3 Error-Resilient Entropy Coding (EREC)	226
	7.2.4 Independent Segment Decoding	228
	7.2.5 Insertion of Intra Blocks or Frames	228
	7.2.6 Scalable Coding	229
	7.2.7 Multiple Description Coding	230
7.3	Channel Coding	232
7.4	Error Concealment	234
	7.4.1 Intra Error Concealment Techniques	234
	7.4.2 Inter Error Concealment Techniques	234
7.5	Error Resilience Features of H.264/AVC	236
	7.5.1 Picture Segmentation	236
	7.5.2 Intra Placement	236
	7.5.3 Reference Picture Selection	237
	7.5.4 Data Partitioning	237
	7.5.5 Parameter Sets	237
	7.5.6 Flexible Macroblock Ordering	238
	7.5.7 Redundant Slices (RSs)	239
References		239

8	**Cross-Layer Optimized Video Delivery over 4G Wireless Networks**	**241**
8.1	Why Cross-Layer Design?	241
8.2	Quality-Driven Cross-Layer Framework	242

8.3   Application Layer   244
8.4   Rate Control at the Transport Layer   244
        8.4.1   Background   244
        8.4.2   System Model   246
        8.4.3   Network Setting   246
        8.4.4   Problem Formulation   248
        8.4.5   Problem Solution   248
        8.4.6   Performance Evaluation   249
8.5   Routing at the Network Layer   252
        8.5.1   Background   252
        8.5.2   System Model   254
        8.5.3   Routing Metric   255
        8.5.4   Problem Formulation   257
        8.5.5   Problem Solution   258
        8.5.6   Implementation Considerations   262
        8.5.7   Performance Evaluation   263
8.6   Content-Aware Real-Time Video Streaming   265
        8.6.1   Background   265
        8.6.2   Background   265
        8.6.3   Problem Formulation   266
        8.6.4   Routing Based on Priority Queuing   267
        8.6.5   Problem Solution   269
        8.6.6   Performance Evaluation   270
8.7   Cross-Layer Optimization for Video Summary Transmission   272
        8.7.1   Background   272
        8.7.2   Problem Formulation   274
        8.7.3   System Model   276
        8.7.4   Link Adaptation for Good Content Coverage   278
        8.7.5   Problem Solution   280
        8.7.6   Performance Evaluation   283
8.8   Conclusions   287
References   287

9   Content-based Video Communications   291

9.1   Network-Adaptive Video Object Encoding   291
9.2   Joint Source Coding and Unequal Error Protection   294
        9.2.1   Problem Formulation   295
                9.2.1.1   System Model   296
                9.2.1.2   Channel Model   297
                9.2.1.3   Expected Distortion   298
                9.2.1.4   Optimization Formulation   298
        9.2.2   Solution and Implementation Details   299
                9.2.2.1   Packetization and Error Concealment   299
                9.2.2.2   Expected Distortion   299
                9.2.2.3   Optimal Solution   300
        9.2.3   Application on Energy-Efficient Wireless Network   301

		9.2.3.1	Channel Model	301
		9.2.3.2	Experimental Results	302
	9.2.4	Application on Differentiated Services Networks		303
9.3	Joint Source-Channel Coding with Utilization of Data Hiding			305
	9.3.1	Hiding Shape in Texture		308
	9.3.2	Joint Source-Channel Coding		309
	9.3.3	Joint Source-Channel Coding and Data Hiding		311
		9.3.3.1	System Model	311
		9.3.3.2	Channel Model	312
		9.3.3.3	Expected Distortion	312
		9.3.3.4	Implementation Details	313
	9.3.4	Experimental Results		315
References				322

10   AVC/H.264 Application – Digital TV   325

10.1	Introduction		325
	10.1.1 Encoder Flexibility		326
10.2	Random Access		326
	10.2.1 GOP Bazaar		327
		10.2.1.1 MPEG-2 Like, 2B, GOP Structure	327
		10.2.1.2 Reference B and Hierarchical GOP structures	330
		10.2.1.3 Low Delay Structure	331
		10.2.1.4 Editable Structure	331
		10.2.1.5 Others	332
	10.2.2 Buffers, Before and After		332
		10.2.2.1 Coded Picture Buffer	332
		10.2.2.2 Decoded Picture Buffer (DPB)	334
10.3	Bitstream Splicing		335
10.4	Trick Modes		337
	10.4.1 Fast Forward		338
	10.4.2 Reverse		338
	10.4.3 Pause		338
10.5	Carriage of AVC/H.264 Over MPEG-2 Systems		338
	10.5.1 Packetization		339
		10.5.1.1 Packetized Elementary Stream (PES)	340
		10.5.1.2 Transport Stream (TS)	340
		10.5.1.3 Program Stream	343
	10.5.2 Audio Video Synchronization		344
	10.5.3 Transmitter and Receiver Clock Synchronization		344
	10.5.4 System Target Decoder and Timing Model		344
References			345

11   Interactive Video Communications   347

| 11.1 | Video Conferencing and Telephony | 347 |
| | 11.1.1 IP and Broadband Video Telephony | 347 |

	11.1.2 Wireless Video Telephony	348
	11.1.3 3G-324M Protocol	348
	11.1.3.1 Multiplexing and Error Handling	349
	11.1.3.2 Adaptation Layers	350
	11.1.3.3 The Control Channel	350
	11.1.3.4 Audio and Video Channels	350
	11.1.3.5 Call Setup	350
11.2	Region-of-Interest Video Communications	351
	11.2.1 ROI based Bit Allocation	351
	11.2.1.1 Quality Metric for ROI Video	351
	11.2.1.2 Bit Allocation Scheme for ROI Video	353
	11.2.1.3 Bit Allocation Models	354
	11.2.2 Content Adaptive Background Skipping	356
	11.2.2.1 Content-based Skip Mode Decision	357
	11.2.2.2 ρ Budget Adjustment	360
References		366

12   Wireless Video Streaming   **369**

12.1	Introduction	369
12.2	Streaming System Architecture	370
	12.2.1 Video Compression	370
	12.2.2 Application Layer QoS Control	372
	12.2.2.1 Rate Control	372
	12.2.2.2 Rate Shaping	373
	12.2.2.3 Error Control	374
	12.2.3 Protocols	374
	12.2.3.1 Transport Protocols	375
	12.2.4 Video/Audio Synchronization	376
12.3	Delay-Constrained Retransmission	377
	12.3.1 Receiver-Based Control	378
	12.3.2 Sender-Based Control	378
	12.3.3 Hybrid Control	379
	12.3.4 Rate-Distortion Optimal Retransmission	379
12.4	Considerations for Wireless Video Streaming	382
	12.4.1 Cross-Layer Optimization and Physical Layer Consideration	383
12.5	P2P Video Streaming	384
References		385

Index   **389**

Foreword

4G Wireless Video Communications is a broad title with a wide scope, bridging video signal processing, video communications, and 4G wireless networks. Currently, 4G wireless communication systems are still in their planning phase. The new infrastructure is expected to provide much higher data rates, lower cost per transmitted bit, more flexible mobile terminals, and seamless connections to different networks. Along with the rapid development of video coding and communications techniques, more and more advanced interactive multimedia applications are emerging as 4G network "killer apps".

Depending on the reader's background and point of view, this topic may be considered and interpreted with different perspectives and foci. Colleagues in the multimedia signal processing area know firsthand the effort and ingenuity it takes to save 1 bit or increase the quality of the reconstructed video during compression by 0.5 dB. The increased bandwidth of 4G systems may open the door to new challenges and open issues that will be encountered for first time or were neglected before. Some topics, such as content interactivity, scalability, and understandability may be reconsidered and be given higher priority, and a more aggressive approach in investigating and creating new applications and service concepts may be adopted. On the other hand, for colleagues with communications and networking backgrounds, it is important to keep in mind that in the multimedia content delivery scenario, every bit or packet is not equal from the perspective of power consumption, delay, or packet loss. In other words, media data with certain semantic and syntactical characteristics may impact the resulting user experience more significantly than others. Therefore it is critical to factor this in to the design of new systems, protocols, and algorithms. Consequently, the cross-layer design and optimization for 4G wireless video is expected to be an active topic. It will gather researchers from various communities resulting in necessary inter-disciplinary collaborations.

In this book, the authors have successfully provided a detailed coverage of the fundamental theory of video compression, communications, and 4G wireless communications. A comprehensive treatment of advanced video analysis and coding techniques, as well as the new emerging techniques and applications, is also provided. So far, to the best of my knowledge, I have seen very few books, journals or magazines on the market focusing on this topic, which may be due to the very fast developments in this field. This book has revealed an integrated picture of 4G wireless video communications to the general audience. It provides both a treatment of the underlying theory and a valuable practical reference for the multimedia and communications practitioners.

Aggelos K. Katsaggelos
Professor of Electrical Engineering and Computer Science
Northwestern University, USA

Preface

This book is one of the first books to discuss the video processing and communications technology over 4G wireless systems and networks. The motivations of writing this book can be traced back to year 2004, when Haohong Wang attended IEEE ICCCN'04 conference at Chicago. During the conference banquet, Prof. Mohsen Guizani, the Editor-in-Chief of Wiley journal of Wireless Communications and Mobile Computing (WCMC), invited him to edit a special Issue for WCMC with one of the hot topics in multimedia communications area. At that time, 4G wireless systems were just in very early stage of planning, but many advanced interactive multimedia applications, such as video telephony, hand-held games, and mobile TV, have been widely known in struggling with significant constraints in data rate, spectral efficiency, and battery limitations of the existing wireless channel conditions. "It might be an interesting topic to investigate the video processing, coding and transmission issues for the forthcoming 4G wireless systems", Haohong checked with Lisimachos Kondi then at SUNY Buffalo and Ajay Luthra at Motorola, both of them were very excited at this idea and willing to participate, then they wrote to around 50 world-class scientists in the field, like Robert Gray (Stanford), Sanjit Mitra (USC), Aggelos Katsaggelos (Northwestern), Ya-Qin Zhang (Microsoft), and so on, to exchange ideas and confirm their visions, the answers were surprisingly and unanimously positive! The large volume paper submissions for the special Issue later also confirms this insightful decision, it is amazing that so many authors are willing to contribute to this Issue with their advanced research results, finally eight papers were selected to cover four major areas: video source coding, video streaming, video delivery over wireless, and cross-layer optimized resource allocation, and the special Issue was published in Feb. 2007.

A few months later after the special Issue of "Recent Advances in Video Communications for 4G Wireless Systems" was published, we were contacted by Mark Hammond, the Editorial Director of John Wiley & Sons, who had read the Issue and had a strong feeling that this Issue could be extended to a book with considerable market potentials around the world. Highly inspired by the success of the earlier Issue, we decided to pursue it. At the same timeframe, Song Ci at University of Nebraska-Lincoln joined in this effort, and Sherman Shen of University of Waterloo invited us to make this book a part of his book series of "Wiley Series on Wireless Communications and Mobile Computing", thus this book got launched.

Video Meets 4G Wireless

4G wireless networks provide many features to handle the current challenges in video communications. It accommodates heterogeneous radio access systems via flexible core IP networks, thus provide the end-users more opportunities and flexibilities to access video and multimedia contents. It features very high data rate, which is expected to relieve the burden of many current networking killer video applications. In addition, it provides QoS and security supports to the end-users, which enhances the user experiences and privacy protection for future multimedia applications.

However, the development of the video applications and services would never stop; on the contrary, we believe many new 4G "killer" multimedia applications would appear when the 4G systems are deployed, as an example mentioned in chapter 1, the high-reality 3D media applications and services would bring in large volume of data transmissions onto the wireless networks, thus could become a potential challenge for 4G systems.

Having said that, we hope our book helps our audiences to understand the latest developments in both wireless communications and video technologies, and highlights the part that may have great potentials for solving the future challenges for video over 4G networks, for example, content analysis techniques, advanced video coding, and cross-layer design and optimization, etc.

Intended Audience

This book is suitable for advanced undergraduates and first-year graduate students in computer science, computer engineering, electrical engineering, and telecommunications majors, and for students in other disciplines who is interested in media processing and communications. The book also will be useful to many professionals, like software/firmware/algorithm engineers, chip and system architects, technical marketing professionals, and researchers in communications, multimedia, semiconductor, and computer industries.

Some chapters of this book are based on existing lecture courses. For example, parts of chapter 3 and chapter 7 are based on the course "EE565: Video Communications" taught at the State University of New York at Buffalo, parts of chapters 2 and 4 are used in lectures at University of Nebraska-Lincoln, and chapters 6, 8 and 9 are used in lectures at University of Nebraska-Lincoln as well as in many invited talks and tutorials to the IEEE computer, signal processing, and communications societies. Parts of chapter 3, chapter 5 and chapter 7 are based on many tutorials presented in various workshops, conferences, trade shows and industry consortiums on digital video compression standards and their applications in Digital TV.

Organization of the book

The book is organized as follows. In general the chapters are grouped into 3 parts:

- Part I: which covers chapters 2–4, it focuses on the fundamentals of wireless communications and networking, video coding and communications, and 4G systems and networks;

- Part II: which covers chapters 5–9, it demonstrates the advanced technologies including advanced video coding, video analysis, error resilience and concealment, cross-layer design and content-based video communications;
- Part III: which covers chapters 10–12, it explores the modern multimedia application, including digital TV, interactive video communications, and wireless video streaming.

Chapter 1 overviews the fundamentals of 4G wireless system and the basics of the video compression and transmission technologies. We highlight the challenges and opportunities for 4G wireless video to give audiences a big picture of the future video communications technology development, and point out the specific chapters that are related to the listed advanced application scenarios.

Chapter 2 introduces the fundamentals in wireless communication and networking, it starts with the introduction of various channel models, then the main concept of adaptive modulation and coding (AMC) is presented; after that, the OFDM and MIMO systems are introduced, followed by the cross-layer design of AMC at physical layer and hybrid automatic repeat request (HARQ) at data link layer; at the end, the basic wireless networking technologies, including network architecture, network services, multiplexing, mobility management and handoff, are reviewed.

Chapter 3 introduces the fundamental technology in video coding and communications, it starts with the very basics of video content, such as data format, video signal components, and so on, and then it gives a short review on the information theory which is the foundation of the video compression; after that, the traditional block-based video coding techniques are overviewed, including motion estimation and compensation, DCT transform, quantization, as well as the embedded decoding loop; at the end, the wavelet-based coding scheme is introduced, which leads to a 3D video coding framework to avoid drift artifacts.

In Chapter 4, the current status of the 4G systems under development is introduced, and three major candidate standards: LTE (3GPP Long Term Evolution), WiMax (IEEE 802.16m), and UMB (3GPP2 Ultra Mobile Broadband), have been briefly reviewed, respectively.

In Chapter 5, a high level overview of the Emmy award winning AVC/H.264 standard, jointly developed by MPEG of ISO/IEC and VCEG of ITU-T, is provided. This standard is expected to play a critical role in the distribution of digital video over 4G networks. Full detailed description of this standard will require a complete book dedicated only to this subject. Hopefully, a good judgment is made in deciding how much and what detail to provide within the available space and this chapter gives a satisfactory basic understanding of the coding tools used in the standard that provide higher coding efficiency over previous standards.

Chapter 6 surveys the content analysis techniques that can be applied for video communications, it includes low-level feature extraction, image segmentation, video object segmentation, video structure understand, and other methods conducted in compressed domain. It also covers the rest part of the content processing lifecycle from representation, coding, description, management to retrieval, as they are heavily dependent on the intelligence obtained from the content analysis.

Chapter 7 discusses the major components for robust video communications, which are error resilience, channel coding and error concealment. The error resilience features in H.264 are summarized at the end to conclude this chapter.

After all the 4G wireless systems and advanced video technologies have been demonstrated, Chapter 8 focuses on a very important topic, which is how to do the cross-layer design and optimization for video delivery over 4G wireless networks. It goes through four detailed applications scenario analysis that emphasize on different network layers, namely, rate control at transport layer, routing at network layer, content-aware video streaming, and video summary transmissions. Chapter 9 discusses more scenarios but emphasizes more on the content-aware video communications, that is, how to combine source coding with the unequal error protection in wireless networks and how to use data hiding as an effective means inside the joint source-channel coding framework to achieve better performance.

The last 3 chapters can be read in almost any order, where each is open-ended and covered at a survey level of a separate type of application. Chapter 10 presents digital TV application of the AVC/H.264 coding standard. It focuses on some of the encoder side flexibilities offered by that standard, their impacts on key operations performed in a digital TV system and what issues will be faced in providing digital TV related services on 4G networks. It also describes how the coded video bitstreams are packed and transmitted, and the features and mechanisms that are required to make the receiver side properly decode and reconstruct the content in the digital TV systems. In Chapter 11, the interactive video communication applications, such as video conferencing and video telephony are discussed. The Region-of-Interest video communication technology is highlighted here which has been proved to be helpful for this type of applications that full of head-and-shoulder video frames. In Chapter 12, the wireless video stream technology is overviewed. It starts with the introduction of the major components (mechanisms and protocols) of the streaming system architecture, and follows with the discussions on a few retransmission schemes, and other additional consideration. At the end, the P2P video streaming, which attracts many attention recently, is introduced and discussed.

Acknowledgements

We would like to thank a few of the great many people whose contributions were instrumental in taking this book from an initial suggestion to a final product. First, we would like to express our gratitude to Dr. Aggelos K. Katsaggelos of Northwestern University for writing the preceding forward. Dr. Katsaggelos has made significant contributions to video communication field, and we are honored and delighted by his involvement in our own modest effort. Second, we would like to thank for Mr. Mark Hammond of John Wiley & Sons and Dr. Sherman Shen of University of Waterloo for inviting us to take this effort.

Several friends provided many invaluable feedback and suggestions in various stages of this book, they are Dr. Khaled El-Maleh (Qualcomm), Dr. Chia-Chin Chong (Docomo USA Labs), Dr. Hideki Tode (Osaka Prefecture University), Dr. Guan-Ming Su (Marvell Semiconductors), Dr. Junqing Chen (Aptina Imaging), Dr. Gokce Dane (Qualcomm), Dr. Sherman (Xuemin) Chen (Broadcom), Dr. Krit Panusopone (Motorola), Dr. Yue Yu (Motorola), Dr. Antonios Argyriou (Phillips Research), Mr. Dalei Wu (University of Nebraska-Lincoln), and Mr. Haiyan Luo (University of Nebraska-Lincoln). We also would like to thank Dr. Walter Weigel (ETSI), Dr. Junqiang Chen (Aptina Imaging) and Dr. Xingquan Zhu (Florida Atlantic University), IEEE, and 3GPP, International Telecommunication Union, for their kind permissions to let us use their figures in this book.

We acknowledge the entire production team at John Wiley & Sons. The editors, Sarah Tilley, Brett Wells, Haseen Khan, Sarah Hinton, and Katharine Unwin, have been a pleasure to work with through the long journey of the preparation and production of this book. I am especially grateful to Dan Leissner, who helped us to correct many typos and grammars in the book.

At the end, the authors appreciate the many contributions and sacrifices that our families have made to this effort. Haohong Wang would like to thank his wife Xin Lu and the coming baby for their kind encouragements and supports. Lisimachos P. Kondi would like to thank his parents Vasileios and Stephi and his brother Dimitrios. Ajay Luthra would like to thank his wife Juhi for her support during many evenings and weekends spent on working on this book. He would also like to thank his sons Tarang and Tanooj for their help in improving the level of their dad's English by a couple of notches. Song Ci would like to thank his wife Jie, daughter Channon, and son Marvin as well as his parents for their support and patience.

The dedication of this book to our families is a sincere but inadequate recognition of all of their contributions to our work.

About the Authors

 **Haohong Wang** received a BSc degree in Computer Science and a M.Eng. degree in Computer & its Application, both from Nanjing University, he also received a MSc degree in Computer Science from University of New Mexico, and his PhD in Electrical and Computer Engineering from Northwestern University. He is currently a Senior System Architect and Manager at Marvell Semiconductors at Santa Clara, California. Prior to joining Marvell, he held various technical positions at AT&T, Catapult Communications, and Qualcomm. Dr Wang's research involves the areas of multimedia communications, graphics and image/video analysis and processing. He has published more than 40 articles in peer-reviewed journals and International conferences. He is the inventor of more than 40 U.S. patents and pending applications. He is the co-author of *4G Wireless Video Communications* (John Wiley & Sons, 2009), and *Computer Graphics* (1997).

Dr Wang is the Associate Editor-in-Chief of the *Journal of Communications*, Editor-in-Chief of the *IEEE MMTC E-Letter*, an Associate Editor of the *Journal of Computer Systems, Networks, and Communications* and a Guest Editor of the *IEEE Transactions on Multimedia*. He served as a Guest Editor of the *IEEE Communications Magazine, Wireless Communications and Mobile Computing*, and *Advances in Multimedia*. Dr Wang is the Technical Program Chair of IEEE GLOBECOM 2010 (Miami). He served as the General Chair of the 17th IEEE International Conference on Computer Communications and Networks (ICCCN 2008) (US Virgin Island), and the Technical Program Chair of many other International conferences including IEEE ICCCN 2007 (Honolulu), IMAP 2007 (Honolulu), ISMW 2006 (Vancouver), and the ISMW 2005 (Maui). He is the Founding Steering Committee Chair of the annual International Symposium on Multimedia over Wireless (2005–). He chairs the TC Promotion & Improvement Sub-Committee, as well as the Cross-layer Communications SIG of the IEEE Multimedia Communications Technical Committee. He is also an elected member of the IEEE Visual Signal Processing and Communications Technical Committee (2005–), and IEEE Multimedia and Systems Applications Technical Committee (2006–).

Lisimachos P. Kondi received a diploma in electrical engineering from the Aristotle University of Thessaloniki, Greece, in 1994 and MSc and PhD degrees, both in electrical and computer engineering, from Northwestern University, Evanston, IL, USA, in 1996 and 1999, respectively. He is currently an Assistant Professor in the Department of Computer Science at the University of Ioannina, Greece. His research interests are in the general area of multimedia communications and signal processing, including image and video compression and transmission over wireless channels and the Internet, super-resolution of video sequences and shape coding. Dr Kondi is an Associate Editor of the *EURASIP Journal of Advances in Signal Processing* and an Associate Editor of *IEEE Signal Processing Letters*.

Ajay Luthra received his B.E. (Hons) from BITS, Pilani, India in 1975, M.Tech. in Communications Engineering from IIT Delhi in 1977 and PhD from Moore School of Electrical Engineering, University of Pennsylvania in 1981. From 1981 to 1984 he was a Senior Engineer at Interspec Inc., where he was involved in digital signal and image processing for bio-medical applications. From 1984 to 1995 he was at Tektronix Inc., where from 1985 to 1990 he was manager of the Digital Signal and Picture Processing Group and from 1990 to 1995 Director of the Communications/Video Systems Research Lab. He is currently a Senior Director in the Advanced Technology Group at Connected Home Solutions, Motorola Inc., where he is involved in advanced development work in the areas of digital video compression and processing, streaming video, interactive TV, cable head-end system design, advanced set top box architectures and IPTV.

Dr Luthra has been an active member of the MPEG Committee for more than twelve years where he has chaired several technical sub-groups and pioneered the MPEG-2 extensions for studio applications. He is currently an associate rapporteur/co-chair of the Joint Video Team (JVT) consisting of ISO/MPEG and ITU-T/VCEG experts working on developing the next generation of video coding standard known as MPEG-4 Part 10 AVC/H.264. He is also the USA's Head of Delegates (HoD) to MPEG. He was an Associate Editor of *IEEE Transactions on Circuits and Systems for Video Technology* (2000–2002) and a Guest Editor for its special issues on the H.264/AVC Video Coding Standard, July 2003 and Streaming Video, March 2001. He holds 30 patents, has published more than 30 papers and has been a guest speaker at numerous conferences.

Song Ci is an Assistant Professor of Computer and Electronics Engineering at the University of Nebraska-Lincoln. He received his BSc from Shandong University, Jinan, China, in 1992, MSc from the Chinese Academy of Sciences, Beijing, China, in 1998, and a PhD from the University of Nebraska-Lincoln in 2002, all in Electrical Engineering. He also worked with China Telecom (Shandong) as a telecommunications engineer from 1992 to 1995, and with the Wireless Connectivity Division of 3COM Cooperation, Santa Clara, CA, as a R&D Engineer in 2001. Prior to joining the University of Nebraska-Lincoln, he was an Assistant Professor of Computer Science at the University of Massachusetts Boston and the University of Michigan-Flint. He is the founding director of the Intelligent Ubiquitous Computing Laboratory (iUbiComp Lab) at the Peter Kiewit Institute of the University of Nebraska. His research interests include cross-layer design for multimedia wireless communications, intelligent network management, resource allocation and scheduling in various wireless networks and power-aware multimedia embedded networked sensing system design and development. He has published more than 60 research papers in referred journals and at international conferences in those areas.

Dr Song Ci serves currently as Associate Editor on the Editorial Board of *Wiley Wireless Communications and Mobile Computing* (WCMC) and Guest Editor of *IEEE Network Magazine Special Issue on Wireless Mesh Networks: Applications, Architectures and Protocols*, Editor of *Journal of Computer Systems, Networks, and Communications* and an Associate Editor of the *Wiley Journal of Security and Communication Networks*. He also serves as the TPC co-Chair of IEEE ICCCN 2007, TPC co-Chair of IEEE WLN 2007, TPC co-Chair of the Wireless Applications track at IEEE VTC 2007 Fall, the session Chair at IEEE MILCOM 2007 and as a reviewer for numerous referred journals and technical committee members at many international conferences. He is the Vice Chair of Communications Society of IEEE Nebraska Section, Senior Member of the IEEE and Member of the ACM and the ASHRAE.

About the Series Editors

 Xuemin (Sherman) Shen (M'97-SM'02) received the BSc degree in Electrical Engineering from Dalian Maritime University, China in 1982, and the MSc and PhD degrees (both in Electrical Engineering) from Rutgers University, New Jersey, USA, in 1987 and 1990 respectively. He is a Professor and University Research Chair, and the Associate Chair for Graduate Studies, Department of Electrical and Computer Engineering, University of Waterloo, Canada. His research focuses on mobility and resource management in interconnected wireless/wired networks, UWB wireless communications systems, wireless security, and ad hoc and sensor networks. He is a co-author of three books, and has published more than 300 papers and book chapters in wireless communications and networks, control and filtering.

Dr Shen serves as a Founding Area Editor for *IEEE Transactions on Wireless Communications*; Editor-in-Chief for *Peer-to-Peer Networking and Application*; Associate Editor for *IEEE Transactions on Vehicular Technology*; *KICS/IEEE Journal of Communications and Networks*, *Computer Networks*; *ACM/Wireless Networks*; and *Wireless Communications and Mobile Computing* (Wiley), etc. He has also served as Guest Editor for *IEEE JSAC, IEEE Wireless Communications*, and *IEEE Communications Magazine*. Dr Shen received the Excellent Graduate Supervision Award in 2006, and the Outstanding Performance Award in 2004 from the University of Waterloo, the Premier's Research Excellence Award (PREA) in 2003 from the Province of Ontario, Canada, and the Distinguished Performance Award in 2002 from the Faculty of Engineering, University of Waterloo. Dr Shen is a registered Professional Engineer of Ontario, Canada.

Dr Yi Pan is the Chair and a Professor in the Department of Computer Science at Georgia State University, USA. Dr Pan received his B.Eng. and M.Eng. degrees in Computer Engineering from Tsinghua University, China, in 1982 and 1984, respectively, and his PhD degree in Computer Science from the University of Pittsburgh, USA, in 1991. Dr Pan's research interests include parallel and distributed computing, optical networks, wireless networks, and bioinformatics. Dr Pan has published more than 100 journal papers with over 30 papers published in various IEEE journals. In addition, he has published over 130 papers in refereed conferences (including IPDPS, ICPP, ICDCS, INFOCOM, and GLOBECOM). He has also co-edited over 30 books. Dr Pan has served as an editor-in-chief or an editorial board member for 15 journals including five IEEE Transactions and has organized many international conferences and workshops. Dr Pan has delivered over 10 keynote speeches at many international conferences. Dr Pan is an IEEE Distinguished Speaker (2000–2002), a Yamacraw Distinguished Speaker (2002), and a Shell Oil Colloquium Speaker (2002). He is listed in Men of Achievement, Who's Who in America, Who's Who in American Education, Who's Who in Computational Science and Engineering, and Who's Who of Asian Americans.

1

Introduction

1.1 Why 4G?

Before we get into too much technical jargon such as 4G and so on, it would be interesting to take a moment to discuss iPhone, which was named *Time* magazine's Inventor of the Year in 2007, and which has a significant impact on many consumers' view of the capability and future of mobile phones. It is amazing to see the enthusiasm of customers if you visit Apple's stores which are always crowded. Many customers were lined up at the Apple stores nationwide on iPhone launch day (the stores were closed at 2 p.m. local time in order to prepare for the 6 p.m. iPhone launch) and Apple sold 270 000 iPhones in the first 30 hours on launch weekend and sold 1 million iPhone 3G in its first three days. There are also many other successful mobile phones produced by companies such as Nokia, Motorola, LG, and Samsung and so on.

It is interesting to observe that these new mobile phones, especially smart phones, are much more than just phones. They are really little mobile PCs, as they provide many of the key functionalities of a PC:

- A keyboard, which is virtual and rendered on a touch screen.
- User friendly graphical user interfaces.
- Internet services such as email, web browsing and local Wi-Fi connectivity.
- Built-in camera with image/video capturing.
- Media player with audio and video decoding capability.
- Smart media management tools for songs, photo albums, videos, etc..
- Phone call functionalities including text messaging, visual voicemail, etc.

However, there are also many features that some mobile phones do not yet support (although they may come soon), for example:

- Mobile TV support to receive live TV programmes.
- Multi-user networked 3D games support.
- Realistic 3D scene rendering.

4G Wireless Video Communications Haohong Wang, Lisimachos P. Kondi, Ajay Luthra and Song Ci
© 2009 John Wiley & Sons, Ltd

- Stereo image and video capturing and rendering.
- High definition visuals.

The lack of these functions is due to many factors, including the computational capability and power constraints of the mobile devices, the available bandwidth and transmission efficiency of the wireless network, the quality of service (QoS) support of the network protocols, the universal access capability of the communication system infrastructure and the compression and error control efficiency of video and graphics data. Although it is expected that the mobile phone will evolve in future generations so as to provide the user with the same or even better experiences as today's PC, there is still long way to go. From a mobile communication point of view, it is expected to have a much higher data transmission rate, one that is comparable to wire line networks as well as services and support for seamless connectivity and access to any application regardless of device and location. That is exactly the purpose for which 4G came into the picture.

4G is an abbreviation for Fourth-Generation, which is a term used to describe the next complete evolution in wireless communication. The 4G wireless system is expected to provide a comprehensive IP solution where multimedia applications and services can be delivered to the user on an 'Anytime, Anywhere' basis with a satisfactory high data rate and premium quality and high security, which is not achievable using the current 3G (third generation) wireless infrastructure. Although so far there is not a final definition for 4G yet, the International Telecommunication Union (ITU) is working on the standard and target for commercial deployment of 4G system in the 2010–2015 timeframe. ITU defined IMT-Advanced as the succeeding of IMT-2000 (or 3G), thus some people call IMT-Advanced as 4G informally.

The advantages of 4G over 3G are listed in Table 1.1. Clearly 4G has improved upon the 3G system significantly not only in bandwidth, coverage and capacity, but also in many advanced features, such as QoS, low latency, high mobility, and security support, etc.

Table 1.1 Comparison of 3G and 4G

	3G	4G
Driving force	Predominantly voice driven, data is secondary concern	Converged data and multimedia services over IP
Network architecture	Wide area networks	Integration of Wireless LAN and Wide area networks
Bandwidth (bps)	384K–2M	100 M for mobile 1 G for stationary
Frequency band (GHz)	1.8–2.4	2–8
Switching	Circuit switched and packet switched	Packet switch only
Access technology	CDMA family	OFDMA family
QoS and security	Not supported	Supported
Multi-antenna techniques	Very limited support	Supported
Multicast/broadcast service	Not supported	Supported

1.2   4G Status and Key Technologies

In a book which discusses multimedia communications across 4G networks, it is exciting to reveal part of the key technologies and innovations in 4G at this moment before we go deeply into video related topics, and before readers jump to the specific chapters in order to find the detail about specific technologies. In general, as the technologies, infrastructures and terminals have evolved in wireless systems (as shown in Figure 1.1) from 1G, 2G, 3G to 4G and from Wireless LAN to Broadband Wireless Access to 4G, the 4G system will contain all of the standards that the earlier generations have implemented. Among the few technologies that are currently being considered for 4G including 3GPP LTE/LTE-Advanced, 3GPP2 UMB, and Mobile WiMAX based on IEEE 802.16 m, we will describe briefly two of them that have wider adoption and deployment, while leaving the details and other technologies for the demonstration provided in Chapter 4.

1.2.1   3GPP LTE

Long Term Evolution (LTE) was introduced in 3GPP (3rd Generation Partnership Project) Release 8 as the next major step for UMTS (Universal Mobile Telecommunications System). It provides an enhanced user experience for broadband wireless networks.

LTE supports a scalable bandwidth from 1.25 to 20 MHz, as well as both FDD (Frequency Division Duplex) and TDD (Time Division Duplex). It supports a downlink peak rate of 100 Mbps and uplink with peak rate of 50 Mbps in 20 MHz channel. Its spectrum efficiency has been greatly improved so as to reach four times the HSDPA (High Speed Downlink Packet Access) for downlink, and three times for uplink. LTE also has a low latency of less than 100 msec for control-plane, and less than 5 msec for user-plane.

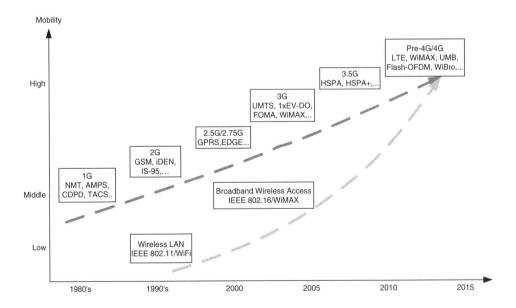

Figure 1.1   Evolving of technology to 4G wireless

It also supports a seamless connection to existing networks, such as GSM, CDMA and HSPA. For multimedia services, LTE provides IP-based traffic as well as the end-to-end Quality of Service (QoS).

LTE Evolved UMTS Terrestrial Radio Access (E-UTRA) has the following key air-interface technology:

- *Downlink based on OFDMA.* The downlink transmission scheme for E-UTRA FDD and TDD modes is based on conventional OFDM, where the available spectrum is divided into multiple sub-carriers, and each sub-carrier is modulated independently by a low rate data stream. As compared to OFDM, OFDMA allows for multiple users to access the available bandwidth and assigns specific time-frequency resources to each user, thus the data channels are shared by multiple users according to the scheduling. The complexity of OFDMA is therefore increased in terms of resource scheduling, however efficiency and latency is achieved.
- *Uplink based on SC-FDMA.* Single-Carrier-Frequency Domain Multiple Access (SC-FDMA) is selected for uplink because the OFDMA signal would result in worse uplink coverage owing to its weaker peak-to-average power ratio (PAPR). SC-FDMA signal processing is similar to OFDMA signal processing, so the parameterization of downlink and uplink can be harmonized. DFT-spread-OFDM has been selected for E-UTRA, where a size-M DFT is applied to a block of M modulation symbols, and then DFT transforms the modulation symbols into the frequency domain; the result is then mapped onto the available sub-carriers. Clearly, DFT processing is the fundamental difference between SC-FDMA and OFDMA signal generation, as each sub-carrier of an OFDMA signal only carries information related to a specific modulation system while SC-FDMA contains all of the transmitted modulation symbols owing to the spread process by the DTF transform.

LTE uses a MIMO (multiple input multiple output) system in order to achieve high throughput and spectral efficiency. It uses 2×2 (i.e., two transmit antennae at the base station and two receive antennae at the terminal side), 3×2 or 4×2 MIMO configurations for downlink. In addition, LTE supports MBMS (multimedia broadcast multicast services) either in single cell or multi-cell mode.

So far the adoption of LTE has been quite successful, as most carriers supporting GSM or HSPA networks (for example, AT&T, T-Mobile, Vodafone etc) have been upgrading their systems to LTE and a few others (for example, Verizon, China Telecom/Unicom, KDDI, and NTT DOCOMO, etc) that use different standards are also upgrading to LTE. More information about LTE, such as its architecture system, protocol stack, etc can be found in Chapter 4.

1.2.2 Mobile WiMAX

Mobile WiMAX is a part of the Worldwide Interoperability for Microwave Access (WiMAX) technology, and is a broadband wireless solution that enables the convergence of mobile and fixed broadband networks through a common wide area broadband radio access technology and flexible network architecture.

In Mobile WiMAX, a scalable channel bandwidth from 1.25 to 20 MHz is supported by scalable OFDMA. WiMAX supports a peak downlink data rate of up to 63 M bps and a peak uplink data rate of up to 28 Mbps in the 10 MHz channel with MIMO antenna techniques and flexible sub channelization schemes. On the other hand, the end-of-end QoS is supported by mapping the service flows to the DiffServ code points of MPLS flow labels. Security is protected by EAP-based authentication, AES-CCM-based authentication encryption and CMAC and HCMAC based control message protection schemes. In addition, the optimized handover schemes are supported in Mobile WiMAX by latencies of less than 50 milliseconds.

In the Physical layer, the OFDMA air interface is adopted for downlink and uplink, along with TDD. FDD has the potential to be included in the future in order to address specific market opportunities where local spectrum regulatory requirements either prohibit TDD or are more suitable for FDD deployment. In order to enhance coverage and capacity, a few advanced features, such as AMC (adaptive modulation and coding), HARQ (hybrid automatic repeat request), and CQICH (fast channel feedback), are supported.

In Mobile WiMAX, QoS is guaranteed by fast air link, asymmetric downlink and uplink capability, fine resource granularity and flexible resource allocation mechanism. Various data scheduling services are supported in order to handle bursty data traffic, time-varying channel coditions, frequency-time resource allocation in both uplink and downlink on per-frame basis, different QoS requirements for uplink and downlink, etc.

The advanced features of Mobile WiMAX may be summarized as follows:

- A full range of smart antenna technologies including Beamforming, Space-Time Code, and Spatial Multiplexing is supported. It uses multiple-antennae to transmit weighted signals using Beamforming in order to reduce the probability of outage and improve system capacity and coverage; On the other hand, spatial diversity and fade margin reduction are supported by STC, and SM helps to increase the peak data rate and throughput.
- Flexible sub-channel reuse is supported by sub-channel segmentation and permutation zone. Thus resource reuse is possible even for a small fraction of the entire channel bandwidth.
- The Multicast and Broadcast Service (MBS) is supported.

So far more than 350 trials and deployments of the WiMAX networks have been announced by the WiMAX Forum, and many WiMAX final user terminals have been produced by Nokia, Motorola, Samsung and others. On the other hand, WiMAX has been deployed in quite a number of developing countries such as India, Pakistan, Malaysia, Middle East and Africa countries, etc.

1.3 Video Over Wireless

As explained in the title of this book, our focus is on the video processing and communications techniques for the next generation of wireless communication system (4G and beyond). As the system infrastructure has evolved so as to provide better QoS support and higher data bandwidth, more exciting video applications and more innovations in video processing are expected. In this section, we describe the big picture of video over wireless from video compression and error resilience to video delivery over wireless channels. More information can be found in Chapters 3, 5 and 7.

1.3.1 Video Compression Basics

Clearly the purpose of video compression is to save the bandwidth of the communication channel. As video is a specific form of data, the history of video compression can be traced back to Shannon's seminal work [1] on data compression, where the quantitative measure of information called *self-information* is defined. The amount of self-information is associated with the probability of an event to occur; that is, an unlikely event (with lower probability) contains more information than a high probability event. When a set of independent outcome of events is considered, a quantity called *entropy* is used to count the average self-information associated with the random experiments, such as:

$$H = \sum p(A_i)i(A_i) = -\sum p(A_i) \log p(A_i)$$

where $p(A_i)$ is the probability of the event A_i, and $i(A_i)$ the associated self-information.

Entropy concept constitutes the basis of data compression, where we consider a source containing a set of symbols (or its alphabet), and model the source output as a discrete random variable. Shannon's first theorem (or source coding theorem) claims that no matter how one assigns a binary codeword to represent each symbol, in order to make the source code uniquely decodable, the average number of bits per source symbol used in the source coding process is lower bounded by the entropy of the source. In other words, entropy represents a fundamental limit on the average number of bits per source symbol. This boundary is very important in evaluating the efficiency of a lossless coding algorithm.

Modeling is an important stage in data compression. Generally, it is impossible to know the entropy of a physical source. The estimation of the entropy of a physical source is highly dependent on the assumptions about the structure of the source sequence. These assumptions are called the model for the sequence. Having a good model for the data can be useful in estimating the entropy of the source and thus achieving more efficient compression algorithms. You can either construct a physical model based on the understanding of the physics of data generation, or build a probability model based on empirical observation of the statistics of the data, or build another model based on a set of assumptions, for example, aMarkov model, which assumes that the knowledge of the past k symbols is equivalent to the knowledge of the entire past history of the process (for the kth order Markov models).

After the theoretical lower bound on the information source coding bitrate is introduced, we can consider the means for data compression. Huffman coding [2] and arithmetic coding [3] are two of the most popular lossless data compression approaches. The Huffman code is a source code whose average word length approaches the fundamental limit set by the entropy of a source. It is optimum in the sense that no other uniquely decodable set of code words has a smaller average code word length for a given discrete memory less source. It is based on two basic observations of an optimum code: 1) symbols that occur more frequently will have shorter code words than symbols that occur less frequently; 2) the two symbols that occur least frequently will have the same length [4]. Therefore, Huffman codes are variable length code words (VLC), which are built as follows: the symbols constitute initially a set of leaf nodes of a binary tree; a new node is created as the father of the two nodes with the smallest probability, and it is assigned the sum of its offspring's probabilities; this new node adding procedure is repeated until the root of the tree is reached. The Huffman code for any symbol can be obtained by traversing the tree from the root node to the leaf corresponding to the symbol, adding a *0* to the code word every time the traversal passes a left branch and a *1* every time the traversal passes a right branch.

Arithmetic coding is different from Huffman coding in that there are no VLC code words associated with each symbol. Instead, the arithmetic code generates a floating-point output based on the input sequence, which is described as follows: first, based on the order of the N symbols listed in the alphabet, the initial interval [0, 1] is divided into N ordered sub-intervals with their lengths proportional to the probabilities of the symbols. Then, the first input symbol is read and its associated sub-interval is further divided into N smaller sub-intervals. In the similar manner, the next input symbol is read and its associated smaller sub-interval is further divided, and this procedure is repeated until the last input symbol is read and its associated sub-interval is determined. Finally, this sub-interval is represented by a binary fraction. Compared to Huffman coding, arithmetic coding suffers from a higher complexity and sensitivity to transmission errors. However, it is especially useful when dealing with sources with small alphabets, such as binary sources, and alphabets with highly skewed probabilities. Arithmetic coding procedure does not need to build the entire code book (which could be huge) as in Huffman coding, and for most cases, using arithmetic coding can get rates closer to the entropy than using Huffman coding. More detail of these lossless data coding methods can be found in section 3.2.3. We consider video as a sequence of images. A great deal of effort has been expended in seeking the best model for video compression, whose goal is to reduce the spatial and temporal correlations in video data and to reduce the perceptual redundancies.

Motion compensation (MC) is the paradigm used most often in order to employ the temporal correlation efficiently. In most video sequences, there is little change in the content of the image from one frame to the next. MC coding takes advantage of this redundancy by using the previous frame to generate a prediction for the current frame, and thus only the motion information and the residue (difference between the current frame and the prediction) are coded. In this paradigm, motion estimation [5] is the most important and time-consuming task, which directly affects coding efficiency. Different motion models have been proposed, such as block matching [6], region matching [7] and mesh-based motion estimation [8, 9]. Block matching is the most popular motion estimation technique for video coding. The main reason for its popularity is its simplicity and the fact that several VLSI implementations of block matching algorithms are available. The motion vector (MV) is searched by testing different possible matching blocks in a given search window, and the resulting MV yields the lowest prediction error criterion. Different matching criteria, such as mean square error (MSE) and mean absolution difference (MAD) can be employed. In most cases, a full search requires intolerable computational complexity, thus fast motion estimation approaches arise. The existing popular fast block matching algorithms can be classified into four categories: (1) Heuristic search approaches, whose complexities are reduced by cutting the number of candidate MVs tested in the search area. In those methods, the choices of the positions are driven by some heuristic criterion in order to find the absolute minimum of cost function; (2) Reduced sample matching approaches [10, 11], whose complexities are reduced by cutting the number of points on which the cost function is calculated; (3) Techniques using spatio-temporal correlations [10, 12, 13], where the MVs are selected using the vectors that have already been calculated in the current and in the previous frames; (4) Hierarchical or multi-resolution techniques [12], where the MVs are searched in the low-resolution image and then refined in the normal resolution one.

Transform coding is the most popular technique employed in order to reduce the spatial correlation. In this approach, the input image data in the spatial domain are transformed into another domain, so that the associated energy is concentrated on a few decorrelated coefficients instead of being spread over the whole spatial image. Normally, the efficiency of a transform depends on how much energy compaction is provided by it. In a statistical sense, Karhunen-Loeve Transform (KLT) [4] is the optimal transform for the complete decorrelation of the data and in terms of energy compaction. The main drawback of the KLT is that its base functions are data-dependent, which means these functions need to be transmitted as overhead for the decoding of the image. The overhead can be so significant that it diminishes the advantages of using this optimum transform. Discrete Fourier Transform (DFT) has also been studied, however it generates large number of coefficients in the complex domain, and some of them are partly redundant owing to the symmetry property. The discrete cosine transform (DCT) [14] is the most commonly used transform for image and video coding. It is substantially better than the DFT on energy compaction for most correlated sources [15]. The popularity of DCT is based on its properties. First of all, the basis functions of the DCT are data independency, which means that none of them needs to be transmitted to the decoder. Second, for Markov sources with high correlation coefficient, the compaction ability of the DCT is very close to that of the KLT. Because many sources can be modeled as Markov sources with a high correlation coefficient, this superior compaction ability makes the DCT very attractive. Third, the availability of VLSI implementations of the DCT makes it very attractive for hardware-based real time implementation.

For lossy compression, Quantization is the most commonly used method for reducing the data rate, which represents a large set of values with a much smaller set. In many cases, scalar quantizers are used to quantize the transformed coefficients or DFD data in order to obtain an approximate representation of the image. When the number of reconstruction levels is given, the index of the reconstructed level is sent by using a fixed length code word. The Max-Lloyd quantizer [16, 17] is a well-known optimal quantizer, which results in the minimum mean squared quantization error. When the output of the quantization is entropy coded, a complicated general solution [18] is proposed. Fortunately, at high rates, the design of optimum quantization becomes simple because the optimum entropy-coded quantizer is a uniform quantizer [19]. In addition, it has been shown that the results also hold for low rates [18]. Instead of being quantized independently, pixels can be grouped into blocks or vectors for quantization, which is called vector quantization (VQ). The main advantage of VQ over scalar quantization stems from the fact that VQ can utilize the correlation between pixels.

After the data correlation has been reduced in both the spatial and temporal and the de-correlated data are quantized, the quantized samples are encoded differentially so as to further reduce the correlations as there may be some correlation from sample to sample. Thus we can predict each sample based on its past and encode and transmit only the differences between the prediction and the sample value. The basic differential encoding system is known as the differential pulse code modulation or DPCM system [20], which is an integration of quantizing and differential coding methods, and a variant of the PCM (Pulse code modulation) system.

It is very important to realize that the current typical image and video coding approaches are a hybrid coding that combines various coding approaches within the same framework.

For example, in JPEG, the block-based DCT transform, DPCM coding of the DC coefficient, quantization, zig-zag scan, run-length coding and Huffman coding are combined in the image compression procedure. In video coding, the hybrid motion-compensated DCT coding scheme is the most popular scheme adopted by most of the video coding standards. In this hybrid scheme, the video sequence is first motion compensated predicted, and the resulted residue is transformed by DCT. The resulted DCT coefficients are then quantized, and the quantized data are entropy coded. More information about the DCT-based video compression can be found in Chapter 3.

1.3.2 Video Coding Standards

The work to standardize video coding began in the 1980s and several standards have been set up by two organizations, ITU-T and ISO/IEC, including H.26x and the MPEG-x series. So far, MPEG-2/H.262 and MPEG-4 AVC/H.264 have been recognized as the most successful video coding standards. Currently, MPEG and VCEG are looking into the requirements and the feasibility of developing the next generation of video coding standards with significant improvement in coding efficiency over AVC/H.264. We will now review briefly the major standards, more detailed information can be found in Chapter 5.

H.261 [21] was designed in 1990 for low target bit rate applications that are suitable for the transmission of video over ISDN at a range from 64 kb/s to 1920 kb/s with low delay. H.261 adopted a hybrid DCT/DPCM coding scheme where motion compensation is performed on a macroblock basis. In H.261, a standard coded video syntax and decoding procedure is specified, but most choices in the encoding methods, such as allocation of bits to different parts of the picture are left open and can be changed by the encoder at will.

MPEG-1 is a multimedia standard with specifications for the coding, processing and transmission of audio, video and data streams in a series of synchronized multiplexed packets. It was targeted primarily at multimedia CD-ROM applications, and thus provided features including frame based random access of video, fast forward/fast reverse searches through compressed bit streams, reverse playback of video, and edit ability of the compressed bit stream.

Two years later, MPEG-2 was designed so as to provide the coding and transmission of high quality, multi-channel and multimedia signals over terrestrial broadcast, satellite distribution and broadband networks. The concept of 'profiles' and 'levels' was first introduced in MPEG-2 in order to stipulate conformance between equipment that does not support full implementation. As a general rule, each Profile defines a set of algorithms, and a Level specifies the range of the parameters supported by the implementation (i.e., image size, frame rate and bit rates). MPEG-2 supports the coding and processing of interlaced video sequences, as well as scalable coding. The intention of scalable coding is to provide interoperability between different services and to support receivers with different display capabilities flexibly. Three scalable coding schemes, SNR (quality) scalability, spatial scalability and temporal scalability, are defined in MPEG-2.

In 2000, the H.263 [22] video standard was designed so as to target low bit rate video coding applications, such as visual telephony. The target networks are GSTN, ISDN and wireless networks, whose maximum bit rate is below 64 kbit/s. H.263 considers network-related matters, such as error control and graceful degradation, and specific requirements for video telephony application such as visual quality, and low coding delay, to be its main responsibility. In H.263, one or more macroblock rows are organized into a group of blocks (GOP) to enable quick resynchronization in the case of transmission error. In encoding, a 3D run-length VLC table with triplet (LAST, RUN, LEVEL) is used to code the AC coefficients, where LAST indicates if the current code corresponds to the last coefficient in the coded block, RUN represents the distance between two non zero coefficients, and LEVEL is the non zero value to be encoded. H.263 adopts half-pixel motion compensation, and provides advanced coding options including unrestricted motion vectors that are allowed to point outside the picture, overlapped block motion compensation, syntax-based arithmetic coding and a PB-frame mode that combines a bidirectionally predicted picture with a normal forward predicted picture.

After that, MPEG-4 Visual was designed for the coding and flexible representation of audio-visual data in order to meet the challenge of future multimedia applications. In particular, it addressed the need for universal accessibility and robustness in an error-prone environment, high interactive functionality, coding of nature and synthetic data, as well as improved compression efficiency. MPEG-4 was targeted at a bit rate between 5–64 kbits/s for mobile and PSTN video applications, and up to 4 Mbit/s for TV/film applications. MPEG-4 is the first standard that supports object-based video representation and thus provides content-based interactivity and scalability. MPEG-4 also supports Sprite coding technology, which allows for the efficient transmission of the background scene where the changes within the background are caused mainly by camera motion. More information about MPEG-4 and object-based video coding can be found in section 6.4.

H.264/MPEG-4 AVC is the latest video standard developed jointly by ITU and ISO. It is targeted at a very wide range of applications, including video telephony, storage, broadcast and streaming. Motion prediction ability is greatly improved in H.264/AVC by the introduction of directional spatial prediction for intra coding, various block-size motion compensation, quarter sample accurate motion compensation and weighted prediction, etc. A 4×4 integer transform was adopted in H.264/AVC so as to replace the popular 8×8 DCT transform, which will not cause inverse-mismatch; smaller size transform seldom causes ringing artifacts and requires less computation. The details of H.264/AVC are provided in Chapter 5.

1.3.3   Error Resilience

Although the video compression algorithms mentioned above can achieve a very high coding efficiency, the resulted compressed video streams are very vulnerable to errors in error-prone communications networks owing to the coding scheme that used. For example, the desynchronization errors caused by VLC coding and propagation errors caused by predictive coding make error handling very difficult. In lossy wireless networks, error resilient techniques [23] can significantly increase the system's robustness by using one of the following methods: encoder error resilience tools, decoder error concealment, as well as techniques that require cooperation bewteen encoder, decoder and the network. Figure 1.2 illustrates the simplified architecture of a video communication system, where the input video is compressed on the transmitter side and the generated bit stream is channel encoded so as to make it more robust against error-prone channel transmission. On the receiver side, the inverse operations are performed in order to obtain the reconstructed video for displaying.

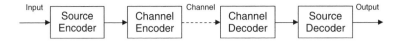

Figure 1.2 Simplified architecture of a video communication system

It is natural to increase the robustness of a compressed video bit stream by optimizing the source coding. The most popular approaches are increasing the amount of synchronization data and using special coding schemes such as RVLC (Reversible Variable Length Codes) [24]. Synchronization markers are bit sequences which have been designed especially so that they can be easily distinguished from other code words or variations of these code words with small perturbations. By inserting a synchronization marker periodically inside the compressed bit stream, any bit error will only affect those data between any two markers, which effectively prevents abusive error flooding. In this way, the decoder can resume proper decoding upon the detection of a resynchronization marker. Forward error correcting (FEC) is another approach, in which the FEC encoder adds redundant information to the bit stream which enables the receiver to detect or even correct transmission errors. This procedure is also known as channel coding. In a case where the data contains various portions of different importance, unequal error protection (UEP) becomes very useful, that is, the encoder uses stronger FEC codes on those important portions, while saving bits from having to protect unimportant portions. More detail on algorithms in this category can be found in sections 7.2 and 7.3.

Decoder error concealment refers to minimizing the negative impact of transmission error on the decoded image. The decoder recovers or estimates lost information by using the available decoded data or the existing knowledge of target applications. The current error concealment approaches can be divided into three types: spatial error concealment, temporal error concealment and adaptive error concealment. As the name indicates, spatial error concealment uses neighboring pixels to recover the corrupted area by using interpolations. It is very useful for high motion images in which frequency and temporal concealment do not yield good results, and it works particularly well for homogenous areas. Temporal error concealment makes use of the motion vectors and the data from previous time instants in order to recreate a missing block. As shown in Figure 1.3, the motion vectors are obtained by interpolation of the motion vectors of the macroblocks

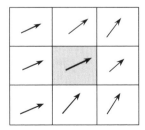

Figure 1.3 Motion vector interpolations in time domain

next to the one lost. This approach works well under the hypothesis that adjacent macroblocks move in the same direction to that of the lost macroblock. For those scenarios presenting high and irregular motion levels and scene changes, spatial concealment gives better results than temporal concealment. So as to take advantage of both the spatial and temporal approaches, adaptive error concealment selects a method according to the characteristics of the missing block, its neighbors and the overall frame, in order to perform the concealment. In this way, some parts of the image might be concealed using spatial concealment, while others might use temporal or other concealment methods, or even a combination of these approaches.

In the system, if there exist mechanisms to provide the encoder with some knowledge of the characteristics of the transmission channel – for example, if a feedback channel is set up from the decoder to the encoder so that the decoder can inform the encoder about which part of the transmitted information is corrupted by errors – then the encoder can adjust its operation correspondingly in order to suppress or eliminate the effect of such error. This type of approach is called network adaptive encoding. Of course, the encoder can also decide if retransmission of the whole video frame or some extra data would be helpful to the decoder so as to recover the packet from error. This is called ARQ (automatic repeat request). ARQ is typically not suitable in real-time communications owing to the intolerable round-trip delay. In summary, the network adaptive encoding approach optimizes the source (and channel) encoder by considering transmission factors, such as error control, packetization, packet scheduling and retransmission, routing and error concealment. It can significantly improve the system's performance (see section 9.4.1 for an example).

In H.264/AVC, many error resilience features have been adopted:

- Slice structured coding, in which slices provide resynchronization points (slice header) within a video frame and the encoder can determine the location of these points on any macroblock boundary.
- Arbitrary slice ordering, in which the decoding order of the slices may not follow the constraint that the address of the first macroblock within a slice is increasing monotonically within the NAL unit stream for a picture.
- Slice data partition, in which the slice data is partitioned into three parts: header, intra and inter-texture data.
- Redundant slices, which are provided as a redundant data in case the primary coded picture is corrupted.
- Flexible macroblock ordering, in which the macroblocks in a frame can be divided into a number of slices in a flexible way without considering the raster scan order limitation.
- Flexible reference frames, by which the reference frames can be changed on the macroblock basis;
- H.264/AVC defines IDR pictures and intra-pictures, so that the intra-picture does not have to provide a random access point function, while the IDR picture plays such a role.

1.3.4 Network Integration

RTP/UDP/IP is the typical protocol stack for video transmission. UDP and TCP are transport protocols supporting functions such as multiplexing, error control and congestion control. Since TCP retransmission introduces delays that are not acceptable for many video applications, UDP (User Data Protocol) is usually employed although it does not guarantee packet delivery. Therefore, the receiver has to rely on the upper layers in order to detect packet loss. RTP (Real-time transport protocol) is the protocol designed to provide end-to-end transport functionality, however, it does not guarantee QoS or reliable delivery. The RTP header contains important information such as timestamp, sequence number, payload type identification and source identification. RTCP (Real Time Control Protocol) is a companion protocol to RTP, which provides QoS feedback through the use of a Sender Report and Receiver Report at the source and destination, respectively. Thus, a video transmission procedure can be described as the following. At the sender side, the video data are compressed and packed into RTP packets and the feedback control information is transferred to the RTCP generator. The resulting RTP and RTCP packets go down to the UDP/IP layer for transport over the Internet. On the receiver side, the received IP packet are first unpacked by the IP and then the UDP layer and are then dispatched by the filter and dispatcher to the RTP and RTCP packets. The RTP packet is unpacked by the RTP analyzer and tested for loss detection. When packet loss is detected, the message will be sent to the error concealment module for further processing. On the other hand, the RTCP packets are unpacked and the message containing feedback information will be processed. It is important to observe that the feedback information exchange mechanism and feedback control algorithms at the end system provide QoS guarantees. For example, when network congestion occurs, the receiver can catch it by detecting symptoms such as packet loss or packet delay. The receiver then sends feedback RTCP packets to the source in order to inform it about the congestion status. Thus, the sender will decrease its transmission rate once it receives the RTCP packet. This way, the source can always keep up with network bandwidth variation and the network is therefore utilized efficiently.

For H.264/AVC, the elementary data unit for encapsulation by transport protocols (for example RTP) is called the network abstraction layer (NAL) unit. The NAL formats the video content data and provides header information in a manner appropriate for conveyance by particular transport layers. Each NAL unit consists of a one-byte header and the payload byte string, and the header indicates the type of the NAL unit and the relative importance of the NAL unit for the decoding process. During packetization, the RTP packet can contain either one NAL unit or several NAL units in the same picture.

As introduced in [25], in a 3GPP multimedia service the compressed IP/UDP/RTP packet (with RoHC) is encapsulated into a single PDCP packet, and then segmented into smaller pieces of RLC-PDU units. The RLC layer can operate in both the unacknowledged mode and the acknowledged mode, in which the former does not guarantee the data delivery while the later uses ARQ for error correction. In the physical layer, the FEC is added to the RLC-PDU depending on the coding schemes in use.

On the other hand, there are many other protocol stacks for video delivery, for example, the PSS (packet-switching streaming service) supports both the IP/UDP/RTP and IP/TCP stack, and the MBMS (multimedia broadband/multicast service) supports both the IP/UDP/RTP and IP/UDP/LCT/ALC/FLUTE stacks. In PSS, an RTCP extension is standardized so as to support packet retransmissions for RTP applicable to unicast and multicast groups.

In [26], the protocol stacks and the end-to-end architecture of a video broadcast system over WIMAX are introduced. As shown in Figure 1.4, the MBS (multicast/broadcast service) server at the controller side handles coding/transcoding, RTP packetization, mapping video channel ID to multicast connection ID (CID), shaping and multiplexing, encryption, FEC coding, constructing MBS-MAC-PDU for transmission over WiMAX PHY/MAC, burst scheduling and allocating OFDMA data region for each MBS-MAC-PDU. The packets then go through UDP/IP/L2/L1 layers for header encapsulation and are transmitted in the wireless PHY medium. When a packet reaches the Base station, the header de-capsulation is conducted in layers L1/L2/IP/UDP, the obtained MBS-MAC-PDU is then encapsulated with headers by WiMAX MAC/PHY layers, the PHY channel coding is conducted to each MBS-MAC-PUD and they are then mapped to the corresponding OFDMA data region that is determined by the MBS server for transmission. At the MSS (Mobile Subscriber Station) side, the MBS client handles error correction, decryption and constructing RTP video packet and video decoding. It is important to emphasize that the MBS client determines multicast CID according to the selected video channel ID so that only those MBS-MAC-PDUS associated with the selected multicast CID will be decoded.

1.3.5 Cross-Layer Design for Wireless Video Delivery

It is natural to consider whether the interactions between the different network protocol layers can be optimized jointly in end-to-end system design in order to achieve better performance; this is called cross-layer design. Content-aware networking and network-adaptive media processing are two widely-used approaches in wireless video delivery, and they are considered to be two sub-sets of the cross-layer design. In the former approach,

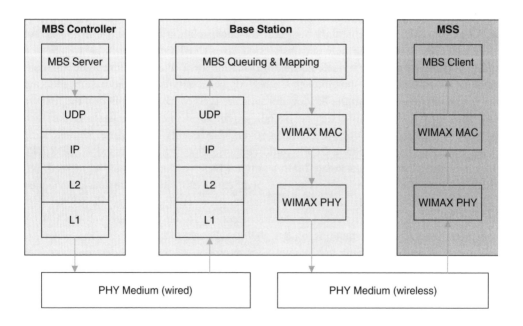

Figure 1.4   Video Broadcasting over MIMAX proposed in [26]

the resource management and protection strategies in the lower layers (i.e. PHY, MAC, network and transport layers) are optimized by considering the specific characteristics of the multimedia applications. The latter approach conducts the media compression and streaming algorithms after taking into account the mechanisms provided by the lower layers for error control and resource allocation. The principal difference between them occurs where the central controller locates: in the content-aware networking scenario, the controller locates at the lower layer, thus the application layer passes its control information, for example rate-distortion table, and requirements to the lower layer, so that after optimization the controller notifies the best strategy or parameters to the application layer for compression and transmission; and in the meantime, the controller also determines the parameters used in the PHY/MAC/IP and transport layers. On the other hand, in the network-adaptive media processing scenario, the controller locates at the application layer, which determines the optimal parameters for all layers given the information provided by the lower layers.

In Chapter 8, a quality-driven cross-layer optimization framework is introduced, which make the quality, in other words the user's experience of the system, the most important factor and the highest priority of design concern. Quality degradation is general is caused by factors such as limited bandwidth, excessive delay, power constraints and computational complexity limitation. Quality is therefore the backbone of the system and connects all other factors. In the optimization framework, the goal of design is to find an optimal balance within an N-dimensional space with given constraints, in which the dimensions include distortion, delay, power, complexity, etc. This chapter introduces an integrated methodology for solving this problem, which constructs a unified cost-to-go function in order to optimize the parameters in the system. It is an approximated dynamic programming (DP) approach, which handles global optimization over time with non-linearity and random distributions very well. The method constructs an optimal cost-to-go function based on the extracted features by using a significance measure model derived from the non-additive measure theory. The unique feature of the significance measure is that the non linear interactions among state variables in the cost-to-go function can be measured quantitatively by solving a generalized non linear Choquet integral.

1.4 Challenges and Opportunities for 4G Wireless Video

As shown in Figure 1.5, the world has become a content producer. People create and upload their own pieces of art onto the network while enjoying other people's masterpieces. It may be expected that in 4G the communication networks will continue to expand so as to include all kinds of channels with various throughputs, quality of services and protocols, and heterogeneous terminals with a wide range of capabilities, accessibilities and user preference. Thus the gap between the richness of multimedia content and the variation of techniques for content access and delivery will increase dramatically. Against such a background of expectations of universal multimedia access (UMA) will become a challenge for the 4G wireless network. The major concept of UMA is universal or seamless access to multimedia content by automatic selection or adaptation of content following user interaction. In Chapter 6, this topic is discussed in detail and the related content analysis techniques and standards are introduced.

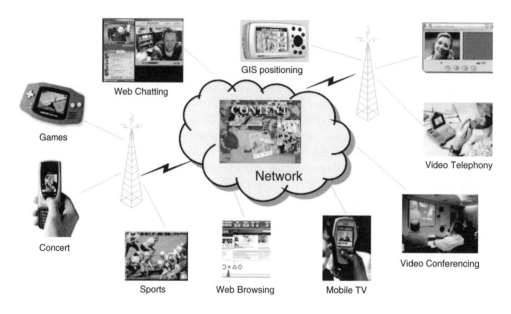

Figure 1.5   Content has become the center of the network communications

Content-based interactivity is highly connected to UMA but imposes higher expectations and requirements on content understanding supports. In the 4G high-speed communication systems, the mobile TV user may customize the content by manipulating the text, image, audio, video and graphics, and may even access information that is not available in the current system (for example, the player's information in a motorcycling racing game, or even the dynamic speed of the motorcycle that he/she is currently riding). Consequently, the 4G wireless system may face the challenge that watching TV or movie has been changed from a passive activity to a new user experience with much higher interactivity. In Chapters 6 and 11, this topic is discussed in detail and the related content-based video representation and communications techniques are demonstrated.

Mobile TV has been a network killer application for a long time, the supported resolution of TV clips by mobile devices has increased dramatically from QCIF (176 × 144) to VGA (640 × 480) in the past four or five years; nVidia showed off their APX2500 solution that can support even 720 p (1280 × 720 resolution) video in April 2008. It is expected that high-definition (HD) TV programmes will soon be delivered to and played in mobile devices in 4G networks, although there are still many challenges to be faced. In Chapter 10, the topics of digital entertainment techniques and mobile TV are discussed in detail.

On the other hand, video on TV will not be flat for much longer; that is, the next step forward is destined to be 3D video or 3D TV services over 4G wireless networks with various representation formats. As shown in Figure 1.6, the expected road map for reality video over wireless was predicted by the Japanese wireless industry in 2005, and it is interesting that the expected deployment of stereo/multi-view/hologram is around the same time period as that of 4G. Currently, stereoscopic and multi-view 3D videos are more developed than other 3D video representation formats, as their coding approaches

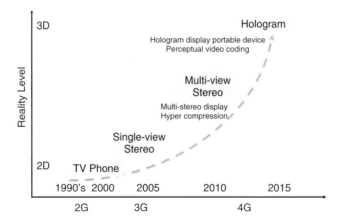

Figure 1.6   Estimated reality video over wireless development roadmap

are standardized in MPEG 'video-plus-depth' and JVT MVC standards, respectively. It is also claimed that coded 3D video only takes a 1.2 bit rate as compared to monoscopic video (i.e., the traditional 2D video). Clearly, higher reality requirements will bring in a larger volume of data to be delivered over the network, and more service and usage scenarios to challenge the 4G wireless infrastructures and protocols. In Chapter 5, some related techniques on the topic of multi-view video are discussed.

Recently, P2P (peer-to-peer) live video streaming has become very popular as it needs much less deployment cost and is able to take advantage of the distributed storage and computing resources in the network. The basic idea of P2P is to treat every node in the network as a peer and the peer can help to pass the video packets to others. However, in the 4G system, the large scale P2P wireless video streaming capability of supporting many viewers would still be very challenging, considering the difficulties of effective incentive mechanisms for peers willing to collaborate, peers' media relay capability, peers' system topology discovery capability, QoS control, etc. In Chapter 12, this topic is discussed in detail and the related video streaming techniques are introduced.

Finally, as so many fancy applications and services are expected to be deployed in 4G networks, the cross-layer design mechanism for content delivery with satisfactory quality of user experience becomes critical. In Chapters 8 and 9, quality-driven cross-layer design and optimization methodology and its applications for various content-based video communication scenarios are addressed in order to resolve this issue.

References

1. C. E. Shannon, A mathematical theory of communication, *Bell System Technical Journal*, Vol. 27, pp. 379–423, 1948.
2. D. A. Huffman, A method for the construction of minimum redundancy codes, *Proceedings of the IRE*, Vol. 40, pp. 1098–1101, 1951.
3. J. Rissanen, G. G. Langdon, Arithmetic coding, *IBM Journal on Research and Development*, Vol. 23, No. 2, pp. 149–62, March 1979.
4. K. Sayood, *Introduction to Data Compression*, Morgan Kaufmann Publishers, Inc. 1996.

5. B. Furht, J. Greenberg, R. Westwater, *Motion Estimation Algorithms for Video Compression*, Kluwer Academic Publishers, 1997.

6. A. K. Jain, Image data compression: a review, *Proceedings of the IEEE*, Vol. 69, pp. 349–89, March 1981.

7. Y. Nakaya and H. Harashima, Motion compensation based on spatial transformations, *IEEE Transactions on Circuits and Systems*, Vol. 4, No. 3, pp. 339–56, June 1994.

8. P. J. L. van Beek, and A. M. Tekalp, Object-based video coding using forward tracking 2-D mesh layers, in *Proc. SPIE Visual Comm. Image Proc.*, Vol. 3024, pp. 699–710, San Jose, California, Feb. 1997.

9. M. Dudon, O. Avaro, and C. Roux, Triangular active mesh for motion estimation, *Signal Processing: Image Communication*, Vol. 10, pp. 21–41, 1997.

10. K. Lengwehasatit and A. Ortega, A novel computationally scalable algorithm for motion estimation, in *Proc. VCIP'98*, 1998.

11. B. Natarajan, V. Bhaskaran, and K. Konstantinides, Low-complexity block-based motion estimation via one-bit transforms, *IEEE Trans. Circuits Syst. Video Technol.*, Vol. 7, pp. 702–6, Aug. 1997.

12. J. Chalidabhongse and C. C. Kuo, Fast motion vector estimation using multiresolution-spatio-temporal correlations, *IEEE Trans. Circuits Syst. Technol.*, Vol. 7, pp. 477–88, June 1997.

13. A. Chimienti, C. Ferraris, and D. Pau, A complexity-bounded motion estimation algorithm, *IEEE Trans. Image Processing*, Vol. 11, No. 4, pp. 387–92, April 2002.

14. K. R. Rao and P. Yip, *Discrete Cosine Transform: algorithms, advantages, applications*, Academic Press, Boston, 1990.

15. N. S. Jayant and P. Noll, *Digital Coding of Waveforms*, Englewood Cliffs, NJ: Prentice Hall, 1984.

16. S. P. Lloyd, Least squares quantization in PCM, *IEEE Trans. Information Theory*, Vol. 28, pp. 129–37, March 1982.

17. J. Max, Quantizing for minimum distortion, *IRE Trans. Information Theory*, Vol. 6, pp. 7–12, March 1960.

18. N. Farvardin and J. W. Modestino, Optimum quantizer performance for a class of non-Gaussian memoryless sources, *IEEE Trans. Information Theory*, IT-30, pp. 485–97, May 1984.

19. H. Gish and J. N. Pierce, Asymptotically efficient quantization, *IEEE Trans. Information Theory*, Vol. 14, pp. 676–83, Sept. 1968.

20. C. C. Cutler, "*Differential quantization for television signals*", U.S. Patent, 2,605,361. July 29, 1952.

21. ITU-T Recommendation H.261, *Video codec for audiovisual services at px64kbps*, 1993.

22. ITU-T Recommendation H.263, *Video coding for low bitrate communication*, 1998.

23. Y. Wang, S. Wenger, J. Wen, and A. K. Katsaggelos, Review of error resilient techniques for video communications, *IEEE Signal Processing Magazine*, Vol. 17, No. 4, pp. 61–82, July 2000.

24. Y. Takishima, M. Wada, H. Murakami, Reversible variable length codes, *IEEE Trans. Communications*, Vol. 43, Nos. 2/3/4, pp. 158–62, Feb./March/April 1995.

25. T. Stockhammer, M. M. Hannuksela, H.264/AVC Video for Wireless Transmission, *IEEE Wireless Communications*, Vol. 12, Issue 4, pp. 6–13, August 2005.

26. J. Wang, M. Venkatachalam, and Y. Fang, System architecture and cross-layer optimization of video broadcast over WiMAX, *IEEE Journal of Selected Areas In Communications*, Vol. 25, No. 4, pp. 712–21, May 2007.

2

Wireless Communications and Networking

In the past decades, wireless communications and networking have enjoyed huge commercial success, thanks to many new wireless technologies and industrial standards. The fundamental difference between wired and wireless technologies resides in the physical channel. The unique features of wireless propagation such as reflection, refraction, and shadowing result in various channel impairments including multi-path fading, path loss, and doppler frequency, making it very challenging for high-speed information delivery over the wireless channel. Furthermore, the time-varying nature of wireless media poses new design challenges for wireless networks, since the existing Internet protocols were developed based on the assumption that all packet losses are caused by network congestions, which is no longer the case in the wireless environment. In this chapter, we will provide a brief overview of the fundamentals of wireless communications and networking, especially the technologies adopted by 4G wireless systems such as cross-layer design, adaptive modulation and coding, hybrid ARQ, MIMO and OFDM.

2.1 Characteristics and Modeling of Wireless Channels

2.1.1 Degradation in Radio Propagation

Errors which occur in data transmission through wireless channels are mainly caused by three types of degradation, i.e. *interference*, *frequency-nonselective fading* and *frequency-selective fading*. Note that although there are other kinds of degradation in wireless networks such as path loss, doppler effect and log-normal distributed shadowing loss, they are approximately constant when a mobile user moves slowly within urban areas. Interferences can be generated either in nature such as additive white Gaussian noise (AWGN) or from other users whose RF transmission occupies the same frequency band without any coordination.

Frequency-selective fading results from delay spread τ_d in channel impulse response (CIR) caused by multi-path radio propagation. Delay spread could cause inter-symbol interference (ISI) at the receiver. If τ_d is smaller than the symbol duration $1/R$, where R is the symbol rate, there is little frequency-selective degradation. In other words, under

4G Wireless Video Communications   Haohong Wang, Lisimachos P. Kondi, Ajay Luthra and Song Ci
© 2009 John Wiley & Sons, Ltd

a certain τ_d, increasing the transmission symbol rate R will cause more serious ISI. Therefore, a channel equalizer should be adopted in order to alleviate the averse impact of ISI on throughput performance. According to [1, 2], in indoor and micro-cellular channels, delay spread is usually small (less than several hundred nanoseconds). Moreover, most of the current research on adaptive modulation and coding techniques is based on narrow-band wireless channels, where bit duration is sufficiently larger than the inter-arrival time of reflected waves. Therefore, in this case, inter-symbol interference is small. For high speed data communications over wireless channel, inter-symbol interference will degrade significantly the throughput performance. In this case, orthogonal frequency division multiplexing (OFDM) provides an effective solution to this problem where a broadband channel will be divided into many narrow-band sub-channels. Therefore, inter-symbol interference will be minimized for a higher throughput.

In wireless mobile networks, owing to user mobility, signal waveforms may experience substantial fading distortion over time which is usually a function of Doppler frequency. This kind of fading is called frequency-nonselective fading or flat fading. During the duration of flat fading, burst errors may occur. In wireless mobile networks the degradation of channel quality is caused mainly by *Rayleigh fading channel*, and most adaptive modulation and coding techniques are based on the Rayleigh fading channel model. Thus, in the following subsection, more detail about the Rayleigh fading channel will be provided.

2.1.2 Rayleigh Fading Channel

For narrow-band signals without a line-of-sight (LOS) component between the transmitter and the receiver, Rayleigh distribution is commonly used in order to characterize the statistics of the time-variant signal envelope with flat fading [3, 4]. If a line-of-sight (LOS) component exists, Rician distribution will be adopted in order to model the received fading signal. Furthermore, it is shown that using Rayleigh distribution when the Rician factor is unknown will represent the worst scenario of the fading channel. For a narrow-band signal $r(t)$ can be expressed as

$$r(t) = a(t)e^{-j(w_c t + \theta(t))} \tag{2.1}$$

Here, $a(t)$ is the time-varying signal envelope, which follows the Rayleigh distribution. $\theta(t)$ is the time-varying phase which is uniformly distributed from $[0, 2\pi]$. Both the in-phase and quadrature components are i.i.d Gaussian random numbers with zero mean and σ^2 variance. Thus, the probability density function (PDF) of $a(t)$ is:

$$f(a) = \frac{a}{\sigma^2} e^{-\frac{a^2}{2\sigma^2}} \tag{2.2}$$

The PDF of the received SNR s, which is proportional to the square of the signal envelope, following the Exponential distribution [5]. It can be written as:

$$f(s) = \frac{1}{\rho} e^{-\frac{s}{\rho}} \tag{2.3}$$

For $s \geq 0$, we have $\rho = E[s]$. To characterize the channel variation, it is very important to know the distribution of SNR and the autocorrelation function, which are widely used in adaptive modulation and coding. For Rayleigh fading channels, the autocorrelation of the channel power gain over time can be written as [6]:

$$R(\tau) = J_0^2 \left(\frac{2\pi v \tau}{\lambda} \right) \qquad (2.4)$$

where v is the velocity of the mobile user and λ is the wavelength of RF. There are two classical channel models, the *Jakes model* and *Markov channel model*, which are widely adopted in the research work on adaptive modulation and coding.

In [6], narrow-band Rayleigh fading is generated by adding a set of six or more sinusoidal signals with predetermined frequency offsets. These frequencies are chosen as to approximate the typical U-shaped Doppler spectrum, and N frequency components are taken at

$$\omega_i = \omega_m \cos \frac{2\pi i}{2(2N+1)} \qquad 1 \leq i \leq N \qquad (2.5)$$

Simulation results have shown that when N is larger than 6, this model can describe Rayleigh fading channels appropriately. Measurements over non-line-of-sight (NLOS) paths at UHF frequencies in urban environments confirmed the accuracy of this model.

Another commonly-used fading channel model is the Markov chain model. Gilbert [7] originally proposed a model based on a two-state Markov model inspired by the fact that burst errors occurred in poor telephone channels. In his work, one state represents the channel in inter-burst state and the other represents the channel in burst state. In the good state, there is error-free; in the bad state, the channel performs as a binary symmetric channel (BSC). Elliott [8] modified this model by changing the good state from error-free BSC to a nonzero crossover probability. This is the well-known Gilbert-Elliot (G-E) bursty channel model [9], illustrated in Figure 2.1.

Fritchman proposed a Markov model with a finite number N of states and derived a transition probability matrix for this model. Other researchers were inspired by Fritchman's model to propose several simplified Fritchman models. Wang *et al.* proposed a finite-state Markov channel [10] for pursuing the higher accuracy of a Rayleigh fading channel. Garcia-Frias *et al.* proposed the hidden Markov model [11] for a more general channel model not only for characterizing Rayleigh fading channels but also for characterizing frequency-selective fading channels. Swarts *et al.* [10] and Wang *et al.* [12, 13] evaluated and validated their finite-state Markov models in order to characterize the Rayleigh fading channel.

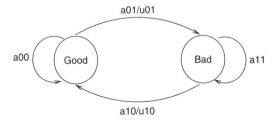

Figure 2.1   Gilbert-Elliot bursty channel model

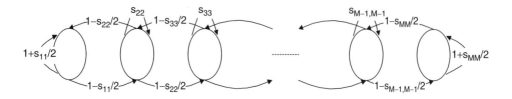

Figure 2.2   Finite-Markov channel model

Figure 2.2 shows a typical finite-Markov model characterizing the Rayleigh fading channel used in adaptive modulation and coding research [14]. In this channel model, each state represents a stationary BSC channel with a constant bit error rate. The states are ordered according to the decreasing values of bit error rates. The switching process between states is described by the transition probability matrix. Slow variations of bit error rate which occur in real channels make it possible for each state to transmit only between its two neighboring states. The transition probability matrix is given by:

$$[\mathbf{S}] = \begin{pmatrix} \frac{1+s_{11}}{2} & \frac{1-s_{11}}{2} & 0 & \cdots & 0 & 0 \\ \frac{1-s_{22}}{2} & s_{22} & \frac{1-s_{22}}{2} & \cdots & 0 & 0 \\ \cdots & \cdots & \cdots & \cdots & \cdots & \cdots \\ 0 & 0 & 0 & \cdots & \frac{1-s_{M,M-1}}{2} & \frac{1+s_{M,M}}{2} \end{pmatrix} \tag{2.6}$$

The matrix of the state probabilities is defined and computed by

$$[W] = [W_1, W_2, \ldots, W_M] \tag{2.7}$$

$$[W][S] = [W] \tag{2.8}$$

$$[W][E] = [I] \tag{2.9}$$

where $[E]$ is a column matrix whose entries are 1s and $[I]$ is the identity matrix.

In the research on adaptive modulation and coding [15], the Nakagami-m probability density function is often used to characterize the multi-path fading environment because this fading channel model can represent a range of multi-path channels via the parameter m, which can be interpreted as the amount of fading on the channel. As the value of m increases, the amount of fading on the channel decreases. In particular, the Nakagami-m distribution includes the one-sided Gaussian distribution ($m = 1/2$ which corresponds to the worst fading) and the Rayleigh distribution (when $m = 1$). Furthermore, the Nakagami-m distribution approximates to the Nakagami-n (Rice) distribution, which often gives the best fit to the land- and indoor-mobile multi-path fading channels as well as satellite radio links. Hence, the channel fading amplitude α is given by:

$$p_\alpha(\alpha) = 2 \left(\frac{m}{\Omega} \right)^m \frac{\alpha^{2m-1}}{\Gamma(m)} e \left(\frac{\alpha^2}{\Omega} \right) \qquad \alpha \geq 0 \tag{2.10}$$

where $\Omega = E(\alpha^2)$ is the average received power, m is the Nakagami fading parameter ($m \geq \frac{1}{2}$) and $\Gamma(\cdot)$ is the gamma function defined as:

$$\Gamma(z) = \int_0^{+\infty} t^{z-1} e^{-t} dt \qquad z \geq 0 \qquad (2.11)$$

2.2 Adaptive Modulation and Coding

With the diffusion of multimedia services in all kinds of wireless networks, Quality of Service (QoS) should be provided in these networks. Among the parameters of quality of services, reliable communication link is the primary requirement. One common metric used to evaluate the reliability of a channel is the signal-to-interference-plus-noise-ratio (SINR). Because the wireless channel is time-varying caused by fading, interference, shadow and path loss, in a multiuser wireless network, a user will perceive different channel quality in terms of SINR from time to time. However, wireless communication system parameters such as data rate, modulation scheme and transmission power have less flexibility. When a user perceived SINR is lower than the minimum SINR that is required to provide a specific service data rate, the current connection will be terminated, leading to channel outage.

In order to cope with channel outage, increasing the transmission power appears to be an effective solution to this problem. But in a multiuser communication system, simply increasing the transmission power of one user will cause stronger interference for other nearby users. Therefore, for non-adaptive communications systems, it requires a fixed link margin in order to maintain acceptable performance even when the channel quality is good. However, this will result in an insufficient utilization of channel capacity. Therefore, adaptive transmission has been proposed in order to maintain a predefined SINR at receiver by dynamically changing system parameters such as transmission power [2], symbol transmission rate [16], constellation size [17], channel coding rate/scheme [14], multi-carrier transmission [18, 19], or any combination of these parameters [15, 20]. In this section, we will address issues related to *adaptive modulation and coding (AMC)*. The motivation behind this idea is to provide an acceptable bit-error-rate (BER) performance while keeping transmit power constant under the time-varying wireless channel quality. Through this technique, spectral efficiency and flexible data rate can be achieved. AMC techniques were first used in V.34 modems to combat the poor channel quality of telephone lines. This idea was extended to Rayleigh fading channels [15] in time-division duplex (TDD) systems. In frequency-division duplex (FDD) systems, a feedback channel should be used in order to make an estimation of channel fading fed back to the transmitter from the receiver. The main purpose of this section is to introduce AMC techniques and to compare the performance of some other adaptive schemes in terms of transmission power, modulation and channel coding rate.

2.2.1 Basics of Modulation Schemes

Most modern communication systems are digital devices, whose output information is coded into bits, represented by a pulse train. Thus, the fundamental question is how much channel bandwidth is needed to transmit the coded information for a given communication system. According to the Fourier analysis, the coefficient of k_{th} harmonic of a given periodic signal $s(t)$ can be represented as:

$$S[k] = \frac{1}{T} \int_0^T s(t) \cdot e^{-i2\pi \frac{n}{T} t} dt$$

where T is the period of signal. Therefore, the minimum bandwidth can be calculated based on how much energy will be kept when the signal passes through a bandwidth-limited channel. Since a coefficient at the corresponding frequency is proportional to the energy contained in that frequency, we can derive the channel bandwidth B by determining how many harmonics need to be passed through the physical channel.

When the input signal is continuous, the Nyquist theorem is used to determine the sampling frequency to generate the digital signal. Then the source data rate R_s of a signal is:

$$R_s = 2B log_2(V)$$

where V is the number of quantization levels adopted to code each discrete data sample. If the channel is noisy, the Shannon theorem is used to determine the maximum data rate under a certain channel condition in terms of signal-to-noise ratio (SNR), which is:

$$R = B log_2(1 + \gamma)$$

where γ is the given SNR.

Theoretically, any time-limited signal has the infinite spectrum in the frequency domain which will suffer a significant amount of attenuation and delay distortion. Therefore, modulation and coding schemes need to be adopted in order to overcome these problems. Modulation is the process for using the information being transmitted in order to change the waveform attributes of the sinusoid wave carrier. For a given sinusoid waveform, there are three waveform attributes: amplitude, phase, and frequency. Therefore, there are three major modulation schemes: Amplitude Shift-Keying (ASK), Phase Shift-Keying (PSK) and Frequency Shift-Keying (FSK).

In ASK, the appearance of a carrier waveform represents a binary '1' and its absence indicates a binary '0':

$$s_0(t) = 0 \quad and \quad s_1(t) = \cos(2\pi f_c t)$$

where $s_0(t)$ is the modulated signal for bit '0' and $s_1(t)$ is for bit '1' and f_c is the carrier frequency. ASK usually works with some other modulation scheme such as PSK in order to form a new modulation scheme called quadrature amplitude modulation (QAM) for higher channel utilization.

In PSK, the phase of the carrier waveform is changed by the binary information. The simplest PSK modulation scheme is Binary PSK (BPSK), which can be written as:

$$s_0(t) = -\cos(2\pi f_c t) \quad and \quad s_1(t) = \cos(2\pi f_c t)$$

There are many other PSK modulation schemes which are often used for higher bandwidth efficiency such as quadrature PSK (QPSK) and 8-PSK. In FSK, the carrier frequency is changed by the binary information:

$$s_0(t) = \cos(2\pi(f_c + f_0)t) \qquad \text{and} \qquad s_1(t) = \cos(2\pi(f_c + f_1)t)$$

where f_0 is corresponding to the frequency offset for bit '0' and f_1 is for bit '1'. Modulation schemes have been discussed in literature. For more detailed information, the reader can refer to [5] and [2].

2.2.2   System Model of AMC

Figure 2.3 describes a general system model of AMC, which is composed of transmitter, channel and receiver. In this figure, **r** is the input information bit vector. **x** is the encoded and interleaved symbol vector. **y** is the output fading vector. **g** is the estimation of current channel power gain which will be sent back through the feedback channel. Note that in TDD systems, it is not necessary to use the feedback channel.

The function of the transmitter can be divided into two units: one is the modulation, encoding and interleaving unit and the other is the power control unit. For most AMC techniques, the first part is necessary. The function of interleaving is to remove the channel memory effect among the received symbols. Whether or not using power control in a transmitter depends on different AMC schemes. For example, in [21, 22], variable-rate and variable-power adaptive schemes are proposed and evaluated. In AMC, the wireless channel is usually assumed to be a discrete-time channel with stationary and ergodic time-varying channel power gain **g** and AWGN noise **n** with noise density $N_0/2$. Assume that an estimation $\hat{g}$ of the channel power gain **g** is available at the receiver after an estimation time delay τ_e with estimation error ε. This estimation is then used by the demodulation and decoding unit to generate the output information bit stream **r**. The

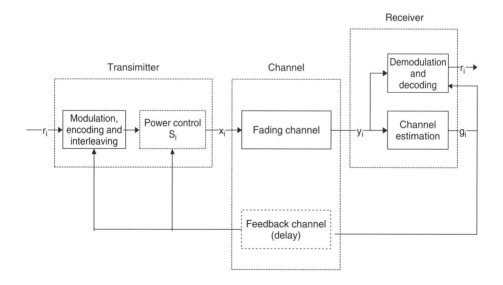

Figure 2.3   System model of adaptive modulation and coding

estimated channel power gain is also sent back to the transmitter through the feedback channel with feedback delay τ_f.

2.2.3 Channel Quality Estimation and Prediction

In order to make AMC techniques respond to the channel variation efficiently, reliable and accurate estimation of the channel quality in terms of SINR or frame error rate is very important to AMC performance. The channel quality estimation includes two main components: channel quality predictor and channel quality indication metric.

The metric of channel quality indication should have the following features [23]:

(1) It should be independent of fading rates, i.e. the metric values should be consistent across a wide range of mobility patterns.
(2) It should be independent of the size of signal constellation in use.
(3) It should be estimated accurately by using a small number of samples at the receiver.
(4) It should provide reliable indicators in both noise- and interference-limited conditions.

In the literature, there are several different channel quality indication metrics in link adaptation such as SINR [21, 24], BER [17], Euclidian Distance (ED) [23], fading depth [25] and channel capacity or throughput [21, 22, 26].

In [21, 24], SINR is used as the channel quality metric. Given a AWGN channel $n(t)$, a stationary ergodic channel gain $\sqrt{g(t)}$, and a constant transmit power $\overline{S}$, the instantaneous received SINR is $\gamma(t) = \overline{S}g(t)/(N_0 B)$ and the average received SINR is $\overline{\gamma} = \overline{S}/(N_0 B)$. The average data rate of an AMC scheme is thus given by:

$$R = \int_{\gamma_0}^{\infty} \frac{1}{T(\gamma)} (\log_2 M(\gamma) - r) p(\gamma) d\gamma \qquad (2.12)$$

where $T(\gamma)$ is the symbol period, $M(\gamma)$ is the corresponding signal set, r is the redundant information used for the channel coding and $p(\gamma)$ is the fading distribution of γ. All of these variables are functions of the channel SINR γ.

There has been a lot of research done on AMC by using BER as the channel quality metric. In general, given a certain SINR, BER can be easily derived according to corresponding modulation schemes. In [14], the bit error probability for BPSK modulation is given by:

$$P_b = \int_{-\infty}^{+\infty} Q\left(\sqrt{\frac{2E_b}{N_0}}\right) p(a) da \qquad (2.13)$$

where $Q(\cdot)$ is the complementary Gaussian error function, E_b is the energy per bit and N_0 is the one-side AWGN power spectral density. a is the fading amplitude at the receiver. The assumption behind this equation is that the channel is a AWGN channel during one symbol period, which is also used in symbol-by-symbol adaptation methods.

In [17, 23], received signal strength indicator (RSSI) is adopted as a channel quality metric because in this research work, block-by-block adaptation methods are used. The

RSSI can be calculated by:

$$RSSI = \frac{1}{K} \sum_{k=0}^{K-1} |r_k|^2 \qquad (2.14)$$

where K is the length of received block.

Euclidean distance is used as a channel quality metric [23]. It has been proved that when coherent demodulation is employed, the normalized expected value of the cumulative path metric may be considered as an accurate approximation of the interference plus noise power spectral density $(I + N)$ per symbol. The Euclidean distance can be written as:

$$\Lambda_K(\hat{s}_k) = \sum_{k=0}^{K-1} |\alpha_k(s_k - \hat{s}_k) + (\gamma_k i_k + n_k)|^2 \qquad (2.15)$$

where Λ_K is the Euclidean distance of the block of length K. $\hat{s}_k$ is the estimation of the symbol sequence s_k at the decoder. α_k and γ_k are the fading channel gain of the transmitted symbol and the interference, respectively. n_k is the sample of the AWGN channel with variance of N_0. Then, the expectation of the Euclidean distance can be approximated as:

$$\frac{1}{K} E \left\{ \sum_{k=0}^{K-1} |r_k i_k + n_k|^2 \right\} = (I + N) \qquad (2.16)$$

Thus, the SNR can be expressed as:

$$SNR = \frac{RSSI}{\mu} \qquad (2.17)$$

where μ is the short-term moving average of the average scaled Euclidean distances.

Fading amplitude and autocorrelation function [25] are employed as a channel quality metric. The idea behind this approach is that the current fading amplitude obeys the Rician distribution, when conditioned on the previous fading amplitudes. The conditional probability density function is given by:

$$p_{Y|\hat{X}}(y|\mathbf{x}) = \frac{y}{\sigma^2} e^{(-y^2 + s^2/2\sigma^2)} I_0 \left(\frac{ys}{\sigma^2} \right) \qquad y \geq 0 \qquad (2.18)$$

where $Y = |X(kT_s)|$ is the amplitude of the fading that multiplies the kth transmitted symbol. $I_0(\cdot)$ is the zeroth-order modified Bessel function and the noncentrality parameter is given by s^2 which is the sum of the squares of the MMSE prediction of the real and imaginary parts of $X(t)$. σ^2 is the mean square error of the MMSE predictor of the in-phase or quadrature fading value of interest. When $N = 1$, i.e. just one outdated fading estimate $\rho = R_X(\tau_1)$ is used, then the number of signals in the signal set $(\tilde{M}(h))$ employed when $|\hat{X}(kT_s - \tau_1)| = h$ at the received SNR $\left(\frac{E_s}{N_0} \right)$ is:

$$\tilde{M}(h) = \max \left\{ M : \ \tilde{P}_M \left(\frac{E_s}{N_0}, h, \rho \right) \leq P_b \right\} \qquad \rho_{\min} \leq \rho \leq 1 \qquad (2.19)$$

where P_b is the target BER. ρ_{min} is the minimum value of $R_X(\tau_1)$ which is approximated to $J_0(2\pi 6\tau_1)$ for indoor channel with walking speed mobility. Through this approach, a guaranteed performance across all possible autocorrelation functions can be ensured [23].

Another channel quality metric is maximum spectral efficiency [21, 24]. The motivation behind this approach is inspired by the analysis of the AWGN channel capacity. The capacity of a communication channel is limited by available transmit power and channel bandwidth. Then the maximum spectral efficiency can be written as:

$$\frac{C}{B} = \int_{\gamma_0}^{\infty} B \log_2(\frac{\gamma}{\gamma_0}) p(\gamma) d\gamma \tag{2.20}$$

where C is the fading channel capacity and B is the wireless channel bandwidth. γ is the instant received SNR and γ_0 is the cut-off SNR. $p(\gamma)$ is the probability distribution of the received SNR.

Most of aforementioned channel quality metrics are developed for narrow-band channels which are not adequate for wide-band channel quality metrics because more transmission bursts experience ISI. Therefore, pseudo-SNR [27] at the output of the decision-feedback equalizer (DFE) is proposed as a channel quality metric, which is defined as the desired signal power divided by residual ISI power plus effective noise power. So the pseudo-SNR can be written as:

$$\gamma_{\text{DFE}} = \frac{E\left[|S_k \sum_{m=0}^{N_f-1} C_m h_m|^2\right]}{\sum_{q=-(N_f-1)}^{-1} E[|d_q S_{k-q}|^2] + N_0 \sum_{m=0}^{N_f-1} |C_m|^2} \tag{2.21}$$

where $d_q = \sum_{m=0}^{N_f-1} C_m h_{m+q}$. S_k is the transmitted signal at time k which is assumed to be uncorrelated to each other. N_f and $q \in [1, N_b]$ represent the number of taps in the forward filter and the backward filter, respectively. N_0 is the single-side AWGN power spectral density. C_m is the optimum coefficient for the forward filter and h_i denotes the ith path of the channel impulse response (CIR).

For the channel quality predictor, the Lagrange equation is widely adopted in symbol-by-symbol adaptation schemes under the assumption that the channel fading will be constant during one symbol. Other kinds of channel quality predictors are also adopted. For example, a third-order optimal one-step linear predictor is used in [23] for predicting the SNR over the next time slot from the past SNR values. In general, prediction accuracy and estimation delay are two very important factors which can have a significant effect on the performance of AMC techniques. Therefore, accuracy and speed are primary criteria when choosing a channel quality predictor. Moreover, optimal prediction filters can be designed for further improvement in performance.

2.2.4 Modulation and Coding Parameter Adaptation

Modulation and coding parameters such as the constellation size, coding rate of Trellis code and transmission power can be adapted to various anticipated channel conditions. Adapting these modulation and coding parameters in response to the predicted local SINR can be employed in order to accommodate a wide range of trade-offs between the received data integrity and link throughput. Generally speaking, methods of parameter adaptation used in different adaptive modulation and coding schemes are similar. Given a predetermined performance requirement in terms of packet loss rate and throughput, during each adaptation period, channel quality is predicted and used so as to select the most appropriate modulation and coding mode from a set of candidate modulation and coding modes. In most cases, the forward error control (FEC) coding is used at the transmitter for stronger error control capability. The adaptable channel coding parameters include code rate, interleaving and puncturing for convolutional and turbo codes, and block lengths for block codes. Depending on whether channel coding is combined with modulation, the adaptive modulation techniques can be divided into adaptive uncoded modulation and adaptive coded modulation. The issue of pure adaptive channel coding is beyond the scope of this book, more detail can be found in [14, 19, 20, 28].

A spectrally efficient M-ary quadrature amplitude modulation (MQAM) is adopted in the research work [17, 21, 22, 24, 25, 29], and different MQAM constellations are employed by different researchers. For example, in [17], a circular star constellation in conjunction with differential channel coding is adopted rather than the conventional square constellation. However, it suffers from the difficulty of carrier recovery in a fading environment. In addition to MQAM, M-ary phase-shift modulation (MPSK) is another choice of modulation scheme [20, 30]. In [30], a new nonuniform M-ary phase-shift modulation is proposed for supporting multimedia services in CDMA systems. In [18, 19], adaptive OFDM is proposed and evaluated. The idea behind adaptive OFDM is to adapt the data rate of each carrier of OFDM in response to the different channel quality perceived by each frequency carrier.

Furthermore, all modulation schemes can also work with Trellis coded modulation (TCM) in order to achieve a better tradeoff between data integrity and high link throughput [20, 23–25, 30]. Figure 2.4 illustrates the basic structure of adaptive Trellis coded modulation [25], where b_i is the input bit stream, $\underline{X}$ is the predicted channel quality metric, and z_i is the output symbol stream. The number of subsets 2^n is kept constant

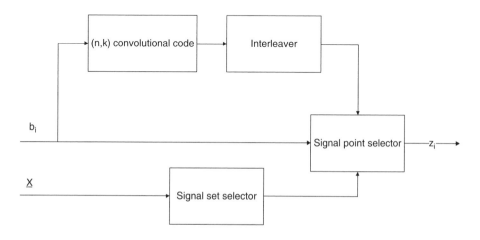

Figure 2.4   The adaptive trellis-coded modulation

across all signal sets and only the number of signals per subset is adapted. At the encoder, sets of i information bits are taken into the convolutional encoder, which produces n bit subset selectors that are interleaved. The signal constellation is chosen from the valid signal sets $\zeta_1, \zeta_2, \ldots, \zeta_L$ based on the predicted channel quality metric. Finally, the appropriate number of information bits in a symbol is determined by the logarithm with the base two of the size of a subset ζ. The ith transmitted symbol z_i is then chosen by ith output of interleaver (i.e. subset selector) from $\tilde{\zeta}_i$. There are two kinds of subset partitioning methods widely used in current research, one is Gray mapping [23] and the other is the standard Ungerboeck mapping [20, 24, 25, 30]. In this figure, interleaving subset selectors does not exist in some schemes [23] because this will introduce more delay which may curtail the performance of adaptive modulation and coding schemes.

2.2.5 Estimation Error and Delay in AMC

It has been widely shown that some issues such as estimation error, estimation and feedback delay and adaptation rate will affect the performance of AMC dramatically [14, 17, 21, 23, 31, 32]. Since most of these issues are optimization problems, the solutions to them may vary scheme by scheme. In this section, several important issues that could greatly affect the performance of adaptive modulation and coding are discussed.

Channel estimation error and delay are the most important factors affecting system performance [14, 16, 21, 22, 31]. Channel estimation error is caused mainly by delay and accuracy of the channel quality metric predictor as well as loss or quantization error of feedback information. Since the fading channel is time-varying, delay caused by estimation and feedback will make instantaneous channel estimation outdated. In general, there are two main types of errors in channel estimation. First, a wrong decision can be made by using an overestimated channel quality metric. Second, an erroneous conclusion is drawn using an underestimated channel quality metric. The first type of error will degrade the BER of the system; and the second type of error will degrade the throughput of the system. If reliability is more important than throughput in a given system, it is essential to avoid the first type of error in channel estimation. However, accurate channel quality predictors are in practice constrained by the total delay bound and implementation complexity.

In order to improve the accuracy of channel estimation, several measures have been proposed in the literature. In [21–24], pilot-symbol-assisted modulation (PSAM) in Rayleigh fading channel is proposed. Through using pilot-symbol, the estimation delay and the accuracy of channel estimation are greatly improved, thus the performance of proposed adaptive modulation schemes can be achieved. In [25], the concept of strongly robust autocorrelation is proposed in order to combat feedback delay, the problem of accuracy predictor and the time-varying Doppler frequency effect. In [16], the effect of feedback information quantization is discussed. In general, the feedback channel is usually assumed to be perfect owing to the fact that the probability of packet loss for very short packets is very small.

2.2.6 Selection of Adaptation Interval

The link adaptation interval should be selected so as to be long enough to provide an accurate channel estimation. On the other hand, it should be short enough to prevent long delay in channel estimation. In general, the adaptation interval is shorter than the expected fading duration which may affect tens or hundreds of symbols [14, 17, 23, 24, 31]. To reflect the time-varying characteristics of a Rayleigh fading channel, many adaptation schemes adopt the symbol-by-symbol method. But symbol-by-symbol adaptation is difficult to achieve in practice owing to feedback delay and feedback bandwidth constraints [23]. In addition, these schemes result in highly wide range signals, which may drive power amplifiers to inefficient operating points. On the other hand, symbol-by-symbol adaptation may provide valuable information in the form of achievable upper bounds. Therefore, how to determine the fading duration is of interest since it determines the tradeoff between the number of regions and the adaptation rate of power, modulation and coding. There has some research focusing on this topic [2, 21, 33].

2.3 Orthogonal Frequency Division Multiplexing

2.3.1 Background

The principle of OFDM is to transmit data by dividing the data stream into multiple parallel bit streams with much lower bit rates, which can be further modulated onto a set of low frequency carriers. Although OFDM has been studied since the 1960s, only recently has it been recognized as an outstanding method for high speed, bi-directional wireless data communications.

In OFDM, the frequency band of interest is divided up into a number of low frequency carriers, or subcarriers. These subcarriers are normally orthogonal which means that at the receiver each subcarrier can be detected without having interference from other subcarriers. This is made possible by the mathematical property of orthogonal waveforms, which ensures that the integral of the product of any two subcarriers is zero. Therefore, by dividing the frequency band into a large number of narrow-band carriers, wireless channel impairments are significantly reduced, since fading only impacts on a very limited number of the subcarriers. Therefore, OFDM provides superior link quality and robustness of high-speed data communications over the wireless channel.

2.3.2 System Model and Implementation

The OFDM implementation is shown in Figure 2.5. The input data stream is modulated by a modulator, resulting in a complex symbol stream $\{X[0], X[1], \ldots\}$. This symbol steam is passed through a serial-to-parallel converter, whose output is multiple sets of N parallel modulation symbols. Each set is defined as $X := \{X[0], \ldots, X[N-1]\}$ with each symbol of X being transmitted by each of the subcarriers. Thus, the N-symbol output from the serial-to-parallel converter are the discrete frequency components of the OFDM modulator output $s(t)$. In order to generate $s(t)$, the frequency components are converted into time samples by performing an inverse DFT on these N symbols, which is implemented efficiently using the IFFT algorithm. The IFFT yields the OFDM symbol consisting of the sequence $x[n] = x[0], \ldots, x[N-1]$ of length N, where:

$$x[n] = \frac{1}{\sqrt{N}} \sum_{i=0}^{N-1} X[i] e^{j2\pi ni/N}, \quad 0 \le n \le N-1 \tag{2.22}$$

This sequence corresponds to samples of the multicarrier time signal $s(t)$.

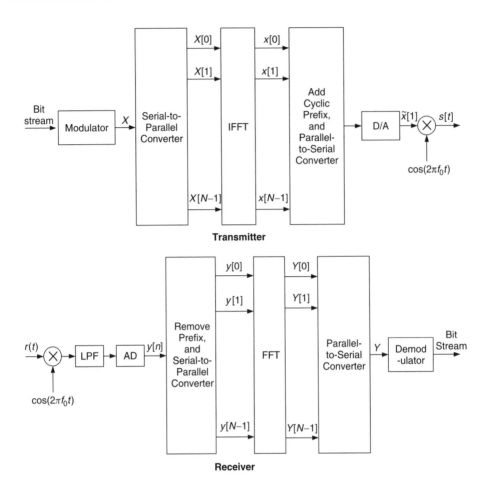

Figure 2.5 OFDM with IFFT/FFT implementation

When $x[n]$ is sent through a linear time-invariant discrete-time channel $h[n]$, the output $y[n]$ is the discrete-time linear convolution of the input and the channel impulse response:

$$y[n] = h[n] * x[n] = x[n] * h[n] = \sum_k h[k]x[n-k] \qquad (2.23)$$

In contrast to the above linear convolution, the N-point *circular convolution* of $x[n]$ and $h[n]$ is defined as:

$$y[n] = h[n] \circledast x[n] = x[n] \circledast h[n] = \sum_k h[k]x[n-k]_N \qquad (2.24)$$

where $[n-k]_N$ denotes $[n-k]$ modulo N. Moreover, according to the definition of DFT, the circular convolution of two signals in time leads to their multiplication in frequency, i.e.:

$$\text{DFT}\{x[n] \circledast h[n]\} = X[i]H[i], \quad 0 \leq i \leq N - 1 \tag{2.25}$$

where $H[i]$ is the N-point DFT of $\{h[n]\}$. Unfortunately, the channel output $y[n]$ in equation (2.23) is not a circular convolution but a linear convolution. However, the linear convolution can be turned into a circular convolution by adding a special prefix to the input called a *cyclic prefix*. The cyclic prefix for $x[n]$ is defined as $\{x[N - \mu], \ldots, x[N - 1]\}$ consisting of the last μ values of the $x[n]$ sequence.

After the cyclic prefix is added to the OFDM symbol $x[n]$ in equation (2.22), the resulting time samples $\tilde{x}[n] = \{\tilde{x}[-\mu], \ldots, \tilde{x}[N - 1]\} = \{x[N - \mu], \ldots, x[0], \ldots, x[N - 1]\}$ are ordered by the parallel-to-serial converter and are passed through a D/A converter, and the baseband OFDM signal $\tilde{x}(t)$. $\tilde{x}(t)$ is then upconverted to frequency f_0 forming $s(t)$.

At the receiver side, the received signal $r(t)$ is downconverted to base band by removing the high-frequency component. The A/D converter samples the baseband signal and obtains $y[n] = \tilde{x}[n] * h[n] + v[n]$, $-\mu \leq n \leq N - 1$, where $h[n]$ is the discrete-time equivalent low-pass impulse response of the channel and $v[n]$ is the channel noise. The prefix of $y[n]$ consisting of the first μ samples is then removed. Then the resulting N time samples are serial-to-parallel converted and passed through an FFT. According to equation (2.25), this generates scaled versions of the original symbols $H[i]X[i] =: Y[i]$, where $H[i] = H(f_i)$ is the flat fading channel gain associated with the ith sub-channel. Finally, the FFT output is parallel-to-serial converted and passed through a demodulator in order to recover the original data.

The above analysis indicates that the orthogonality of sub-channels in an OFDM system makes the recovery of the original symbols very easy by just dividing out sub-channel gains from $Y[i]$, i.e., $X[i] = Y[i]/H[i]$.

2.3.3 Pros and Cons

To summarize, the main advantages of OFDM are:

- OFDM deals efficiently with frequency-selective fading without the need for a complicated equalizer.
- OFDM supports a high data rate by dividing an entire channel into many overlapping narrow-band sub-channels by using orthogonal signals.
- Different subcarriers can be allocated to different users in order to provide a flexible multiuser access scheme and to exploit multiuser diversity.
- OFDM has a high degree of flexibility of radio resource management, including the different frequency responses of different channels for various users, data rate adaptation over each subcarrier, dynamic sub-carrier assignment and adaptive power allocation.

The main drawbacks of OFDM are

- OFDM signals have a large peak to average power ratio (PAPR), which increases approximately linearly with the number of subcarriers. Large PAPRs force the transmitter power amplifier to have a large *backoff* in order to ensure the linear amplification of the signal. The techniques to reduce or tolerate the high PAPR of OFDM signals include clipping the OFDM signal above some threshold, peak cancellation with a complementary signal, allowing nonlinear distortion from the power amplifier and special coding techniques.

- OFDM systems are highly sensitive to frequency offsets caused by the oscillator inaccuracies and the Doppler shift due to mobility, which gives rise to inter-carrier interference (ICI). Methods of reducing frequency offset effects include windowing of the transmitted signal, self ICI cancellation schemes and frequency offset estimation methods.

2.4 Multiple-Input Multiple-Output Systems

A MIMO system has multiple antennae at transmitter and receiver, which can be used to increase data rates through multiplexing or to improve performance through spatial diversity. Owing to its significant performance gain, MIMO has been adopted by IEEE 802.11n, 802.16-2004 and 802.16e as well as by 3GPP and 3GPP2.

2.4.1 MIMO System Model

A narrow-band point-to-point communication system of M_t transmit antennae and M_r receive antennae is shown in Figure 2.6. This system can be represented by the following discrete-time model:

$$
\begin{bmatrix} y_1 \\ \vdots \\ y_{M_r} \end{bmatrix} = \begin{bmatrix} h_{11} & \cdots & h_{1M_t} \\ \vdots & \ddots & \vdots \\ h_{M_r 1} & \cdots & h_{M_r M_t} \end{bmatrix} \begin{bmatrix} x_1 \\ \vdots \\ x_{M_t} \end{bmatrix} + \begin{bmatrix} n_1 \\ \vdots \\ n_{M_r} \end{bmatrix}
\tag{2.26}
$$

or simply as $\mathbf{y} = \mathbf{Hx} + \mathbf{n}$. Here $\mathbf{x}$ represents the M_t-dimensional transmitted symbol, $\mathbf{n}$ is the M_r-dimension complex Gaussian noise vector with zero mean and covariance matrix

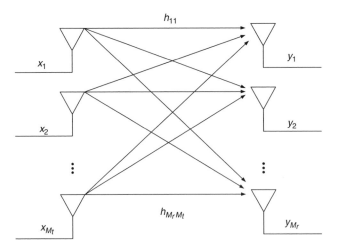

Figure 2.6 MIMO systems

$\sigma^2 \mathbf{I}_{M_r}$, where typically $\sigma^2 \triangleq \mathbf{E}[n_i^2] = N_0/2$, the power spectral density of the channel noise. $\mathbf{H}$ is the $M_r \times M_t$ matrix of channel gain h_{ij} representing the channel gain between the jth transmit antenna and the ith receive antenna. In general, different assumptions about channel side information (CSI) and the distribution of $\mathbf{H}$ lead to different channel capacities, different approaches to space-time signaling and different decoding complexities of the received signals.

2.4.2 MIMO Capacity Gain: Multiplexing

When both the transmitter and the receiver equip multiple antennae, there exists a mechanism to boost the performance gain, which is also called *capacity gain*. The capacity gain of a MIMO system results from the fact that a MIMO channel can be divided into a number $R_{\mathbf{H}}$ of parallel independent channels where $R_{\mathbf{H}}$ is the rank of MIMO channel $\mathbf{H}$. By multiplexing independent data onto these $R_{\mathbf{H}}$ independent channels, we get an $R_{\mathbf{H}}$-fold increase on data rate in comparison to a system with just one pair of antennae between the transmitter and the receiver. This increased data rate is called capacity gain or *multiplexing gain*. The parallel division of channel $\mathbf{H}$ is obtained by defining a transformation on the channel input and output via *transmit precoding* and *receiver shaping* [34]. The representative method of achieving spatial multiplexing is the Bell Labs Layered Space Time (BLAST) architectures for MIMO channels [35]. According to different transmit precoding schemes, BLAST architectures are classified into V-BLAST [36], where the serial data is parallel encoded into M_t independent streams, and D-BLAST [35], where the serial data is parallel encoded with a stream rotation operation.

2.4.3 MIMO Diversity Gain: Beamforming

In contrast to capacity gain, MIMO can also provide another type of performance gain called *diversity gain*. When the same symbol – weighted by a complex scale factor – is sent over each transmit antenna, the diversity gain is obtained via coherent combining of the multiple signal paths at the receiver. This scheme is also referred as *MIMO beamforming*. To perform coherent combining, channel knowledge at the receiver is assumed as known. The diversity gain also depends on whether or not channel is known at the transmitter. When channel matrix $\mathbf{H}$ is known, the maximum obtainable diversity gain is $M_t M_r$. When the channel is not known to the transmitter, other methods have to be adopted. For $M_t = 2$, the Alamouti scheme [37] can be used to extract the maximum diversity gain of $2M_r$ [38]. For $M_t > 2$, full diversity gain can also be obtained using other space-time block codes, which will be introduced later in this chapter.

2.4.4 Diversity-Multiplexing Trade-offs

The discussion above suggests that capacity gain and diversity gain in MIMO systems are obtained by two totally different mechanisms. Either type of performance gain can be maximized at the expense of the other. However, in the design of real communication systems, it is not necessary to use all antennae just for multiplexing or diversity. Thus, some space-time dimensions can be used for diversity gain, and other dimensions can be used for multiplexing gain.

Actually, it has been shown that a flexible trade-off between diversity and multiplexing can be achieved [39]. Let d and r be the diversity gain and multiplexing gain, respectively. d and r can be expressed as:

$$\lim_{\gamma \to \infty} \frac{R(\gamma)}{\log_2 \gamma} = r \tag{2.27}$$

and

$$\lim_{\gamma \to \infty} \frac{P_e(\gamma)}{\log \gamma} = -d \tag{2.28}$$

where $R(\gamma)$ is the data rate (bps) per unit Hertz and $P_e(\gamma)$ is the probability of error, and both are functions of SNR γ. For each r, the optimal diversity gain $d_{opt}(r)$ is the maximum diversity gain that can be achieved by any scheme. Then, the trade-off between r and $d_{opt}(r)$ is:

$$d_{opt}(r) = (M_t - r)(M_r - r), \quad 0 \leq r \leq \min(M_t, M_r) \tag{2.29}$$

Equation (2.29) is plotted in Figure 2.7, which implies (1) the maximum diversity gain and the maximum multiplexing gain cannot be achieved simultaneously in a MIMO system; (2) if all transmit *and* receive antennae are used for diversity then the maximum diversity gain $M_t M_r$ can be obtained; and (3) some antennae can be used to increase data rate at the expense of losing diversity gain.

2.4.5 Space-Time Coding

Space-time coding (STC) has received extensive attention from both academia and industry in recent years. It has the following advantages: (1) STC improves downlink performance without the need for multiple receive antennae at the receiver side; (2) STC can be

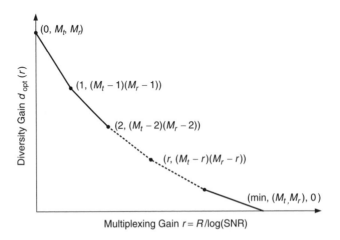

Figure 2.7 The diversity-multiplexing trade-off curve of a MIMO system

easily combined with channel coding in order to achieve a coding gain in addition to the spatial diversity gain; (3) STC does not require channel side information at the transmitter, thus eliminating the need for an expensive system design, especially in the case of a fast time-varying channel; and (4) STC are robust against non-ideal operating conditions such as antenna correlations, channel estimation errors and Doppler effects. To utilize these advantages and design practical space-time codes in order to achieve a performance target, STC design criteria and STC structures have been proposed. For example, *rand criterion* is proposed in order to design the space-time codes that achieve the maximum diversity gain, and *determination criterion* is used to design the space-time codes that obtain high coding gain [34, 40]. Two representative STC structures are space-time trellis codes (STTC) and space-time block codes (STBC).

- *Space-time trellis codes*: the STTC encoder maps the information bit stream into M_t streams of symbols that are transmitted simultaneously. Like conventional trellis codes, STTC uses a trellis and use maximum likelihood (ML) sequence estimation via the Viterbi algorithm. The advantage of STTC is that it can achieve the maximum diversity gain and the maximum coding gain. However, the complexity of decoding is increased exponentially along with the increase of diversity level and transmission rate. Detailed STTC structures for different signal constellations and different numbers of antennae are presented in [41].
- *Space-time block codes*: Alamouti [37] first developed a space-time block coding scheme for transmission with two antennae, based on the assumption that channel gain is constant over two consecutive symbol periods. In Alamouti's scheme, input symbols are grouped into pairs where symbols x_n and x_{n+1} are transmitted over the nth symbol period from the first and second antennae, respectively. Then, over the $(n + 1)$th symbol period, symbol $-x_{n+1}^*$ is transmitted from the first antenna and symbol x_n^* is transmitted from the second antenna, where $*$ denotes the complex conjugate transpose. This imposes an orthogonal spatio-temporal structure on the transmitted symbols which leads to linear processing complexity in decoding. Alamouti's scheme has the following attractive features: (1) achieving full diversity at full transmission rate for any (real or complex) signal constellation; (2) not requiring CSI at the transmitter; and (3) ML decoding involves linear processing complexity at the receiver. This scheme was generalized in [42] for STBCs to achieve full diversity gain with an arbitrary number of transmit antennae.

2.5 Cross-Layer Design of AMC and HARQ

To enhance spectral efficiency and channel utilization in future wireless communications systems, adaptive modulation and coding (AMC) has been advocated at the physical layer. However, since the constellation size and coding rate are chosen based on the corresponding carrier-to-noise ratio (CNR) region, AMC cannot guarantee to achieve maximum spectral efficiency, especially when the number of available AMC modes is limited and the current CNR is low. In this section, we discuss how to combine adaptive modulation and coding (AMC) at the physical layer with hybrid automatic repeat request (HARQ) at the data link layer in a cross-layer fashion, where the upper layer HARQ provides an effective means to fine-tune the performance of the lower layer AMC. Both type-I and type-II HARQ are considered in our discussion.

2.5.1 Background

In recent wireless network standards, such as 3GPP/3GPP2, HIPERLAN/2 and IEEE 802.11/16 [43–46] adaptive modulation and coding (AMC) has been advocated at the physical layer in order to enhance channel utilization and throughput of future wireless communication systems. The most common way of performing AMC adaptation is to simply divide the carrier-to-noise ratio (CNR) range into several fading regions and assign different AMC modes with different constellation sizes and coding rates to the different CNR regions. However, since the constellation size and coding rate are chosen based on the corresponding CNR region, this method cannot achieve maximum spectral efficiency, especially when the number of available AMC modes is limited. Therefore, the chosen AMC parameters cannot keep up with the dynamic change of the instantaneous CNR value in the same CNR region. In addition, since explicit channel estimations and measurements are required by AMC in order to set suitable modulation and coding parameters, the accuracy of AMC parameter choice is affected by possible measurement error and delay.

Hybrid automatic repeat request (HARQ), described and used in 3GPP, 3GPP2 and IEEE 802.16 standards [43, 44, 46], is an alternative way to mitigate channel fading at the data link layer. Hybrid ARQ schemes can be classified into two categories, namely type-I and type-II schemes [47, 48]. A general type-I HARQ scheme uses error detection and correction codes for each transmission and retransmission. Previously received packets with uncorrectable errors are discarded at the receiver. A general type-II HARQ scheme uses a low rate error correction code. An information packet is first transmitted with parity bits for error detection with none or a few parity bits for error correction. Incremental blocks of redundancy bits are transmitted upon retransmission requests. The receiver combines the transmitted and retransmitted blocks together so as to form a more powerful error correction code in order to recover the original information. Obviously, the mechanism of either type of HARQ schemes makes itself adapt autonomously to the instantaneous channel conditions by a different number of retransmissions and by being insensitive to the errors and delays incurred in channel measurements. However, during a long deep fading, a large number of retransmissions will be required for certain packets to be received correctly at the receiver. This, in turn, may lead to unacceptably large delay and buffer sizes. Comparing the aforementioned two types of HARQ, type-I has two major drawbacks: 1) the receiver discards the uncorrectable packet of every transmission, which may be helpful for error-correcting if combined with subsequent received packets; and 2) once the coding rate is fixed, all parity bits for error correction are transmitted even if they are not all needed, thus reducing channel efficiency. These two drawbacks can be overcome partially by type-II HARQ by using incremental redundancy transmissions.

Based on the above analysis of the advantages and drawbacks of AMC and HARQ, cross-layer design of combining AMC and HARQ is proposed in order to achieve maximum spectral efficiency under certain QoS constraints such as PER and transmission delay. To satisfy QoS performance requirements such as delay, packet loss rate and goodput, the maximum number of transmission attempts for an information packet and the targeted packet error rate (PER) should be limited. The rationale of cross-layer design is that AMC provides only a coarse data rate selection by choosing an appropriate constellation size and coding rate based on the CNR region that the current CNR values fall into, then HARQ can be used to provide fine data rate adjustments through autonomously changing numbers of retransmissions according to the instantaneous channel condition. In long deep fading scenarios, to avoid large number of HARQ retransmissions, an AMC mode with smaller constellation size and lower coding rate is chosen. In the following, we will focus on describing the cross-layer design method and performance analysis for combining AMC with type-II HARQ. The same design and analysis methods can be applied to combining AMC with type-I HARQ.

2.5.2 System Modeling

In this section, type-II HARQ with rate-compatible convolutional (RCC) codes [49] is employed. Let $R_1 > R_2 > \cdots > R_M$ denote the M rates offered by a family of RCC codes $C_1, C_2, \cdots, C_M$ which are obtained from a good low rate C_M (e.g., 1/2 or 1/3) code with a puncturing technique. The rate-compatibility restriction to the puncturing rule implies that all the code bits of a high rate punctured code are used by the lower rate codes [50]. As a result, only one single decoder is needed to decode all the received codewords at different rates. Let L_i denote the number of transmitted bits of HARQ at the ith transmission attempt. L_i can be expressed as

$$
L_i = \begin{cases} L\left(\frac{1}{R_1}\right) & i = 1 \\ L\left(\frac{1}{R_i} - \frac{1}{R_{i-1}}\right) & 1 < i \leq M \end{cases} \tag{2.30}
$$

Let $R_1 = 1$, i.e. $L_1 = L$ because uncoded transmission performs well enough in good channel conditions and because there are already channel coding schemes in AMC at the physical layer. Let the number of RCC codes M be equal to the allowed maximum number of transmissions N_t for an information packet. Therefore, if there are still uncorrectable errors after all the bits of yielding M codes are transmitted, the packet is assumed to be lost.

Let N denote the total number of AMC modes available at the physical layer. To illustrate the key idea, in this section, the same $N = 6$ AMC modes are adopted as in [46]. These AMC modes and fitting parameters of their BER performance are listed in Table 2.1, in a rate ascending order as the mode index n increases. Let R_n denote the rate of AMC mode n. Then, L_i / R_n symbols are transmitted at the ith transmission in the physical layer if AMC mode n is used.

The system structure of the cross-layer design is shown in Figure 2.8, which consists of a HARQ module at the data link layer and an AMC module at the physical layer. At the transmitter, an information packet with length of L from the higher layer, including cyclic redundancy check (CRC) bits, is coded and punctured into N_t blocks. The length of the ith block is equal to L_i. For the ith transmission of an information packet, the ith block is processed by the AMC controller with feedback from the receiver. At the receiver, based on the measured CSI, the AMC selector determines the new mode for the next transmission, which will be sent back to the transmitter through a feedback channel. In addition, decoding is performed using code C_i by combining the ith block with the $i - 1$ previously received and buffered blocks. After decoding and error checking, if an error occurs in the decoded packet, the ith block is put into buffer and a retransmission request is generated by the HARQ generator, which is sent to the HARQ controller

Table 2.1   AMC Modes at the physical layer

	Mode1	Mode2	Mode3	Mode4	Mode5	Mode6
Modulation	BPSK	QPSK	QPSK	16-QAM	16-QAM	64-QAM
Coding Rate r_c	1/2	1/2	3/4	9/16	3/4	3/4
R_n (bits/sym.)	0.50	1.00	1.50	2.25	3.00	4.50
a_n	1.1369	0.3351	0.2197	0.2081	0.1936	0.1887
b_n	7.5556	3.2543	1.5244	0.6250	0.3484	0.0871
$\gamma_n^{(1)}(dB)$	1.2632	4.3617	7.4442	11.2882	13.7883	19.7961
$\gamma_n^{(2)}(dB)$	-2.1898	0.1150	2.8227	6.6132	9.0399	15.0211
$\gamma_n^{(3)}(dB)$	-3.4098	-1.6532	0.7272	4.4662	6.8206	12.7749

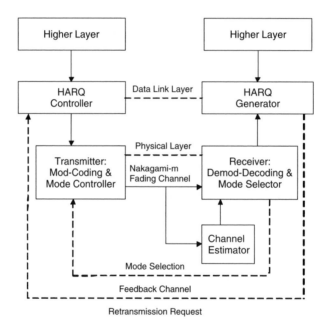

Figure 2.8   The cross-layer structure of combining AMC and HARQ

at the transmitter via a feedback channel. Then, the HARQ controller at the transmitter arranges transmission for the $(i + 1)$th block that is stored in the buffer. If there still exists decoding error after N_t blocks for an information packet are all sent out, packet loss is declared. Then, transmissions for next information packet begin.

The following assumptions are usually needed in this scheme:

- The channel remains time invariant during a block, but it could vary from block to block. Thus, the AMC scheme is updated for every transmission and retransmission attempt.

- Perfect channel state information (CSI) is available at the receiver. The corresponding mode selection is sent back to the transmitter without error and latency.
- Error detection can be handled by CRC.

As in [2], the received CNR γ per block under a Raleigh fading channel is a random variable with a Gamma probability density function (pdf):

$$p(\gamma) = \frac{1}{\overline{\gamma}} \exp\left(-\frac{\gamma}{\overline{\gamma}}\right) \tag{2.31}$$

where $\overline{\gamma} := E\{\gamma\}$ is the average received CNR.

2.5.3 Cross-Layer Design

To satisfy some defined QoS requirements, the following constraints are considered:

C1: The allowed maximum number of transmission attempts for each information packet is N_t.

C2: The probability of packets loss after N_t transmission attempts is no greater than P_{loss}.

C1 and C2 can be extracted from the required QoS in the specific applications. Generally, N_t can be specified by dividing the maximum allowable system delay over the round trip delay required for each transmission. P_{loss} can be specified by the required BER and packet size L with the equation:

$$P_{\text{loss}} = 1 - (1 - \text{BER})^L \tag{2.32}$$

Suppose the average PER after decoding using code C_i is P. As stated before, a packet is dropped if it is not received correctly after N_t transmissions. Considering C2, the following inequality should be satisfied:

$$P^{N_t} \le P_{\text{loss}} \tag{2.33}$$

Specifically:

$$P \le P_{\text{loss}}^{1/N_t} := P_{\text{target}} \tag{2.34}$$

Thus, as long as the P is bounded as in equation (2.34), both C1 and C2 will be satisfied. To satisfy equation (2.34), P_{target} exerts BER requirement on the adopted AMC mode at the physical layer. Since the codeword of C_1 only contains information bits while the codewords of C_i $(i \ge 2)$ contain both information bits and redundancy bits. In the following, the BER requirement by P_{target} on AMC at the physical layer in two cases for C_1 and C_i $(i \ge 2)$ will be discussed respectively.

Case 1: For C_1, each bit inside the output block of after demodulation and decoding of AMC is assumed to have the same BER and bit-errors are uncorrelated. Since C_1 only contains information bits, the PER of C_1 can be related to the BER through:

$$P^{(1)} = 1 - \left(1 - BER^{(1)}\right)^L \tag{2.35}$$

where $P^{(1)}$ denotes the PER of code C_1, and $BER^{(1)}$ is the BER of after demodulation and decoding of AMC. If let $P^{(1)} = P_{target}$, the required target BER on AMC can be obtained by C_1 to reach P_{target} as:

$$BER_{target}^{(1)} = 1 - \left(1 - P_{target}\right)^{1/L_1} \tag{2.36}$$

Case 2: For C_i ($i \geq 2$), since the errors detected the decoder corresponding to C_i are not independent, it is difficult to obtain the exact relationship between PER and BER after C_i decoding. Here a PER upper bound is adopted [51] under the assumption that C_i is a binary convolutional code with hard-decision Viterbi decoding and independent errors at the channel input. For a L long packet coded by C_i, the PER upper bound is:

$$P^{(i)} \leq 1 - (1 - P_u^{(i)})^L := P_{bound}^{(i)} \tag{2.37}$$

where $P^{(i)}$ is the PER after C_i decoding, and $P_u^{(i)}$ is the union bound of the first-event error probability corresponding to code C_i, which can be approximately expressed as [5]:

$$P_u^{(i)} \approx a_{d_f^{(i)}}^{(i)} 2^{d_f^{(i)}} \left(\rho^{(i)}\right)^{d_f^{(i)}/2} \tag{2.38}$$

where $d_f^{(i)}$ is the free distance of the convolutional code C_i; $a_d^{(i)}$ is the total number of error events with weight $d_f^{(i)}$, and $\rho^{(i)}$ is the bit error probability of AMC at the physical layer. Obviously, if:

$$P_{bound}^{(i)} = P_{target}, \tag{2.39}$$

using equations (2.34), (2.37) and (2.38), we can obtain $\rho^{(i)}$ required by code C_i to achieve P_{target}. The required $\rho^{(i)}$ is denoted as:

$$BER_{target}^{(i)} = \left[\frac{1 - \left(1 - P_{target}\right)^{1/L}}{a_{d_f^{(i)}}^{(i)} \cdot 2^{d_f^{(i)}}}\right]^{2/d_f^{(i)}}, \quad i \geq 2 \tag{2.40}$$

Actually, $BER_{target}^{(i)}$ is the required total BER on i received blocks for code C_i decoding using a combination of these blocks to reach P_{target}. To simplify the design, $BER_{target}^{(i)}$ is regarded as the required BER on the ith received block for code C_i decoding by using the combination of i received blocks to achieve P_{target}. This is reasonable since in order to reach the same target PER, BER requirement of code C_i is looser than that of code C_j, $j < i$ due to their different coding rates.

We will now discuss how to determine AMC mode by using equations (2.36) and (2.40) to maximize system spectral efficiency as well as to satisfy C1 and C2.

As in [52, 53], the total CNR range is divided into $N + 1$ continuous and non-overlapping fading regions with the region boundary $\gamma_n, n = 0, 1, \ldots, N + 1$, where N is the number of AMC mode. For each transmission, when the estimated channel CNR γ satisfies $\gamma \in [\gamma_n, \gamma_{n+1})$, AMC mode n is chosen. In the following, we discuss how to determine the boundary point $\gamma_n, n = 0, 1, \ldots, N + 1$.

The boundary points are specified for the given target BER in equations (2.36) and (2.40). To simplify AMC design and facilitate performance analysis, the following approximate BER expression is used as in [52, 53]:

$$\text{BER}_n(\gamma) \approx a_n \exp\left(-b_n \gamma\right) \tag{2.41}$$

where n is AMC mode index and γ is the received CNR. Parameters a_n, b_n can be obtained by fitting equation (2.41) to the exact BER of mode n. The region boundary γ_n for AMC mode n is set to the minimum CNR required to achieve target BER. Using equations (2.36), (2.40) and (2.41), then:

$$\gamma_0^{(i)} = 0,$$

$$\gamma_n^{(i)} = \frac{1}{b_n} \ln\left(\frac{a_n}{\text{BER}_{\text{target}}^{(i)}}\right) \quad n = 1, \cdots, N \tag{2.42}$$

$$\gamma_{N+1}^{(i)} = +\infty$$

where $\gamma_n^{(i)}$ denotes the least boundary CNR of using AMC mode n at the ith transmission. To avoid deep fading, there is no data transmission when CNR falls into $[\gamma_0^{(i)}, \gamma_1^{(i)})$.

With $\gamma_n^{(i)}$ specified in equation (2.42), the AMC mode n will operate with the actual BER $\rho_n^{(i)}$ satisfying

$$\rho_n^{(i)} \leq \text{BER}_{\text{target}}^{(i)} \tag{2.43}$$

Therefore, the system PER P in equation (2.33) will be also guaranteed.

To summarize, cross-layer design are listed as below.

Step 1: Given C1 and C2, determine P_{target} from equation (2.34).

Step 2: For the P_{target} found, determine $BER_{\text{target}}^{(i)}$ for the ith transmission block from equations (2.34), (2.35), (2.36), (2.37), (2.39) and (2.40).

Step 3: For the $BER_{\text{target}}^{(i)}$ found, determine $\gamma_n^{(i)}$ of AMC mode n for the ith transmission from equation (2.42).

It is important to point out that for simplicity performance bounds on PER of RCC codes are used to obtain the approximate performance requirement. Thus, the tighter the bound is in equation (2.37), the more accurate AMC mode selection we can achieve, then the higher spectral efficiency we can obtain. In [51], it has been shown that the PER bound $1 - (1 - P_u)^L$ is strictly tighter than another bound LP_u as developed in [54].

2.5.4 Performance Analysis

(1) Packet Error Rate: Let $P\{F_{n_1,\cdots,n_i}^{(i)}(\gamma^{(1)},\cdots,\gamma^{(i)})\}$ denote the probability of the events of "decoding failure with code C_i after i transmissions using AMC mode $n_1,\cdots,n_i$ under channel CNR values equal to $\gamma^{(1)},\cdots,\gamma^{(i)}$ respectively." Recall that an error packet occurs at the receiver when the number of transmissions reaches the maximum number N_t if there are still detected errors in the packet. The probability of this under channel states $\{\gamma^{(1)},\gamma^{(2)},\cdots,\gamma^{(N_t)}\}$ with AMC modes $\{n_1,n_2,\cdots,n_{N_t}\}$ is:

$$
\begin{aligned}
PER_{n_1,n_2,\cdots,n_{N_t}}&(\gamma^{(1)},\gamma^{(2)},\cdots,\gamma^{(N_t)})\\
&= P\left\{F_{n_1}^{(1)}(\gamma^{(1)}), F_{n_1,n_2}^{(2)}(\gamma^{(1)},\gamma^{(2)}),\right.\\
&\left.\quad\cdots, F_{n_1,n_2,\cdots,n_{N_t}}^{(N_t)}(\gamma^{(1)},\gamma^{(2)},\cdots,\gamma^{(N_t)})\right\}
\end{aligned}
\tag{2.44}
$$

By integrating equation (2.44) over all possible values of CNR vector $\{\gamma^{(1)},\gamma^{(2)},\cdots,\gamma^{(N_t)}\}$, we can obtain the average PER of our cross-layer design as:

$$
\begin{aligned}
PER = \frac{1}{P_{\text{CNR}}} \sum_{n_1=1}^{N}\sum_{n_2=1}^{N}\cdots\sum_{N_t=1}^{N}\int_{\gamma_{n_1}^{(1)}}^{\gamma_{n_1+1}^{(1)}}\int_{\gamma_{n_2}^{(2)}}^{\gamma_{n_2+1}^{(2)}}\cdots\int_{\gamma_{n_{N_t}}^{(N_t)}}^{\gamma_{n_{N_t}+1}^{(N_t)}}\\
P\left\{F_{n_1}^{(1)}(\gamma^{(1)}),\cdots,F_{n_1,\cdots,n_{N_t}}^{(N_t)}(\gamma^{(1)},\cdots,\gamma^{(N_t)})\right\}\\
\cdot p(\gamma^{(1)})\cdots p(\gamma^{(N_t)})\cdot d\gamma^{(1)}\cdots d\gamma^{(N_t)}
\end{aligned}
\tag{2.45}
$$

where P_{CNR} is the joint probability that the CNR value for the ith transmission satisfies $\text{CNR} \geq \gamma_1^{(i)}, i=1,\cdots,N_t$. P_{CNR} can be computed as:

$$
\begin{aligned}
P_{\text{CNR}} = \sum_{n_1=1}^{N}\sum_{n_2=1}^{N}\cdots\sum_{N_t=1}^{N}\int_{\gamma_{n_1}^{(1)}}^{\gamma_{n_1+1}^{(1)}}\int_{\gamma_{n_2}^{(2)}}^{\gamma_{n_2+1}^{(2)}}\cdots\int_{\gamma_{n_{N_t}}^{(N_t)}}^{\gamma_{n_{N_t}+1}^{(N_t)}}\\
p(\gamma^{(1)})\cdots p(\gamma^{(N_t)})\cdot d\gamma^{(1)}\cdots d\gamma^{(N_t)}
\end{aligned}
\tag{2.46}
$$

To simplify computation, we employ $P\{F^{(N_t)}\}$ to approximate the joint probability $P\{F^{(1)}, F^{(2)},\cdots,F^{(N_t)}\}$ in equation (2.45). Here, to compute $P\{F_{n_1,\cdots,n_i}^{(i)}(\gamma^{(1)},\cdots,\gamma^{(i)})\}$, the following equations are used:

$$
\begin{aligned}
P\left\{F_{n_1,\cdots,n_i}^{(i)}(\gamma^{(1)},\cdots,\gamma^{(i)})\right\}\\
\approx 1 - \left\{1 - P_u^{(i)}{}_{n_1,\cdots,n_i}[\gamma^{(1)},\cdots,\gamma^{(i)}]\right\}^L
\end{aligned}
\tag{2.47}
$$

$$P_u^{(i)}{}_{n_1,\cdots,n_i}[\gamma^{(1)},\cdots,\gamma^{(i)}]$$

$$\approx a_{d_f^{(i)}}^{(i)} 2^{d_f^{(i)}} \left(\mathrm{BER}_{n_1,\cdots,n_i}^{(i)}(\gamma^{(1)},\cdots,\gamma^{(i)})\right)^{d_f^{(i)}/2} \tag{2.48}$$

$$\mathrm{BER}_{n_1,\cdots,n_i}^{(i)}(\gamma^{(1)},\cdots,\gamma^{(i)})$$

$$\approx \frac{\sum_{j=1}^{i} L_j \cdot \mathrm{BER}_{n_j}^{(j)}(\gamma^{(j)})}{\sum_{j=1}^{i} L_j} \tag{2.49}$$

$$\mathrm{BER}_{n_j}^{(j)}(\gamma^{(j)}) \approx a_{n_j} \exp\left(-b_{n_j}\gamma^{(j)}\right) \tag{2.50}$$

where $P_u^{(i)}{}_{n_1,\cdots,n_i}[\gamma^{(1)},\cdots,\gamma^{(i)}]$ is the union bound of the first-event error probability with code C_i decoding after i transmissions using AMC mode $n_1,\cdots,n_i$ under channel CNR values equal to $\gamma^{(1)},\cdots,\gamma^{(i)}$ respectively. $\mathrm{BER}_{n_1,\cdots,n_i}^{(i)}(\gamma^{(1)},\cdots,\gamma^{(i)})$ is the approximate total BER of the combined i received blocks which are input to code C_i for decoding. $\mathrm{BER}_{n_j}^{(j)}(\gamma^{(j)})$ is the BER of the jth received block using AMC mode n_j under channel CNR equal to $\gamma^{(j)}$.

(2) Spectral Efficiency: Spectral efficiency η is defined as the average number of accepted information bits per transmitted symbol under constraints C1 and C2, given by:

$$\eta = \frac{L}{\overline{L}} \tag{2.51}$$

where L is information packet size and $\overline{L}$ is the average number of transmitted symbols to transmit an information packet under C1 and C2. For each information packet, after the maximum N_t number of transmissions by using AMC mode $n_1,\cdots,n_{N_t}$ under channel CNR values equal to $\gamma^{(1)},\cdots,\gamma^{(N_t)}$, respectively, the average number of transmitted symbols is:

$$\overline{L}_{n_1,\cdots,n_{N_t}}(\gamma^{(1)},\cdots,\gamma^{(N_t)})$$

$$= \frac{L_1}{R_{n_1}} + \sum_{i=2}^{N_t} \frac{L_i}{R_{n_i}}$$

$$\cdot P\left\{F_{n_1}^{(1)}(\gamma^{(1)}),\cdots,F_{n_1,\cdots,n_{i-1}}^{(i-1)}(\gamma^{(1)},\cdots,\gamma^{(i-1)})\right\} \tag{2.52}$$

Like equation (2.45), through averaging equation (2.52) over all possible CNR values of $(\gamma^{(1)},\cdots,\gamma^{(N_t)})$, the average number of transmitted symbols $\overline{L}$ and η can be obtained.

2.5.5 Performance

We will now discuss the performance of cross-layer design in terms of PER and spectral efficiency. For the scheme combining AMC with type-II HARQ, the set of rates R_1, R_2, R_3 of RCC codes C_1, C_2, C_3 used in the performance evaluation are 1, 1/2 and 1/3, which are generated from a rate 1/3 code with memory $m = 4$. For code C_2, $a_d^{(2)} = 2$ and $d_f^{(2)} = 7$; For C_3, $a_d^{(2)} = 5$ and $d_f^{(2)} = 12$. Let delay constraint in C1 be $N_t = 3$ and performance constraint in C2 be $P_{\mathrm{loss}} = 10^{-4}$. Then, resulting CNR boundary values $\gamma_n^{(i)}$ are listed in Table 2.1.

For the scheme combining AMC with type-I HARQ, considering comparison fairness, the same constraints C1 and C2 are used, i.e. the delay constraint is $N_t = 3$ and the PER performance constraint is $P_{\mathrm{loss}} = 10^{-4}$. Compared with the cross-layer design of using AMC and type-II HARQ, using AMC and type-I HARQ has the following differences: type-I HARQ uses only one code with a fixed coding rate. For an information packet, both information bits and redundancy bits are transmitted at every transmission; there exists $\gamma_n^{(i)} = \gamma_n$, $i = 1, \cdots, N_t$, which means that for N_t transmissions, AMC mode n has the same CNR boundary; and there is no combining decoding at the receiver.

Figure 2.9 shows the average PERs for different cross-layer design schemes combining AMC with type-I or type-II HARQ. Type-I HARQ uses the coding rate 1/2 or 1/3. For all the schemes, information packet size are set to 1024 bits. From the figure, we can observe that all schemes satisfy the PER constraint: $P_{\mathrm{loss}} = 10^{-4}$. In a higher CNR region, the PER of all schemes are much below P_{loss}. In addition, the scheme with AMC and type-I HARQ has a much lower PER than the scheme with AMC and type-II HARQ. This is because redundancy bits are transmitted at every transmission in type-I HARQ.

Figure 2.10 shows the average spectral efficiency comparison between the cross-layer design schemes combining AMC with type-I or type-II HARQ. We can observe that in a

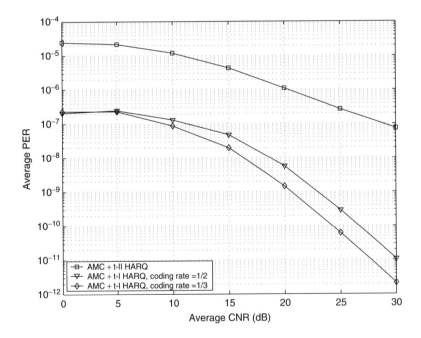

Figure 2.9 Average PER comparison of different cross-layer design schemes with two types of HARQ

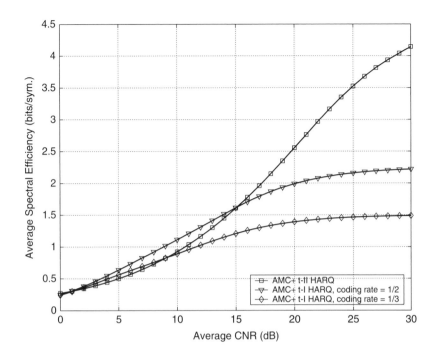

Figure 2.10   Average spectral efficiency for AMC using two types of HARQ repectively

high CNR region the average spectral efficiency of the scheme with type-II HARQ has shown much improvement over the one with type-I HARQ. The higher the average CNR is, the more improvement in spectral efficiency can be obtained by type-II HARQ. This is because unlike type-I HARQ, type-II HARQ transmits no or few unnecessary redundancy bits under good channel conditions. In low CNR region, the scheme with type-I HARQ has a better performance. This implies that sending some redundancy bits at the first transmission is good for spectral efficiency enhancement under poor channel conditions.

The spectral efficiency comparison of the cross-layer design using AMC and type-II HARQ with different packet size L is shown in Figure 2.11. The system with smallest packet size always has the highest spectral efficiency. However, using smaller packet size means more overhead of transmissions in communication systems. Therefore, the results in Figure 2.11 offer us valuable insight into choosing proper packet sizes in real communication system design. Note that in Figure 2.11, the spectral efficiency difference in moderate CNR regions is larger than that in lower and higher CNR regions. This means that the choice of packet size will greatly affect the spectral efficiency achieved by cross-layer design, especially when the channel is in moderate conditions.

2.6   Wireless Networking

The freedom gained from wireless enables tetherless computing. Wireless networks have been one of fastest growing segments of the telecommunications industry and have been

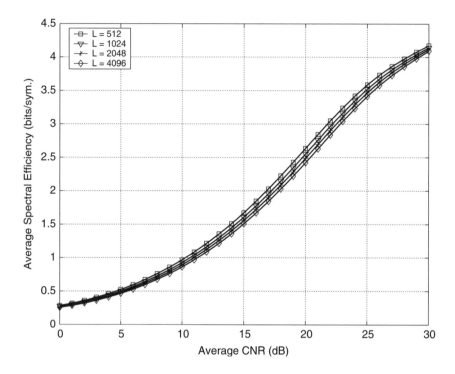

Figure 2.11 Average spectral efficiency for AMC combining with type-II HARQ with different packet sizes

widely used in many applications. For example, wireless networks are widely regarded as an effective and efficient solution to solve the so-called 'last-mile' network access problem, where each node can access the network services through its wireless network interface. Another example is that wireless networks can also be used as one of the backbone network technologies such as microwave backbone networks and wireless mesh networks. Furthermore, the wireless capability of a node implies that the node can roam around for supporting mobile computing applications. Therefore, mobility and location management is necessary. In this section we will review the layering network architectures and the network service models.

2.6.1 Layering Network Architectures

A computer network is defined as a collection of interconnected autonomous computers. However, designing a computer network with high performance is very challenging because a computer network is essentially a complex system – many subsystems interconnected to each other work synergetically. Therefore, modular design methodology has been adopted in the modern computer network design. Among various computer network architectures, the most influential conceptual model is the OSI (Open System Interconnection) 7-layer network model. In the OSI model, network functionality has been broken down into seven layers, and each layer provides a set of functions. Figure 2.12 shows

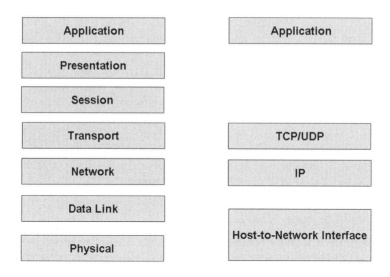

Figure 2.12 OSI model vs. TCP/IP model

the seven layers defined in the OSI model. The basic functions of each layer in the OSI model are summarized as follows:

- The application layer provides various services and applications to the end user such as WWW, remote login, email and so on. Note that an end user is usually an application process.
- The presentation layer provides the formatting and display of application data to the end user. For example, whether a web page encoded and displayed in English or French is decided in the presentation layer.
- The session layer sets up, maintains and terminates the sessions among application processes. An example of session layer function is Internet video conferencing.
- The transport layer provides a reliable bit-stream pipe between two application processes. Each application process can be identified by a socket number, which is composed of an IP address plus a port number assigned by the operating system of an end-user system. Therefore, the transport layer usually provides an ordering delivery service. Also, the transport layer has built-in mechanisms to deal with end-to-end congestion control.
- The network layer provides the routing function to the application process. In other words, it decides how to route a packet through a large-scale interconnected routers. In addition, the network layer needs to monitor the anomaly scenarios of a network.
- The data link layer divides a big chunk of data into frames, to handle point-to-point flow control and to provide error control. In the data link layer, there is an media access sub-layer, which provides multiple channel access methods, especially for shared-media networks such as Ethernet.
- The physical layer provides an adequate physical interface between the application process and the network and provides reliable information delivery schemes between two network nodes.

Although the OSI model demonstrates the most important network design methodology, it is mainly used for pedagogical purposes. The main reasons for its failure have been well summarized in [55], which are bad-timing, bad implementation and bad politics.

Meanwhile, the TCP/IP model has been gaining in popularity due to its efficient design and open source foundation. Actually, there was no TCP/IP model in mind when it was designed and developed, it was just an effective and efficient protocol stack running on top of the operating system. However, the negative side of this is that there are some deeply entrenched implementations based on engineering heuristics. Today, the TCP/IP protocol stack is the cornerstone of the Internet, and the IP protocol provides an effective way to support various data services over heterogeneous networks. A side-by-side comparison of the OSI model and the TCP/IP model is shown in Figure 2.12.

2.6.2 Network Service Models

In general, there are two major network service models: client-server and peer-to-peer. In the client-server model, each client process initiates service requests to the server process, and the server process provides the requested services to the client. Figure 2.13 illustrates the client-server model. Note that in this model the roles of client and server are clearly specified. Generally speaking, the server process needs to have more resources in order to handle multiple service requests, implying that the cost could be very high if it has to satisfy too many service requests. The client-server model has been widely adopted by many applications such as Web, FTP and email. Another major network service model is peer-to-peer model, as shown in Figure 2.14, where there is no clear-cutting role of server and client, meaning that each node could act as server or client. Essentially, the peer-to-peer model reflects a new design methodology: rather than building one powerful server, the workload can be distributed over many less powerful hosts, and each of them takes on a little workload at a time. This will significantly reduce the cost and time for otherwise slow computation and uneven load distribution. For example, the peer-to-peer model has been widely adopted in multimedia streaming applications such as PPlive and Bittorrent, where a big file can be truncated into many small data segments, allowing a user to download different segments from different hosts across a large-scale network. In this way, the downloading speed perceived by each user can be greatly improved without adding extra resources such as hardware and network infrastructure. However, resource allocation in the peer-to-peer model for Quality-of-Service (QoS) provisioning can be very complicated. Also, privacy and security are among the major technical challenges for the peer-to-peer networks.

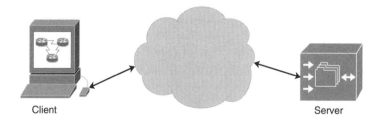

Figure 2.13   The client-server model

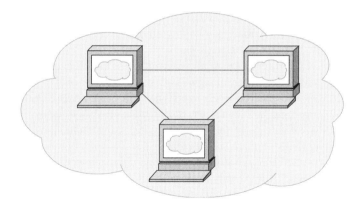

Figure 2.14   The peer-to-peer model

2.6.3   Multiplexing Methods

Multiplexing is one of the most important techniques in telecommunications networks, which is to combine and transmit multiple signals for higher utilization of precious network resources. Usually, the multiplexing procedure occurs at the transmitter, and the multiplexed signal will be demultiplexed at the receiver. Multiplexing and demultiplexing can be done at the transport layer, medica access control sub-layer and physical layer.

   In wireless networks, the major multiplexing methods are time division multiplexing (TDM), frequency division multiplexing (FDM) and code division multiplexing (CDM), as shown in Figure 2.15. In TDM, the resource access time is divided into many small

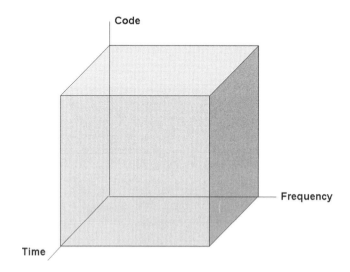

Figure 2.15   The major multiplexing methods

time slots, and each signal will be transmitted over one or multiple time slots. In TDM, each signal can use the entire frequency bandwidth but access to the bandwidth has to be sequential. Likewise, in FDM, all signals can access the channel simultaneously but each of them can only use a portion of the total bandwidth. For CDM, each signal is designated a code, and codes assigned to different signals are usually orthogonal or pseudo-random so as to minimize the interference between two signals. In CDM, each signal can access the total bandwidth at any time but at the cost of complicated system designs such as power control in CDMA networks. Note that spatial multiplexing techniques such as multiple input multiple output (MIMO) have seen a great deal of research in recent years. In MIMO, signals can be transmitted through different antenna elements to greatly enhance the link throughput. In this sense, MIMO can work with all of the aforementioned multiplexing methods for a much higher network capacity.

In mobile wireless networks, radio spectrum is the most precious resource, especially for the licensed radio spectrum. Furthermore, users are mobile, so providing good coverage for high-speed data services poses a big challenge for network design. Therefore, the cellular concept has been developed for network capacity enhancement, where a geographical area will be covered seamlessly by many small cells, usually in the shape of hexagons. Cell size depends on many factors such as the number of users, geographical terrain, bandwidth, transmission power and user applications. The frequency bandwidth efficiency of a cellular network can be evaluated by the frequency reuse factor K, which is the number of neighboring cells not being allowed to use the same frequency. If N directional antennae are used in each cell to further divide each cell into N sectors, the available bandwidth of each cell is NB/K, where B is the total bandwidth of a cellular network. Figure 2.16 shows the basic concept of cellular networks.

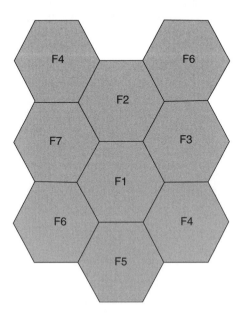

Figure 2.16   A cellular network with K = 7

2.6.4   Connection Management in IP-Based Data Networks

In mobile wireless networks, mobility management is one of the most challenging problems, which includes connection management and location management. Connection management can be realized by using various routing protocols.

Existing routing protocols such as IP protocol do not support host mobility. In the traditional Internet, in order to aggregate routing information and routing decisions at each level of the Internet topology, routing protocols use hierarchical addressing and routing schemes. For example, in the Internet, each IP address is interpreted in two parts: network number and host number. Routers throughout the Internet just need to send the packet to the right network, then that network will be responsible for sending the packet to the right host. Unfortunately, this hierarchy addressing method fails when the mobile host is moving from one network to another network. According to hierarchy addressing and routing, packets are only sent to a mobile host's home network, regardless of where the host is. One possible solution to this problem is to change the IP address of the mobile host to an address of its visiting network, but doing so will trigger a number of configuration files on both the host and the related network routers, which often requires all existing upper layer connections to be restarted and the host to be rebooted.

Mobile IP has been proposed for supporting host mobility without changing its IP address. Actually, this method is very similar with the roaming registration procedure specified in GSM cellular systems. In mobile IP, each mobile host must have a home agent on its home network, which forwards IP packets to the mobile host while it is away from its home network. When visiting another network, each mobile host must acquire a care-of address, which is assigned by a foreign agent. The mobile host needs to register with its home agent in order to let its home agent know about the current care-of address. Furthermore, it has to register with the foreign agent in the visiting network by using the agent discovery protocol. In this way, the mobile host can obtain the mobile IP services provided by the foreign agent.

Usually, a pairing of home address and care-of address is called mobility binding. When sending a packet to a mobile host, it is not necessary for the sender to know the mobile IP of the mobile host. It just sends packets to the mobile host's home address, then the home agent will forward packets to its foreign agent by encapsulating each IP packet with the mobile host's care-of address. After the mobile host receives these encapsulated packets, it strips off the care-of address and gets the right IP packets. The whole procedure described above is called 'IP-in-IP-encapsulation', a.k.a., IP tunneling. Through the IP tunneling technique, host mobility can be supported successfully in current IP-based mobile computing environment. But there are still some technical challenges, such as the triangle routing problem, which need to be studied further.

Moreover, QoS routing has been developed in order to maintain data integrity in terms of packet loss avoidance and delay-bounded delivery. Also, the commitment to QoS should be maintained regardless of host mobility. Resource reservations by using QoS scheduling and buffer management schemes are believed to be effective in providing quick handoff procedure and re-routing, while maintaining the commitment to QoS. Usually, resource reservation schemes are closely coupled with routing because most of them need to know which route is taken by the data flow. However, doing so might introduce a great deal of overhead in practice, especially in mobile wireless networks.

(1) Location management: Location management is one of the key components of mobility management, which deals with how to track a mobile host in mobile wireless networks [56–59]. Host locations are derived by using two basic operations: location update and location paging. Location update, also called registration, is the process by which the network tracks the location of a mobile host not engaged in conversation. A mobile host dynamically updates its location information, where a location area may include one or more cells. Furthermore, when packets arrive, the network will send out a polling signal to a location area for searching for the addressed mobile host. This process is called paging.

Static location management would incur a great deal of overhead in terms of bandwidth and power, since the paging area is fixed and equal to a local area, every time all of the cells of a local area are paged. Moreover, whenever a mobile host moving into a new location area, the location update procedure will be triggered, which is not efficient, especially when a mobile host moves back and forth between two local areas.

Dynamic location management has been proposed in order to solve this problem. In dynamic location management, the size of a local area keeps changing with the change of user mobility and service. So far, there are three types of dynamic location management strategies: time-based update, movement-based update and distance-based update [56]. Their performance varies under different movement patterns. Typically, there are two movement patterns used in analyzing the performance of dynamic location management strategies, one is i.i.d. (independent and identically distributed) movement, and the other is Markovian movement. In the i.i.d. model, during a time slot, the probability that a user is in cell i at the beginning of the slot is independent of the probability that it moves to cell $i + 1$, moves to cell $i - 1$, or remains in cell i. On the other hand, in the Markovian model, during a time slot, user movement during the current slot depends on its state (stationary, right-move state, or left-move state) in the previous slots. Let $X(t)$ be the state during slot t, then $X(t), t = 0, 1, 2 \ldots$ is a Markov chain. The assumptions here are: (1) time is slotted; (2) a user can move into at most one cell during each time slot; (3) location update and paging are done at the beginning of each slot; and (4) paging can be finished within one slot. In practice, these assumptions can be relaxed easily. Readers may refer to [56] for more information about dynamic location management strategies under the i.i.d. model.

2.6.5   QoS Handoff

4G wireless networks should support broadband data services to mobile users while still retaining the commitment to QoS. Unlike its wired counterpart, where relatively there is plenty of bandwidth and reliable channels are available, mobile wireless data networks must face problems such as limited bandwidth, error-prone and time-varying channel and mobility. In general, handoff happens at the cell border, where link quality is significantly less reliable. Besides, link throughput is also crucial in order to lower the handoff latency. Although link adaptation can improve wireless link throughput, handoff re-routing schemes have to be chosen for achieving optimal routes during handoff.

Research on handoff re-routing for connection-oriented wireless networks has originated from two different goals. The first goal is to ensure to establish an optimal route when the mobile host moves to a new location. Although an optimal route can be selected eventually, the handoff latency may be increased significantly owing to routing table calculation and connection establishment. Another goal is to achieve the shortest handoff latency by keeping the original route, and then extending this route hop-by-hop as the mobile host is moving. However, in this way network resources will be wasted by many redundant hops. So far, there have four types of connection-oriented handoff re-routing schemes, which are anchored handoff re-route, multicast handoff re-route, chained handoff re-route and classified handoff re-route.

In anchored handoff re-route [60], an anchor switch or router is designated for each mobile host in the original connection route. During the lifetime of a connection, the path that connects the anchor switches (anchor segments) will never change. Then, when one of the mobile host moves to a new location, only part of the path which connects the mobile host with its own anchor switch needs to be re-routed. This scheme avoids establishing a whole new route for each handoff. However, anchor segments may cause longer routes, looped routes, or higher complexity and packet losses.

In multicasting handoff re-route [61], incoming traffic for a mobile host is multicasted to several surrounding base stations, which will buffer all packets for the mobile host until it enters into a new location. This scheme can significantly reduce handoff latency. But it expends much multicasting bandwidth for each user and requires excessive buffer usage in those locations where the mobile host is not entering.

In chained handoff re-route [62], a temporary route is established by simply extending the original route to the current location. This temporary route will be expired until an optimal route is found. This scheme requires several intermediate routes that consume bandwidth for each movement of the mobile host.

In classified handoff re-route [63], the source switch or router determines the new route to the moving endpoints of an end-to-end connection, regardless of which endpoint has moved. Thus, the number of intermediate routes is reduced by centralizing the establishment of the new routes and preventing the need for establishing and terminating routes while an endpoint is still moving. In this scheme, each endpoint switch or router will divide the connections into two groups: delay sensitive and loss sensitive. Then, when one (or both) of the endpoints of a connection moves, the endpoint switches will be able to re-route the connection based on the traffic type by using different means of re-route. For example, for delay sensitive connections, handoff latency is critical, so chained handoff re-routing will be used; for loss sensitive connections, multicast re-routing can be adopted in order to prevent data losses until an optimal route is calculated.

2.7 Summary

In this chapter, the fundamentals of wireless communications and networking have been reviewed. We started with the characteristics and modeling of wireless channels, and major slowing fading channels were introduced. We then focused on the popular link adaptation techniques such as adaptive modulation and coding and HARQ as well as cross-layer design of link adaptation, which have been widely adopted in wireless communications. A comprehensive review has been provided. Since 4G wireless networks are all IP networks, we discussed the basics of computer networks. Two network design architectures, the OSI model and the TCP/IP model, were covered. Because multiplexing is an important technique for enhancing wireless network capacity, the current major multiplexing methods were reviewed. In wireless networks using licensed frequency bands, the concept of cellular networks has been widely adopted in order to improve frequency utilization. As a result, mobility management and handoff are very important in order to provide QoS to mobile users in cellular networks. The performance of major mobility and location management schemes was also discussed.

References

1. B. S. Publishers, *Wireless Communication*. http://www.baltzer.nl/wicom/, 1999.
2. T. S. Rappaport, *Wireless Communications: Principles and Practice*. Prentice Hall, Inc., 1996.
3. W. Feller, *An Introduction to probability Theory and Its Applications*. John Wiley & Sons, Inc., 1968.
4. H. Stark and J. Woods, *Probability, Random Processes and Estimation Theory for Engineers*. Prentice Hall, Inc., 1994.
5. J. G. Proakis, *Digital Communications*. New York: McGraw-Hill, 1995.
6. W. Jakes, *Microwave Mobile Communications*. John Wiley & Sons, Inc., 1974.
7. E. Gilbert, "Capacity of a Burst-Noise Channel," *The Bell System Technical Journal*, Vol. 39, pp. 1253–63, 1963.
8. E. Elliott, "Estimates of Error Rates for Codes on Burst-Noise Channels," *The Bell System Technical Journal*, Vol. 42, pp. 1977–97, 1963.
9. S. Wilson, *Digital Modulation and Coding*. Prentice Hall, Inc., 1996.
10. F. swarts and H. Ferreira, "Markov Characterization of Digital Fading Mobile VHF Channels," *IEEE Transactions on Vehicular Technology*, Vol. 43, No. 4, pp. 977–85, 1994.
11. J. Garcia-Frias and P. Crespo, "Hidden Markov Models for Burst Error Characterization in Indoor Radio Channels," *IEEE Transactions on Vehicular Technology*, Vol. 46, No. 4, pp. 1106–20, 1997.
12. H. Wang and N. Moayeri, "Finite-Stae Markov Channel-A Useful Model for Radio Communication Channels," *IEEE Transactions on Vehicular Technology*, Vol. 44, No. 1, pp. 163–71, 1995.
13. H. Wang and P. Chang, "On Verifying the First-Order Markovian Assumption for a Rayleigh Fading Channel Model," *IEEE Transactions on Vehicular Technology*, Vol. 44, No. 2, pp. 353–57, 1996.
14. B. Vucetic, "An Adaptive Coding Scheme for Time-Varying Channels," *IEEE Transactions*, Vol. 30, No. 5, pp. 653–63, 1991.
15. M.-S. Alouini, X. Tang, and A. Goldsmith, "An Adaptive Modulation Scheme for Simultaneous Voice and Data Transmission over Fading Channels," *IEEE Journal on Selected Areas in Communications*, Vol. 17, No. 5, pp. 837–50, 1999.
16. J. Cavers, "Variable-Rate Transmission for Rayleigh Fading Channels," *IEEE Transactions on Communications*, Vol. COM-20, No. 1, pp. 15–22, 1972.
17. W. Webb and R. Steele, "Variable Rate QAM for Mobile Radio," *IEEE Transactions on Communications*, Vol. 43, No. 7, pp. 2223–30, 1995.
18. J. Bingham, "Multicarrier Modulation for Data Transmission: An Idea Whose Time Has Come," *IEEE Communications Magzine*, Vol. 28, No. 5, pp. 5–14, 1990.
19. T. Keller and L. Hanzo, "Adaptive multicarrier modulation: a convenient framework for time-frequency processing in wireless communications," *Proceedings of the IEEE*, Vol. 88, No. 5, pp. 611–40, 2000.
20. S. Alamouti and S. Kallel, "Adaptive Trellis-Coded Multiple-Phase-Shift Keying for Rayleigh Fading Channels," *IEEE Transactions on Communications*, Vol. 42, No. 6, pp. 2305–14, 1994.
21. A. Goldsmith and S. Chua, "Variable-Rate Variable-Power MQAM for Fading Channels," *IEEE Transactions on Communications*, Vol. 45, No. 10, pp. 1218–30, 1997.
22. X. Qiu and K. Chawla, "On the Performance of Adaptive Modulation in Cellular Systems," *IEEE Transactions on Communications*, Vol. 47, No. 6, pp. 884–95, 1999.
23. K. Balachandran, S. Kadaba, and S. Nanda, "Channel quality estimation and rate adaptation for cellular mobile radio," *IEEE Journal on Selected Areas in Communications*, Vol. 17, No. 7, pp. 1244–56, 1999.
24. A. Goldsmith and C. S. G., "Adaptive Coded Modulation for Fading Channels," *IEEE Transactions on Communications*, Vol. 46, No. 5, pp. 595–602, 1998.

25. D. Goeckel, "Adaptive Coding for Time-Varying Channels Using Outdated Fading Estimates," *IEEE Transactions on Communications*, Vol. 47, No. 6, pp. 844–55, 1999.

26. F. Babich, "Considerations on adaptive techniques for time-division multiplexing radio systems," *IEEE Transactions on Vehicular Technology*, Vol. 48, No. 6, pp. 1862–73, 1999.

27. C. Wong and L. Hanzo, "Upper-bound performance of a wide-band adaptive modem," *IEEE Transactions on Communications*, Vol. 48, No. 3, pp. 367–9, 2000.

28. C. C. et al., "Adaptive Radio for Multimedia Wireless Links," *IEEE Journal on Selected Areas in Communications*, Vol. 17, No. 5, pp. 793–813, 1999.

29. V. Lau and S. Maric, "Variable rate adaptive modulation for DS-CDMA," *IEEE Transactions on Communications*, Vol. 47, No. 4, pp. 577–89, 1999.

30. M. Pursley and J. Shea, "Adaptive nonuniform phase-shift-key modulation for multimedia traffic in wireless networks," *IEEE Journal on Selected Areas in Communications*, Vol. 18, No. 8, pp. 1394–407, 2000.

31. J. Torrance and L. Hanzo, "Latency and networking aspects of adaptive modems over slow indoors rayleigh fading channels," *IEEE Transactions on Vehicular Technology*, Vol. 48, No. 4, pp. 1237–51, 1999.

32. J. Torrance, L. Hanzo, and T. Keller, "Interference aspects of adaptive modems over slow rayleigh fading channels," *IEEE Transactions on Vehicular Technology*, Vol. 48, No. 5, pp. 1527–45, 1999.

33. L. Chang, "Throughput Estimation of ARQ Protocols for a Rayleigh Fading Channel Using Fade- and Interfade-Duration Statistics," *IEEE Transactions on Vehicular Technology*, Vol. 40, No. 1, pp. 223–29, 1991.

34. A. Goldsmith, Wireless Communications. Cambridge University Press, 2005.

35. G. J. Foschini, "Layered space-time architecture for wireless communication in fading environments when using multi-element antennas," *Bell System Tech. J.*, pp. 41–59, 1996.

36. P. Wolniansky, G. Foschini, G. Golden, and R. Valenzuela, "V-blast: an architecture for realizing very high data rates over the rich-scattering wireless channel," in *Proc. URSI Intl. Symp. Sign. Syst. Electr.*, Oct. 1998, pp. 295–300.

37. S. Alamouti, "A simple transmit diversity technique for wireless communications," *IEEE J. Sel. Areas Commun.*, pp. 1451–1458, Oct. 1998.

38. A. Paulraj, R. Nabar, and D. Gore, *Introduction to Space-Time Wireless Communications*. Cambridge University Press, 2003.

39. L. Zheng and D. N. Tse, "Communication on the grassmann manifold: A geometric approach to the noncoherent multi-antenna channel," *IEEE Trans. Inf. Theory*, Vol. 48, pp. 359–383, Feb. 2002.

40. E. Biglieri, R. Calderbank, A. Constantinides, A. Goldsmith, A. Paulraj, and H. V. Poor, *MIMO Wireless Communications*. Cambridge University Press, 2007.

41. V. Tarokh, N. Seshadri, and A. Calderbank, "Space-time codes for high data rate wireless communications: performance criterion and code construction," *IEEE Trans. Inf. Theory*, Vol. 44, No. 2, pp. 744–765, Mar. 1998.

42. V. Tarokh, H. Jafarkhani, and A. Calderbank, "Space-time block codes from orthogonal designs," *IEEE Trans. Inf. Theory*, Vol. 45, pp. 1456–1467, Jul. 1999.

43. (2004) 3GPP TR 25.848 V4.0.0, Physical Layer Aspects of UTRA High Speed Downlink Packet Access (release 4).

44. (1999) 3GPP2C.S0002-0 Version 1.0, Physical Layer Standard for cdma2000 Spread Spectrum Systems.

45. A. Doufexi, S. Armour, M. Butler, A. Nix, D. Bull, J. McGeehan, and P. Karlsson, "A Comparison of the HIPERLAN/2 and IEEE 802.11a Wireless LAN Standards," *IEEE Communication Magazine*, Vol. 40, pp. 172–180, May 2002.

46. (2002) IEEE Standard 802.16 Working Group, IEEE Standard for Local and Metropolitan Area Networks Part 16: Air Interface for Fixed Broadband Wireless Access Systems.

47. S. Lin and D. Costello, *Error Control Coding: Fundamentals and Applications*. Englewood Cliffs, NJ: Prentice-Hall, 1983.

48. S. Lin, D. Costello, and M. Miller, "Automatic-repeat-request Error-control Schemes," *IEEE Communication Magazine*, Vol. 22, pp. 5–17, Dec. 1984.

49. J. hagenauer, "Rate-compatible punctured convolutional codes (rcpc codes) and their applications," *IEEE Trans. Commun.*, Vol. 36, pp. 389–400, Apr. 1988.

50. Q. Zhang and S. A. Kassam, "Hybrid arq with selective combining for fading channels," *IEEE J. Sel. Areas Commun.*, Vol. 17, pp. 867–880, May 1999.

51. M. B. Pursley and D. J. Taipale, "Error probabilities for spread-spectrum packet radio with convolutional codes and viterbi decoding," *IEEE Trans. Commun.*, Vol. COM-35, pp. 1–12, Jan. 1987.

52. Q. Liu, S. Zhou, and G. B. Giannakis, "Cross-layer combining of adaptive modulation and coding with truncated arq over wireless links," *IEEE Transactions on Wireless Communications*, Vol. 3, pp. 1746–1755, Sep. 2004.

53. M. Alouini and A. J. Goldsmith, "Adaptive modulation over nakagami fading channels," *Kluwer J. Wireless Communications*, Vol. 13, pp. 119–143, May 2000.

54. A. J. Viterbi, "Convolutional codes and their performance in communication systems," *IEEE Trans. Commun.*, Vol. COM-19, pp. 751–772, Oct. 1971.

55. A. S. Tanenbaum, *Computer Networks*. Upper Saddle River, NJ: Prentice Hall, 2003.

56. Bar-Noy, A., Kessler, I., and Sidi, M., "Mobile Users: To Update or not to Update?" *Proceeding of IEEE INFOCOM'94*, Vol. 2, 1994.

57. Liu, T. et al., "Mobility Modeling, Location Tracking, and Trajectory Prediction in Wireless ATM Network," *IEEE JSAC*, Vol. 6, No. 6, 1998.

58. Rose, C. and Yates, R., "Location Uncertainty in Mobile Networks: a theretical framework," *Technical Report*, 1996.

59. SAC 2001, "Analysis of Dynamic Location Management for PCS Networks," *ACM SAC2001*, 2000.

60. Veeraraghavan, M. et al., "Mobility and Connection Management in a Wireless ATM LAN," *IEEE JSAC*, Vol. 15, No. 1, 1997.

61. Ghai, R. and Singh, S., "An Architecture and Communication Protocol for Picocellular Networks," *IEEE Personal Communications Magazine*, Vol. 1, 1994.

62. Rajagopalan, B, "Mobility Management in Integrated Wireless-ATM Networks," *ACM-Baltzer Journal of Mobile Networks and Applications (MONET)*, Vol. 1, No. 3, 1996.

63. McNair, J. et al., "Handoff Rerouting Scheme for Multimedia Connections in ATM-based Mobile Networks," *Proceeding of IEEE Fall VTC'00*, 2000.

3

Video Coding
and Communications

3.1 Digital Video Compression – Why and How Much?

Uncompressed standard definition (SD) video, captured by following the ITU-R 601 4:2:2 standard [1] with 10 bits per pixel, consists of a 270 Mbits/sec bit rate. The typical channel capacity that is available and is used for the transmission of a TV channel is less than 6 Mbits/sec and averages around 3 Mbits/sec. There is a strong push to further reduce that capacity due to economic reasons. This implies that one is required to remove more than 98% of the original information and compress the digital video to less than 2% of the initial bit rate. It requires more than $50 \times$ compression factor. It is not possible to achieve that amount of compression without loss of information. The loss of information adds distortion to the decoded video which appears as a degradation of visual quality in the displayed video. Therefore, one of the key goals of a compression algorithm is, at a given bit rate, to add as little distortion in the decoded video as possible and to lose the information so that the degradation is as invisible as possible. To minimize the perceptual distortion some processing steps are taken before encoding the video. The typical pre-compression steps taken for consumer applications are to compress only the visible part of the video, reduce the number of bits per pixel from 10 to 8 bits and reduce the color resolution before compressing the video.

The visible part of a frame of video contains between 704 and 720 pixels horizontally and 483 lines vertically in the 525-line system used in North America (the other commonly used system outside North America is the 625-line system with 704 to 720 pixels horizontally and 576 active lines vertically). Owing to the inherent block processing nature of the most commonly used compression schemes, the number of pixels used to represent the visible part of the video needs to be a multiple of 16 and is either 704 or 720. Vertically, the number of lines is required to be a multiple of 16 for progressively scanned video and a multiple of 32 for interlaced scanned video. Therefore, 480 lines are commonly used to represent the visible portion of an SD video in the 525-line system. Thus a video picture at SD consists of maximum of $720 \times 480 = 345\ 600$ pixels. Assigning

8 bits for the luminance of each of the pixels provides $720 \times 480 \times 8 = 2\ 764\ 800$ bits for the luminance part of a frame of a SD video. In the ITU-R 601 4:2:2 standard the color resolution is reduced horizontally by a factor of 2. The color information is represented by two difference signals (R-Y, B-Y) and is assigned to only every other pixel horizontally. Therefore, an SD video frame contains $360 \times 480 \times 8 \times 2 = 2\ 764\ 800$ bits for the chrominance part. Counting both luminance and chrominance parts, it requires $2 \times 2\ 764\ 800 = 5\ 529\ 600$ bits to represent a SD video frame. In North America, standard definition video is sent at a rate of 30 (or 29.97) frames per sec (the 625-line system uses a rate of 25 frames/sec). Therefore, the bandwidth required for the visible part of the video following the above mentioned format is $5\ 529\ 600 \times 30$ bits/sec which is approximately 165 Mbits/sec. For consumer applications, one more compromise is made in the representation of a video frame – color information is decimated by a factor of 2 vertically, i.e. color information is kept only for every other pixel (line) vertically. This is also known as 4:2:0 chroma format. In this format, uncompressed SD video requires 1.5 $\times 2\ 764\ 800 \times 30$ which is about 124 Mbits/sec. This is a very large number in comparison to the bandwidth available for the transmission of digital television. Similarly, the storing of uncompressed video also requires an uneconomically large storage capacity. As an example, a 2 hour long movie at a rate of 124 Mbit/sec requires more than 110 GBytes of storage capacity. It is much larger than what is available and practical today. Hence there has been a strong desire to improve video coding efficiency.

In High Definition TV (HDTV), two video formats are commonly used in North America – 1920×1080 size frames at 30 frames/sec and 1280×720 size frames at 60 frames/sec. In uncompressed 4:2:0 formats, they require about 750 Mbits/sec and 660 Mbits/sec respectively. It will require about 675 GBytes to store one 2+ hrs movie. Recently, many users have expressed an interest in using 1920×1080 size frames at a rate of 60 frames/sec. In the uncompressed 4:2:0 format, this requires around 1.5 Gbits/sec of bandwidth or a storage capacity of more than 1.2 Terra Bytes for a 2+ hrs long movie! In addition, with the availability of high quality large TV displays 10 bits and/or 4:4:4 format and/or higher than 1920×1080 resolution of video may also be used in digital TV in the future. Thus, the introduction of HDTV, and the resolution higher than what is used in HDTV today, puts further pressure on channel and storage capacity and emphasizes the need for having a compression algorithm with very high coding efficiency.

3.2   Basics

Now that we have described the need for video compression, we will discuss the basics of video representation, including scanning and color representation.

3.2.1   Video Formats

3.2.1.1   Scanning

Video pictures consist of a certain number of scanned lines. Each line is scanned from left to right and various standards specify different numbers of scanned lines in the vertical dimension. In the 525 line standard for SDTV, commonly used in North America, there are total of 525 lines of which the active visual area consists of 483 lines. Another commonly used standard consists of a total of 625 lines out of which the active visual area consists of 576 lines. When these standards were developed, it was very expensive to develop the devices with that many lines at the specified frame rates – 30 (or 29.97) frames/sec for 525 line systems and 25 frames/sec for 625 line systems. Therefore, in order to reduce the bandwidth required to capture and display a frame, it was scanned in an interlaced format where a frame consists of two fields. In an interlaced format, each field consists of half the number of lines – one with odd numbered lines (line numbers 1, 3, 5 and so on) and another with even numbered lines (line numbers 2, 4, 6 and so on), as shown in Figure 3.1. In this figure the solid lines correspond to field 1 and the dotted lines correspond to field 2.

Field 1 is also called the top field and field 2 the bottom field. It should also be noted that the top field (field 1) is not always displayed first. The syntaxes of the compression standards allow one to specify which field needs to be displayed first.

In HDTV there are two most commonly used formats: one with 1080 lines and other one with 720 lines in the visual area of a frame. In the 1080 line format frames are captured at 24 frames/sec or 30 frames/sec. The 24 frames/sec rate is used for movie material. In the 30 frames/sec format, each frame is scanned in the interlaced format consisting of two fields (top and bottom) at a rate of 60 fields/sec. In the 720 line format the frames are captured with progressive scanning at 60 frames/sec rate. The 1080 line format consists of 1920 pixels horizontally in each line and the 720 line format consists of 1280 pixels horizontally. These numbers are such that the picture aspect ratio (1920:1080 and 1280:720) is 16:9 and the pixel rate is approximately similar – 62 208 000 pixels/sec (1920 × 1080 × 30) and 55 296 000 pixels/sec (1280 × 720 × 60).

There has been a heated discussion in the technical community as to which scanning format – interlaced or progressive – is better. They both have their plusses and minuses. Without going into that argument, the AVC/H.264 standard provides tools for both, interlaced and progressive, formats. However, with the advancement in display and video capturing technology, the progressive format is expected to dominate and the interlaced format is expected to fade away in the future.

3.2.1.2 Color

Color TV displays and many video capture devices are based on representing pixels in a video frame in terms of three primary colors Red (R), Green (G) and Blue (B) [2, 3]. However, before video compression, RGB values are transformed into Y, Cr, Cb where

Figure 3.1 Interlaced scanned frame

Y, also known as luminance or luma, corresponds to luminance values, Cr and Cb, also known as chrominance or chroma, correspond to the two color difference signals R-Y and B-Y. When there are chroma values present for each pixel, it is called 4:4:4 sampling. The roots of this nomenclature lie in the analog video domain. When analog composite video was sampled, it was sampled at four times the frequency of the color sub-carrier ($4 \times f_{sc}$). There were two dominant standards for analog video: NTSC (National Television System Committee) and PAL (Phase Alternating Line). The sub-carriers for the NTSC and PAL systems were at different frequencies. Therefore, as a compromise, ITU-R (known as CCIR before changing its name to ITU-R) developed a universal international standard to sample the luminance part of analog video at 13.5 Msamples/sec rate [1]. The Human Visual System is relatively less sensitive to chroma resolution than to luma resolution. Therefore, many applications do not use the same chroma resolution as luma resolution. In some systems it was sampled at 1/4th the rate of Luma sampling rate. To signify that luma and chroma are sampled at different rate, the format with the chroma sampling rate of 3.375 Msamples/sec was called 4:1:1 (luma is sampled at four times the rate of chroma). In the ITU-R 601 4:2:2 format, chroma is sampled at half the rate of luma. The nomenclature 4:2:2 signified that chroma was sampled at half the rate ($13.5/2 = 6.75$ Msamples/sec). In the early days the number 4 was closely tied to the 13.5 Msamples/sec rate. So, in the systems where the luma was sampled at twice the rate, it was denoted as 8:4:4 or 8:8:8 depending upon the relative sampling rates of luma and chroma. However, as High Definition Tele Vision (HDTV) became real the same nomenclature 4:4:4 and 4:2:2 was used even though the sampling rate of each luma and chroma component is much higher than for SDTV. So, currently 4:2:2 merely signifies that chroma sampling rate is half of the luma sampling rate.

In the early digital video applications, chroma was sub-sampled only horizontally. Vertically it had the same number of samples in a picture as luma samples. However, in digital video compression applications, in addition to horizontal decimation, the chrominance is also decimated vertically by a factor of 2 before encoding. This format is called 4:2:0 where last digit 0 does not have any real significance. Conversion from 4:4:4 to 4:2:0 is done as a pre-processing step to the digital video encoding. As this step also reduces the number of bits representing digital video, it can also be seen as the first step of digital video compression.

Figure 3.2(a) illustrates the 4:4:4 sampling structure. Here X marks represent the locations of luma samples and circles represent the locations of chroma samples. Figure 3.2(b) shows the 4:2:2 sampling structure. In this format, chroma samples are present with every other luma sample horizontally. Figure 3.2(c) depicts the 4:2:0 sampling structure used in MPEG-2 [4] and AVC/H.264 [5] and Figure 3.2(d) shows the 4:2:0 sampling structure used in MPEG-1 [6]. In MPEG-1 the chroma sample is centered horizontally as well as vertically between the luma samples. Note that, as shown in Figure 3.3, for interlaced video, only one out of four frame-lines in a field has chroma information in 4:2:0 sampling format. This causes color smearing. It is generally not objectionable in consumer applications but becomes objectionable in studio and video production applications. For this reason, 4:2:2 is more popular in a studio and video production environment. In progressively scanned video 4:2:0 sampling does not introduce significant color spreading and provides symmetrical horizontal and vertical chroma resolution. However, 4:2:2 remains a very popular digital video capturing and storing standard for a progressively scanned

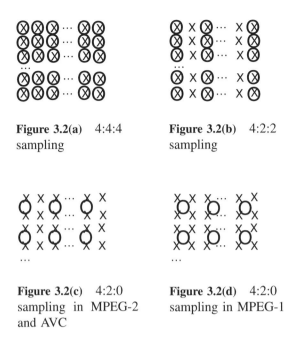

Figure 3.2(a)   4:4:4
sampling

Figure 3.2(b)   4:2:2
sampling

Figure 3.2(c)   4:2:0
sampling in MPEG-2
and AVC

Figure 3.2(d)   4:2:0
sampling in MPEG-1

video also. Therefore, conversion to 4:2:0 is required before encoding the video. This introduces some artifacts, especially if it is done on multiple occasions.

The AVC/H.264 standard is defined for all three 4:4:4, 4:2:2 and 4:2:0 color formats. As 4:2:2 is a very popular video capturing format, a natural question arises – should one apply the digital video encoding algorithm directly in the 4:2:2 domain as opposed to first going down to the 4:2:0 domain. Both digital video encoding and chroma sub sampling introduce artifacts but they look visually different. Therefore, the answer depends on the type of sequences, the bit rates, the compression standard and the pre-processing technology used to do the conversion. With current pre-processing and video coding technology typically used in the products, generally, at higher bit rates (for example, above 12 Mbits/sec for MPEG-2) one obtains better video quality by staying in the 4:2:2 domain and at lower bit rates (such as below 6 Mbits/sec for MPEG-2) one obtains less objectionable artifacts by first going from 4:2:2 to 4:2:0 before encoding.

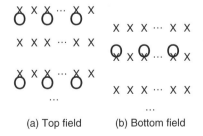

(a) Top field   (b) Bottom field

Figure 3.3   Chroma locations in 4:2:0 format for Interlace scanned video

3.2.1.3 Luminance, Luma, Chrominance, Chroma

The Cathode Ray Tube (CRT) of television does not have linear electrical to optical transfer characteristics. Generally, the optical output is proportional to the input voltage raised to some power. That power is represented by the Greek character gamma (γ) [7]. The value of γ depends on the display type. For CRTs it is around 2.2. Therefore, in order to compensate for that non-linear behavior, cameras go through a gamma correction process where in the optical to electrical conversion step the inverse of gamma is used. To distinguish between the pre-gamma corrected (linear) stage and the post gamma corrected stage, the brightness and color difference signals before correction are called luminance and chrominance signals respectively and after the gamma correction those signals are called luma and chroma signals respectively. As all the video compression and processing in an encoder and decoder happens in the gamma corrected domain, the recognition of that difference is acknowledged in the AVC/H.264 video coding standard (see Chapter 5) and the terms Luma and Chroma are used.

3.3 Information Theory

So far we have discussed the basics of video representation. Before continuing with video compression, we will review some of the basic aspects of information theory. Information theory is the fundamental mathematical theory that governs compression and sets its theoretical limits. More information can be found in [8–10]. Let $\mathbf{F} = \{F_n\}$ represent an *information source* consisting of a number of samples. Each sample F_n of the source is a random variable. In this discussion, we are considering *discrete sources*. Thus, F_n is a discrete random variable and can only take values from a discrete set (alphabet) of symbols $\mathbf{A} = \{a_1, a_2, \ldots, a_L\}$. In the context of video coding, a discrete source may be, for example, the output of a quantizer. A discrete source can alternatively be seen as a discrete-state random process [11]. In this chapter, we only consider *stationary* sources (processes). A source is stationary if the distribution of F_n does not depend on the index n, and the joint distribution of a group of N samples is invariant with respect to a common shift in the index. Let $p_{F_n}(f)$ be the *probability mass function* (pmf) of discrete random variable (sample of the source) F_n. Also, let $p_{F_{n+1}, F_{n+2}, \ldots, F_{n+N}}(f_1, f_2, \ldots, f_N)$ be the joint pmf of N consecutive samples of the source $\mathbf{F}$. Furthermore, let $p_{F_n|F_{n-1}, F_{n-2}, \ldots, F_{n-M}}(f_{M+1}|f_M, f_{M-1}, \ldots, f_1)$ be the conditional pmf of sample F_M given the previous M samples. For simplicity, we will use interchangeably the notations $p(f)$, $p(f_1, f_2, \ldots, f_N)$ and $p(f_{M+1}|f_M, f_{M-1}, \ldots, f_1)$ to denote the preceding functions when there is no risk of confusion.

A stationary source is called *independent identically distributed (i.i.d.)* if:

$$p(f_1, f_2, \ldots, f_N) = p(f_1)p(f_2) \cdots p(f_N) \tag{3.1}$$

or equivalently,

$$p(f_{M+1}|f_M, f_{M-1}, \ldots, f_1) = p(f_{M+1}) \tag{3.2}$$

3.3.1 Entropy and Mutual Information

The entropy of a discrete random variable F with alphabet $\mathbf{A}$ is defined as

$$H(F) = -\sum_{f \in \mathbf{A}} p_F(f) \log_2 p_F(f) \tag{3.3}$$

Since a base-2 logarithm is used in the above definition, the unit for entropy is *bits/ sample*.

Entropy is a measure of the uncertainty about a random variable F. Let us consider the example of a random variable F that corresponds to a coin toss and takes on values $\mathbf{A} = \{\text{heads, tails}\}$. Let p be the probability of 'heads'. Then, $1 - p$ is the probability of 'tails'. The entropy of such a random variable is:

$$H(F) = -p \log_2 p - (1 - p) \log_2 (1 - p) \tag{3.4}$$

Let us assume that the coin is not fair and the outcome is always 'heads'. Then, $p = 1$. Thus, $H(F) = -1 \log_2 1 - 0 \log_2 0$. Assuming that $0 \log_2 0 = \lim_{q \to 0} q \log_2 q = 0$, it is clear that $H(F) = 0$. Similarly, if the outcome is always 'tails', we also have $H(F) = 0$. The entropy of this random variable is maximum when the coin is fair and $p = 0.5$. In that case, $H(F) = 1\text{bit/sample}$. If the coin is unfair and always gives the same result, then, there is no uncertainty in the random variable F. The outcome of the experiment is totally expected. We gain no information when we are told the outcome of such an experiment. On the other hand, when $p = 0.5$, there is maximum uncertainty in the random variable F. Neither 'heads' nor 'tails' is favored against the other outcome. The greater the uncertainty, the greater the amount of information we acquire when we learn the outcome of the experiment. Thus, the uncertainty about a random variable can be considered as the information that is conveyed by it.

The *joint entropy* of two discrete random variables F and G with a joint pmf $p_{F,G}(f, g)$ and alphabets $\mathbf{A}_f$ and $\mathbf{A}_g$, respectively is defined as:

$$H(F, G) = -\sum_{f \in \mathbf{A}_f} \sum_{g \in \mathbf{A}_g} p_{F,G}(f, g) \log_2 p_{F,G}(f, g) \tag{3.5}$$

The *conditional entropy* of discrete random variable F given discrete random variable G is defined as:

$$H(F|G) = -\sum_{g \in \mathbf{A}_g} p_G(g) \sum_{f \in \mathbf{A}_f} p_{F|G}(f, g) \log_2 p_{F|G}(f, g) \tag{3.6}$$

where $p_{F|G}(f, g)$ is the conditional pmf of F given G and $p_G(g)$ is the marginal pmf of G.

We are now ready to extend the entropy definitions from random variables to information sources.

The *Nth order entropy* of a discrete stationary source **F** is the joint entropy of N consecutive samples $F_1, F_2, \ldots, F_N$ of **F**:

$$H_N(\mathbf{F}) = H(F_1, F_2, \ldots F_N) = - \sum_{[f_1, f_2, \ldots, f_N] \in \mathbf{A}^N} p(f_1, f_2, \ldots, f_N) \log_2 p(f_1, f_2, \ldots, f_N)$$

(3.7)

where $p(f_1, f_2, \ldots, f_N)$ is the joint pmf of N consecutive samples of **F** and $\mathbf{A}^N$ is the N-fold Cartesian product of **A**.

The *Mth* order conditional entropy of a discrete stationary source **F** is the conditional entropy of a sample F_{M+1} given its previous M samples $F_M, F_{M-1}, \ldots, F_1$:

$$H_{C,M}(\mathbf{F}) = H(F_{M+1}|F_M, F_{M-1}, \ldots, F_1)$$

$$= \sum_{[f_1, f_2, \ldots, f_N] \in \mathbf{A}^N} p(f_1, f_2, \ldots, f_M) H(F_{M+1}|f_M, F_{M-1}, \ldots, f_1) \qquad (3.8)$$

where

$$H(F_{M+1}|f_M, F_{M-1}, \ldots, f_1) = - \sum_{f_{M+1} \in \mathbf{A}} p(f_{M+1}|f_M, f_{M-1}, \ldots, f_1)$$

$$\times \log_2 p(f_{M+1}|f_M, f_{M-1}, \ldots, f_1)$$

and $p(f_{M+1}|f_M, f_{M-1}, \ldots, f_1)$ is the *Mth* order conditional pmf of **F**.

We will also define the *entropy rate* of a discrete source as:

$$\overline{H}(\mathbf{F}) = \lim_{N \to \infty} \frac{1}{N} H_N(\mathbf{F}) \qquad (3.9)$$

It can be shown that the entropy rate is also equal to

$$\overline{H}(\mathbf{F}) = \lim_{N \to \infty} H_{C,N}(\mathbf{F}) \qquad (3.10)$$

3.3.2 Encoding of an Information Source

Let us assume that we wish to encode a particular realization $\{f_n\}$ of a discrete source **F** with alphabet $\mathbf{A} = \{a_1, a_2, \ldots, a_L\}$. We will be assigning a binary code word c_n to each sample of the realization f_n. We will refer to this procedure as *scalar lossless coding*. We will need to design a one-to-one mapping between each element of the source alphabet a_n and a binary code word c_n. The coding is called lossless because the mapping is such that the original sequence of source symbols can always be recovered without loss from the stream of binary code words. In order for this to happen, the coded sequence must be *uniquely decodable*. This means that the stream of binary code words must correspond to only one possible sequence of source symbols. Let $l(a_i)$ be the length (in bits) of the code word that corresponds to symbol a_i. Then, the *average code word length* for encoding a source realization $\{f_n\}$ is:

$$R = \sum_{a_i \in \mathbf{A}} p(a_i) l(a_i) \tag{3.11}$$

We will also refer to R as the *bit rate*. Its unit is bits per sample.

The following theorem can be proven:

Theorem 3.1

The minimum bit rate $\overline{R}$ required to represent a discrete stationary source $\mathbf{F}$ using scalar lossless coding (that is, by assigning one code word to each sample) satisfies:

$$H_1(\mathbf{F}) \leq \overline{R} \leq H_1(\mathbf{F}) + 1 \tag{3.12}$$

The lower bound of the above inequality cannot always be achieved. It is achievable when the symbol probabilities are negative powers of 2. That is, $p(a_i) = 2^{-m_i}$ for a set of integers $\{m_1, m_2, \ldots, m_L\}$. How far the minimum bit rate is from the lower bound depends on the actual symbol probabilities.

It becomes clear that the optimal scalar lossless encoding for a particular discrete stationary source can have a bit rate that is up to one bit per sample greater than the entropy. It is possible to encode the source so that the average number of bits per sample is arbitrarily close to $H_1(\mathbf{F})$. This can be accomplished by assigning a binary code word to a large number N of consecutive source symbols. Thus, we can treat N source samples as a vector sample that comes from vector source with alphabet $\mathbf{A}^N$. The first order entropy of the vector source is the *Nth* order entropy of the original source:

$$H_N(\mathbf{F}) \leq \overline{R}^N \leq H_N(\mathbf{F}) + 1 \tag{3.13}$$

where $\overline{R}^N$ is the minimum bit rate of the vector source.

It can be proven for an i.i.d. source that $H_N(\mathbf{F}) = N \cdot H_1(\mathbf{F})$. It is clear that by encoding one sample of the vector source, we are encoding N samples of the original source. Then, $\overline{R}_N = \frac{\overline{R}^N}{N}$ will be the average number of bits per original sample that results from the vector sample encoding. Then,

$$N \cdot H_1(\mathbf{F}) \leq N \cdot \overline{R}_N \leq N \cdot H_1(\mathbf{F}) + 1 \tag{3.14}$$

and

$$H_1(\mathbf{F}) \leq \overline{R}_N \leq H_1(\mathbf{F}) + \frac{1}{N} \tag{3.15}$$

Thus, for an i.i.d. source, the average code word length can be made arbitrarily close to the first order entropy of the source by encoding a large number N together. If the source is not necessarily i.i.d., we have:

$$\frac{H_N(\mathbf{F})}{N} \leq \overline{R}_N \leq \frac{H_N(\mathbf{F})}{N} + \frac{1}{N} \tag{3.16}$$

Then, as $N \to \infty$ we have

$$\lim_{N \to \infty} \overline{R}_N = \lim_{N \to \infty} \frac{H_N(\mathbf{F})}{N} = \overline{H}(\mathbf{F}) \tag{3.17}$$

Thus, we can say that the bit rate can always be made arbitrarily close to the source entropy rate by encoding a large number of original source samples together.

Instead of using vector coding, we can improve coding efficiency by using *conditional coding*, also known as *context-based coding*. In *Mth* order conditional coding, the code word for the current sample is selected taking into account the pattern formed by the previous M samples. We will refer to such as pattern as the *context*. We design a separate code book for each of the possible contexts based on the conditional distribution of the output sample given the context. If the size of the source alphabet is L, then, the maximum number of contexts for *Mth* order conditional coding is L^M. Applying the theorem to the conditional distribution under context m, we get:

$$H_{C,M}^m(\mathbf{F}) \leq \overline{R}_{C,M}^m \leq H_{C,M}^m(\mathbf{F}) + 1 \tag{3.18}$$

where $H_{C,M}^m(\mathbf{F})$ is the entropy of the *Mth* order conditional distribution under context m and $\overline{R}_{C,M}^m$ is the minimum bit rate that can be achieved when encoding under context m. If p_m represents the probability of context m, the average minimum bit rate using *Mth* order conditional coding is:

$$\overline{R}_{C,M} = \sum_m p_m \overline{R}_{C,M}^m \tag{3.19}$$

Then, it can be shown that:

$$H_{C,M}(\mathbf{F}) \leq \overline{R}_{C,M} \leq H_{C,M}(\mathbf{F}) + 1 \tag{3.20}$$

When $M \to \infty$, we get:

$$\overline{H}(\mathbf{F}) \leq \lim_{M \to \infty} \overline{R}_{C,M} \leq \overline{H}(\mathbf{F}) + 1 \tag{3.21}$$

Conditional coding can achieve a better coding efficiency than scalar coding because $\overline{H}(\mathbf{F}) < H_1(\mathbf{F})$, except when the source is i.i.d., in which case $\overline{H}(\mathbf{F}) = H_1(\mathbf{F})$. However, whether *Mth* order conditional coding is more efficient than *Nth* order vector coding depends on the actual source statistics.

3.3.3 Variable Length Coding

So far, we have discussed theoretical lower bounds on the bit rate with which an information source can be coded. However, the previously presented theory does not give us a means for encoding the information source in order to achieve these bounds. Binary encoding corresponds to assigning a binary code word to each symbol of an information source. As mentioned previously, we are interested in lossless coding, where there exists a one-to-one mapping between each symbol and a binary code word and the original sequence of source symbols can always be recovered from the binary bit stream. We will also refer to lossless codes as uniquely decodable codes. An important subset of lossless codes is the *prefix codes*. A code is a prefix code if no code word is a prefix of another valid code word in the code. This means that the beginning bits of a code word do not correspond to another code word. Clearly, prefix codes are always uniquely decodable. Another property of prefix codes is that they are instantaneously decodable. This means that we can decode a group of bits into a symbol if it matches a valid code word and we do not have to wait for more bits to arrive, as may be the case for lossless but non-prefix codes. Imposing a code to be a prefix code is a stronger requirement than just being a lossless code. However, it can be proven that no non-prefix uniquely decodable codes can outperform prefix codes in terms of bit rate. Thus, prefix codes are used in all practical applications.

A code is *fixed-length* if every code word has the same number of bits. If the size of the source alphabet is L, then the number of bits that need to be assigned for each code word is $\lfloor \log_2 L \rfloor$. As we discussed earlier, the lower bound for the bit rate is equal to the source entropy rate $\overline{H}(\mathbf{F})$ (or the first order entropy $H_1(\mathbf{F})$ for i.i.d. sources). Since the bit rate (average code word length) is weighted by the probability of each symbol, it is inefficient in most cases to assign the same number of symbols to low-probability and high-probability symbols. It can be shown that fixed-length coding is optimal only in the case when all symbol probabilities are equal. The lower bounds will be achieved if all symbol probabilities are equal and negative powers of 2.

Intuitively, to minimize the average code word length, we will need to assign shorter codes to high-probability symbols and longer codes to low-probability symbols. *Variable-Length Coding* (VLC) assigns code words of different lengths to each symbol. We next describe two popular VLC methods: *Huffman coding* and *arithmetic coding*.

Huffman Coding

Let us assume that we wish to encode a discrete stationary source with alphabet $\mathbf{A} = \{a_1, a_2, \ldots, a_L\}$ and pmf $p(a_l)$. *Huffman coding* is a VLC method that assigns shorter code words to higher probability symbols. Huffman coding assigns prefix codes. It can be shown that, when Huffman coding is applied to individual symbols (scalar lossless coding), the resulting bit rate R is the minimum possible bit rate $\overline{R}$ that can be achieved and satisfies the theorem $H_1(\mathbf{F}) \leq \overline{R} \leq H_1(\mathbf{F}) + 1$. Thus, Huffman coding is optimal for scalar lossless coding in terms of bit rate.

The Huffman coding algorithm is as follows:

1. Arrange the symbols in terms of decreasing probability order and consider them as leaf nodes of a tree.
2. While more than one nodes remain:
   a. Arbitrarily assign '1' and '0' to the two nodes with the smallest probabilities.
   b. Merge the two nodes with the smallest probabilities to form a new node with a probability that is the sum of the probabilities of the two nodes. Go back to step 1.
3. Determine the code word for each symbol by tracing the assigned bits from the corresponding leaf node to the top of the tree. The first bit of the code word is at the top of the tree and the last bit is at the leaf node.

Symbol Probability

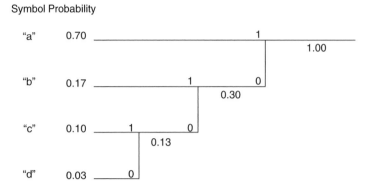

Figure 3.4 A Huffman coding example

Figure 3.4 shows an example of the Huffman coding of a source that emits symbols 'a', 'b', 'c' and 'd' with probabilities 0.70, 0.17, 0.10 and 0.03, respectively. As can be seen from the figure, the following code words are assigned to each symbol:

'a'	1
'b'	01
'c'	001
'd'	000

Let us observe that the resulting code is a prefix code. It is obvious that Huffman codes always produce prefix codes due to the tree structure that they employ. Also, as expected, symbol 'a', the symbol with the highest probability, receives the shortest code word, with a length of one bit. The bit rate (average code word length) for this code is:

$$R = 0.70 \cdot 1 + 0.17 \cdot 2 + 0.10 \cdot 3 + 0.03 \cdot 3 = 1.43 \text{ bits/symbol}$$

The first order entropy of the source is:

$$H_1(\mathbf{F}) = -0.70 \log_2 0.70 - 0.17 \log_2 0.17 - 0.10 \log_2 0.10 - 0.03 \log_2 0.03$$

$$= 0.70 \cdot 0.51 + 0.17 \cdot 2.56 + 0.1 \cdot 3.32 + 0.03 \cdot 5.06 = 1.28 \text{ bits/symbol}$$

As expected, we indeed have $H_1(\mathbf{F}) \le R \le H_1(\mathbf{F}) + 1$. However, the bit rate of 1.43 bits/symbol is not very close to the first order entropy of 1.28 bits/symbol. Clearly, using Huffman coding, the bit rate may be up to one bit/sample higher than the first order entropy.

As we mentioned earlier, it is possible to approach asymptotically the entropy rate $\overline{H}(\mathbf{F})$, which is always less than or equal to the first order entropy, by using vector lossless coding. Thus, we may create new vector symbols by grouping N original symbols together and encode these vector symbols using Huffman coding. As $N \to \infty$, the average bit rate per original symbol will approach asymptotically the entropy rate. This is at the expense of a delay that is introduced from having to wait for N original symbols before we can decode one vector symbol. Another drawback is that we now have to assign Huffman codes to L^N symbols instead of N symbols. Thus, the number of elements in the codebook grows exponentially with N. This limits the vector length N that can be used in practice.

Arithmetic Coding

An alternative VLC method is *arithmetic coding*. The idea behind arithmetic coding is to convert a variable number of symbols into a variable-length code word and represent a sequence of source symbols by an interval in the line segment from zero to one. The length of the interval is equal to the probability of the sequence. Since the sum of the probabilities of all possible symbol sequences is equal to one, the interval corresponding to all possible sequences will be the entire line segment. The coded bitstream that corresponds to a sequence is basically the binary representation of any point in the interval corresponding to the sequence.

Arithmetic coding does not have to wait for the entire sequence of symbols to start encoding the symbols. An initial interval is determined based on the first symbol. Then, the previous interval is recursively divided after each new symbol appears. The upper and lower boundaries of an interval need to be represented in binary. Whenever the Most Significant Bit (MSB) of the lower boundary is the same as the MSB of the upper boundary, this bit is shifted out. At the end of the sequence of symbols, the output of the arithmetic encoder is the binary representation of an intermediate point in the interval corresponding to the sequence. A high probability sequence will correspond to a longer interval in the line segment. Thus, fewer bits will be needed to specify the interval.

A detailed explanation of arithmetic coding is beyond the scope of this chapter. We next compare the performance of arithmetic coding with that of Huffman coding. As mentioned previously, by utilizing Huffman coding as a technique for vector lossless coding, the resulting bit rate satisfies the inequality $\frac{H_N(\mathbf{F})}{N} \leq R \leq \frac{H_N(\mathbf{F})}{N} + \frac{1}{N}$. As the vector length N becomes large, the bit rate asymptotically approaches the entropy rate. For arithmetic coding, the resulting bit rate satisfies the inequality $\frac{H_N(\mathbf{F})}{N} \leq R \leq \frac{H_N(\mathbf{F})}{N} + \frac{2}{N}$, where N is in this case the number of symbols in the sequence to be coded [10, 12]. Clearly, both Huffman coding and arithmetic coding asymptotically approach the entropy rate when N is large, and Huffman coding has a tighter upper bound than arithmetic coding. However, in Huffman coding, N is the number of original symbols in a vector symbol and cannot be made too large due to delay and complexity issues, as discussed previously. In arithmetic coding, N can be as large as the number of symbols in the whole sequence to be coded. Thus, in practice, N can be much larger in arithmetic coding than in Huffman coding and arithmetic coding will approach the entropy rate more closely.

3.3.4 Quantization

So far, we have talked about the lossless coding of a discrete source. However, not all sources are discrete. A source may be *analog* and each of its samples may take values from a continuous set, such as the set of real numbers. It is possible to convert the analog source symbols into discrete symbols using *quantization*. However, as we will see next, quantization is a many-to-one operation and results in loss of information. In video coding, quantization is used to convert the output of the transform into discrete symbols, which can then be coded using VLC. In this chapter, we are interested in *scalar quantization*, where each source sample is quantized separately. It is also possible to employ *vector quantization*, where a group (vector) of samples is quantized together.

Figure 3.5 shows an example of a scalar uniform quantizer. The horizontal axis corresponds to the quantizer input, which is a real-valued sample, and the vertical axis corresponds to the quantizer output, which can only take values from a discrete set. The parameter Q is called the *quantization step size*. As we can see, an interval of input values is mapped into a single output value. Thus, we have a many-to-one mapping. For example, all input values between $-Q/2$ and $Q/2$ are mapped into output value Q. All values between $3Q/2$ and $5Q/2$ are mapped into output value $2Q$, and so on.

What actually needs to be encoded and transmitted for each quantized sample is an index that shows what '*quantization bin*' the sample belongs to. We will refer to these indices as *quantization levels*. The quantization levels for the quantizer of our example are shown in squares in Figure 3.5. The quantization levels are then typically encoded using VLC.

The decoder decodes the VLC and receives the quantization levels. Then, '*inverse quantization*' needs to be performed. The term 'inverse quantization' is used in practice but can be misleading because it might give the impression that quantization is an invertible procedure. Quantization is not invertible since information is lost. In inverse quantization, the received quantization level is converted to a quantized sample value (the

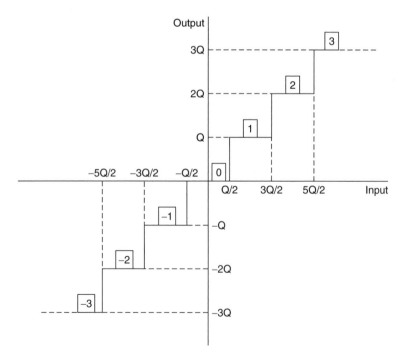

Figure 3.5   An example of a scalar uniform quantizer. The quantization levels are shown in boxes

horizontal axis in Figure 3.5). In the example of Figure 3.5, inverse quantization simply involves the multiplication of the quantization index with the step size Q.

The quantizer of Figure 3.5 is called uniform because a single quantization step size is used for all input value ranges. This mapping clearly results in loss of information. The information loss is smaller when the quantization step size is smaller. However, in that case, there entropy of the quantized data will be higher and a higher bit rate will be required for their encoding. Thus, the quantization parameter can be used to adjust the tradeoff between rate and distortion. A uniform quantizer was used here to introduce the reader to quantization before it is discussed in detail in the context of video coding. A more detailed discussion of quantization is beyond the context of this chapter.

3.4 Encoder Architectures

After reviewing the basics of information theory, we are ready to discuss some basic signal compression architectures.

3.4.1 DPCM

In the early stages of the digital video compression technology DPCM (Differential Pulse Code Modulation) was a commonly used compression technique [13]. Even though it did not provide high coding efficiency, it was easy and economical to implement. In late 1980s, when the implementation technology allowed one to implement more complex algorithms in cost effective ways, hybrid transform-DPCM architecture was used and standardized in H.261. Interestingly, at a high level, fundamental architecture has stayed the same over about two decades from H.261 to MPEG-2 to AVC/H.264. Many proposals with different structures or transforms were received by MPEG and VCEG committees over that period but none of them were able to outperform the hybrid transform-DPCM structure. To get better understanding of basic architecture and some of the issues, let us take a brief look at basic DPCM codec as shown in Figure 3.6. To keep things simpler, let us consider a one-dimensional signal {x(n)}.

Let us assume that the value of nth sample x(n) is predicted from some m number of previous samples (note that the samples in future can also be used for prediction). Let us also assume that the prediction is linear, i.e. the predicted value x'(n) is estimated to be:

$$x'(n) = a_1.x(n-1) + a_2.x(n-2) + \ldots + a_m.x(n-m) \qquad (3.22)$$

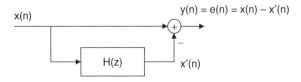

Figure 3.6 Linear prediction

or, in z-domain [14]:

$$X'(z) = H(z).X(z) \tag{3.23}$$

where, $H(z) = a_1.z^{-1} + a_2.z^{-2} + \ldots + a_m.z^{-m}$
The output $y(n)$ of predictor is obtained by:

$$y(n) = x(n) - x'(n) \tag{3.24}$$

or,

$$Y(z) = (1 - H(z)).X(z) \tag{3.25}$$

The output $y(n)$ is sent to decoder (to achieve compression, $y(n)$ is quantized but we will consider quantization later). Decoder, shown in Figure 3.7, implements the transfer function of $1/(1-H(z))$ which is inverse of the transfer function of encoder in equation (3.25).

Let us now consider the effect of quantization. To achieve compression, $y(n)$ is quantized before sending the signal to a receiver. The logical place to put a quantizer is as shown in Figure 3.8.

As explained below, this creates a problem where encoder and decoder start to drift apart. To get a better understanding of this drifting problem and how it is solved, let us consider a simple case where $H(z) = z^{-1}$, i.e. the previous neighboring sample is used as the predictor for $x(n)$. Let us also assume that the quantizer rounds the signal to the nearest integer. Consider an input:

$$x(n) = 0., 3.0, 4.1, 4.9, 5.3, 5.6, 6.0, 6.2, 6.6 \tag{3.26}$$

Output after prediction in this case is:

$$x(n) - x(n-1) = 3, 1.1, 0.8, 0.4, 0.3, 0.4, 0.2, 0.4 \tag{3.27}$$

Output after quantization is 3,1,1,0,0,0,0,0.

When this is used as input to the decoder, the output of the decoder will be 0, 3, 4, 5, 5, 5, 5, 5, 5. It is clear that the input in equation (3.26) and the output of the decoder are drifting apart. The introduction of non-linear block Q has not only destroyed the exact inverse relation between encoder and decoder, it has also created the situation such that for certain inputs the decoder output and the original signal input to the encoder drift far

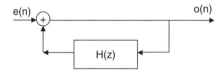

Figure 3.7   Decoder corresponding to DPCM encoder

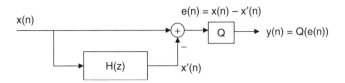

Figure 3.8 Linear prediction with quantization

apart. To avoid this situation, another structure of predictor is used. Consider the structure in Figure 3.9.

At first glance, it does not look like a linear predictor of Figure 3.6. However, it is easy to show by writing signal values at different points that mathematically the transfer function of this structure is also 1-H(z):

$$Y(z) = X(z) - X'(z) \tag{3.28}$$

$$X'(z) = Y'(z).H(z) \tag{3.29}$$

$$Y'(z) = (X'(z) + Y(z)) \tag{3.30}$$

Solving the last two equations:

$$X'(z) = X'(z).H(z) + Y(z).H(z) \tag{3.31}$$

or,

$$X'(z) = Y(z).H(z)/(1 - H(z)) \tag{3.32}$$

Substituting $X'(z)$ in the equation (3.28):

$$Y(z) = X(z) - Y(z).H(z)/(1 - H(z)) \tag{3.33}$$

$$Y(z).(1 - H(z)) = X(z).(1 - H(z)) - Y(z).H(z) \tag{3.34}$$

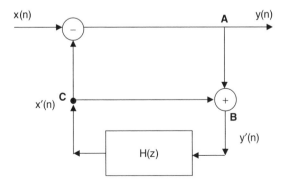

Figure 3.9 Alternative structure of linear predictor

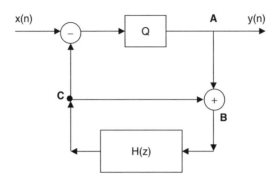

Figure 3.10   DPCM encoder structure

or,

$$Y(z) = (1 - H(z)).X(z) \tag{3.35}$$

which is the same as in equation (3.25).

Even though the structure in Figure 3.9 is mathematically the same as in Figure 3.6, this provides an additional feature – it does not create drift between original input signal and the decoded output in the presence of the quantizer. The quantizer in this structure is introduced as shown in Figure 3.10.

It is easy to confirm that for the input provided in equation (3.26), output does not drift far apart. Indeed, owing to quantization there is error in the decoded signal but it is limited to the quantization step and large drift is not present. A useful way to analyze the structure in Figures 3.9 and 3.10 is to note that the sub-structure between points A, B and C (shown separately in Figure 3.11) is the same as the decoder shown in Figure 3.7.

The signal at point B is the same as the output of the decoder. Hence the prediction $x'(n)$ of $x(n)$ (signal at C) is formed by using the past *decoded* values (at B), as opposed to the past *original* input values (as in equation (3.22) and Figure 3.6). Therefore, whenever the difference between the input and the predicted value starts to drift apart the difference

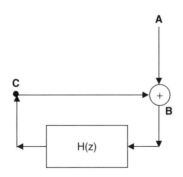

Figure 3.11   Decoder sub-structure inside DPCM encoder shown in Figure 3.9

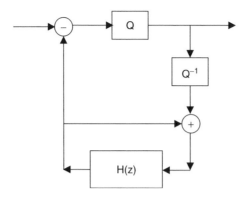

Figure 3.12 DPCM encoder with quantizer and inverse quantizer

gets bigger and corrects the difference for the next sample and thus stops the input and output from drifting. It is critical that this concept is understood.

Generally, scaling is also included as a part of quantization step. Therefore, both the decoder and corresponding decoder sub-structure in the encoder contain the step of inverse quantization (strictly speaking, it is actually an inverse scaling step). Therefore the encoder and decoder structures are as shown in Figures 3.12 and 3.13.

3.4.2 Hybrid Transform-DPCM Architecture

One of the drawbacks of the DPCM style of encoder is that it does not allow one to fully exploit the properties of the human visual system. The human visual system is less sensitive to higher and diagonal frequencies and the DPCM encoder described above works in the spatial domain. Therefore a hybrid transform-DPCM structure was first standardized in H.261. At a high level, the motion compensated hybrid DCT-Transform-DPCM encoder structure is as shown in Figure 3.14.

A two dimensional Discrete Cosine Transform (DCT) is taken before Quantization:

$$F(u, v) = \frac{2}{N} C(u) C(v) \sum_{x=0}^{N-1} \sum_{y=0}^{N-1} f(x, y), \cos\left(\frac{(2x + 1)u\pi}{2N}\right) \cos\left(\frac{(2y + 1)v\pi}{2N}\right) \quad (3.36)$$

where, x and y are the two dimensional locations of the pixels each ranging from 0 to $N-1$, u, v are the two dimensional transform frequencies each ranging from 0 to $N-1$ and:

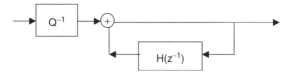

Figure 3.13 DPCM decoder with inverse quantizer

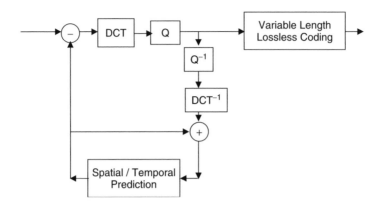

Figure 3.14   Hybrid transform-DPCM structure

$$C(u), C(v) = \begin{cases} \dfrac{1}{\sqrt{2}} & \text{for } u = 0,\, v = 0 \\ 1 & \text{otherwise} \end{cases}$$

The quantization is now done in the transform domain. The characteristics and frequency response of the human visual system are taken into account during the quantization step. Higher and diagonal frequencies are quantized more than the lower frequencies. The chroma component is quantized more heavily than the luma component. Note that the international standards such as MPEG2 and AVC/H.264 (see Chapter 5) specify the decoder behavior. Therefore, only the inverse cosine transform is specified. Equation (3.36) provides the forward transform that an encoder should use to provide the least possible distortion in the decoded video. However, both the forward and inverse transform contain irrational numbers. The standards also specify what level of precision decoders are required to have and how they will convert real numbers to integers within a specified range. Bit stream syntax also specifies the range of coefficients which encoders can use.

After the quantization of two dimensional DCT coefficients, they are scanned out in a zig-zag pattern as a one dimensional signal. This one dimensional signal is passed through the lossless Variable Length Codec (VLC). One dimensionally scanned transform coefficients tend to have long strings of zeroes due to the quantization step. The lossless coders are designed to be very efficient in removing those strings and compress the scanned coefficients with great efficiency.

Overall, the hybrid DCT-transform-DPCM structure is shown in Figure 3.14. This figure also illustrates that the prediction in these codecs is a three dimensional (spatial and temporal) prediction where the estimated value of a current pixel is predicted from the neighboring pixels in the same picture as well as the pixels in other pictures. Figure 3.15 depicts the high level structure of the corresponding decoder.

The basic structure of the encoder and the decoder remained the same in MPEG-1, MPEG-2, MPEG-4 Part 2 (Simple Profile and Advanced Simple Profile) and AVC/H.264. However, the accuracy of the predictions increased significantly.

It should be noted that the other profiles, besides Simple Profile (SP) and Advanced Simple Profile (ASP), of MPEG-4 Part2 included object based coding algorithms and had

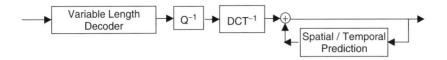

Figure 3.15 Hybrid DCT-transform-DPCM decoder structure

different encoder and decoder structures. However, those algorithms and profiles have not yet been deployed successfully in commercial applications. Those types of algorithms may be adopted in future video coding standards and applications.

Why DCT?

It is well known that the DCT transform of a region of a picture gives coefficients that are centered more around DC than other transforms such as the Discrete Fourier Transform (DFT) and Hadamard Transform [13]. One way to look at the time limited DFT with N samples is as a transform of a signal that is repeated periodically for every N sample. So, if a one dimensional signal is as shown in Figure 3.16 then DFT corresponds to the Fourier transform of a signal as shown in Figure 3.17.

Similarly, N sample DCT is the transform of the signal that is mirrored at the Nth sample and then that mirrored pair is periodically repeated, as shown in Figure 3.18 for the signal in Figure 3.16.

Therefore, for a signal that is not periodic in nature, one side of the signal is at a different level than on the other side and the discontinuity in the repeated signal (Figure 3.18) is smaller than that for DFT. This results in smaller higher frequency components. In the early days of the video compression, the composite NTSC signal was sometimes sampled and compressed directly (this was sometimes also called D-2 format). As the composite signal consisted of sinusoidal chrominance carrier, DFT was used in place of DCT.

3.4.3 A Typical Hybrid Transform DPCM-based Video Codec

We next describe the hybrid transform-DPCM architecture in more detail. This is also known as the *Motion-Compensated Discrete Cosine Transform* (MC-DCT) paradigm for video compression. When specific details are given, they are based on the H.263 video compression standard [15]. H.263 was chosen as a representative example of classic hybrid transform-DPCM video coding. The latest digital video compression standard AVC/H.264 is described in Chapter 5.

In motion-compensated video compression, what we try to do is code the motion that is present in the scene. Figure 3.19 shows an object (a circle) that appears in three consecutive frames, but in different places. In this very simple case where the only

Figure 3.16 Signal to be transformed

Figure 3.17 Underlying assumption about the signal corresponding to DFT

Figure 3.18 Underlying assumption about signal corresponding to DCT

object in the picture is a circle, we only need to code this object in the first frame and just code its motion in subsequent frames. This will lead to a much smaller number of bits to represent the sequence than required if each frame is coded independently. However, most video sequences are not so simple. Thus, more elaborate methods have been invented to code the motion efficiently and compensate for the fact that objects and background can appear and disappear from the scene. The motion estimation problem is a challenging one. Although the motion is three-dimensional, we are only able to observe its two-dimensional projection onto each frame. In addition, most motion estimators assume that pure translation can describe the motion from one frame to the next. Because the estimation of the motion is imperfect, we should also code the error that remains after we compensate for the motion.

The first frame in the video sequence is coded independently, just like a regular image. This is called an *Intra frame*. The compression is done as follows: First, the image is divided into blocks (usually of size 8×8) and the discrete cosine transform (*DCT*) of each block is taken. The DCT maps each block to another 8×8 block of real number

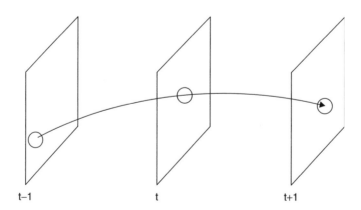

Figure 3.19 Motion-compensated video compression encodes the motion that is present in the scene

coefficients. This is done to decorrelate the pixels. Then, the DCT coefficients are quantized. That is, they are represented by a finite set of values, rather than the whole set of real numbers. Clearly, information is lost at this point. Finally, the quantized coefficients are coded into bits using a lossless variable length coder and the bits are transmitted.

After the first frame, subsequent frames are usually coded as *Inter frames*. That is, interpicture prediction is used. This technique uses the same concept as differential pulse code modulation. A prediction image is formed based on the previously reconstructed frame and some information on the relationship between the two frames, which is provided by the motion vectors. The difference between the actual frame and the prediction is then coded and transmitted.

At the beginning of the transmission of an Inter frame, the picture is divided into blocks (usually of a size 16×16). Unless there is a scene change in the picture, we expect that each of the blocks in the current picture is similar to another 16×16 block in the previously reconstructed frame. Clearly, these two blocks do not have to be in the same spatial position. Thus, the coder uses a technique called *block matching* to identify the block in the previous frame that best matches the block in the current frame. The search is performed in the vicinity of the block in the current frame. Block matching returns a vector that shows the relative spatial location of the best matching block in the previous frame. This vector is called the *motion vector*. Thus, we obtain a motion vector for each 16×16 block in the current frame. Now, we can obtain a prediction of the current frame using the previously reconstructed frame and the motion vectors. The prediction for each block of the current frame will be the block which is pointed to by the corresponding motion vector. Then, the prediction error – that is, the difference between the actual frame and the prediction – is coded using a method similar to the one described above for the Intra frames. The prediction error is transmitted along with the motion vectors. The decoder reconstructs the prediction error and builds the prediction image using the previous frame and the motion vectors. Finally, it produces the reconstructed frame by adding the prediction error to the predicted frame. If the prediction is good, we expect that the coding of the prediction error will require far less bits than the coding of the original picture. However, in some cases, the prediction for some blocks fails. This is the case when new objects are introduced in the picture or new background is revealed. In that case, certain blocks or the entire picture can be coded in Intra mode.

The basic structure of a motion-compensated DCT-based video codec is shown in Figure 3.20. It can be seen that it is very similar in principle to a Differential Pulse Code Modulation (DPCM) system. The switch indicates the choice between Intra and Inter pictures or macroblocks. In the Intra case, the DCT of the image is taken and the coefficients are quantized. In the Inter case, the difference between the predicted and the actual picture is transmitted along with the motion vectors. It can be seen from the picture that the encoder actually includes the decoder. This is because the prediction has to be made using information that is available to the decoder. Thus, every encoded frame has to be decoded by the coder and be used for the prediction of the next frame instead of the original frame, since the actual frame is not available to the decoder. Finally, in both the Intra and Inter cases, the quantized coefficients are variable length coded.

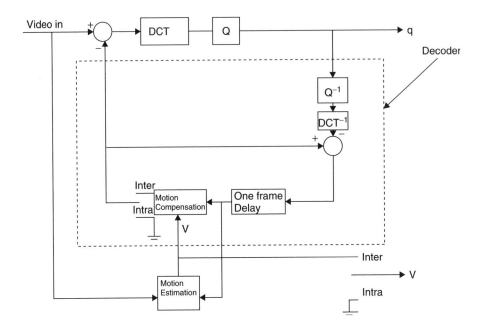

Figure 3.20   The basic structure of a motion-compensated DCT-based video encoder

3.4.4   Motion Compensation

The motion vectors are typically coded differentially due to their redundancy. The difference between the actual motion vector and a predicted motion vector is coded and transmitted. The decoder obtains the actual motion vector by adding the transmitted vector difference to the predicted vector. In H.263, the predicted vector is the median value of three candidate predictors MV1, MV2, and MV3. The candidate predictors are determined as shown in Figure 3.21. Normally, MV1 is the previous motion vector, MV2 is the above motion vector and MV3 is the above right motion vector. However, there are certain exceptions: Candidate predictor MV1 is set to zero if the corresponding macroblock is *outside* the picture. The candidate predictors MV2 and MV3 are set to MV1 if the corresponding macroblocks are outside the picture. Finally, we check if the above right macroblock is outside the picture. In that case, MV3 is set to zero.

Figure 3.21   The candidate predictors for the differential coding of the motion vectors

A positive value of the horizontal or vertical component of a motion vector indicates that the prediction is based on pixels in the referenced frame that are spatially to the right or below the pixels being predicted.

A video compression standard does not specify how the motion vectors are determined. The standard essentially specifies the decoder and a valid bit stream. It is up to the coder to decide on issues such as the determination of the motion vector, mode selection, and so on.

3.4.5 DCT and Quantization

A separable two-dimensional discrete cosine transform (DCT) of size 8×8 is taken for each block using the following formula:

$$F(u, v) = \frac{1}{4} C(u) C(v) \sum_{x=0}^{7} \sum_{y=0}^{7} f(x, y) \cos[\pi (2x + 1)u/16] \cos[\pi (2y + 1)v/16],$$

$$(3.37)$$

with $u, v, x, y = 0, 1, 2, \ldots, 7$, where:

$$C(u) = \begin{cases} 1/\sqrt{2} & \text{if } u = 0 \\ 1 & \text{otherwise,} \end{cases} \qquad (3.38)$$

$$C(v) = \begin{cases} 1/\sqrt{2} & \text{if } v = 0 \\ 1 & \text{otherwise,} \end{cases} \qquad (3.39)$$

x, y are the spatial coordinates in the original block domain and u, v are the coordinates in the transform domain.

For INTRA blocks, the DCT of the original sampled values of Y, Cb and Cr are taken. For INTER blocks, the DCT of the difference between the original sampled values and the prediction is taken.

Up to this point, no information has been lost. However, as noted earlier, in order to achieve sufficient compression ratios, some information needs to be thrown away. Thus, the DCT coefficients have to be quantized. For example, in H.263, the quantization parameter (QP) is used to specify the quantizer. QP may take integer values from 1 to 31. The quantization step size is then $2 \times QP$. QP has to be transmitted along with the quantized coefficients. The following definitions are made:

- /: Integer division with truncation toward zero.
- //: Integer division with rounding to the nearest integer.

Thus, if COF is a transform coefficient to be quantized and LEVEL is the absolute value of the quantized version of the transform coefficient, the quantization is done as follows.

For INTRA blocks (except for the dc coefficient):

$$LEVEL = |COF|/(2 \times QP). \tag{3.40}$$

For INTER blocks (all coefficients, including the dc):

$$LEVEL = (|COF| - QP/2)/(2 \times QP) \tag{3.41}$$

For the dc coefficient of an INTRA block:

$$LEVEL = COF//8 \tag{3.42}$$

The quantized coefficients along with the motion vectors are variable length coded.

3.4.6 Procedures Performed at the Decoder

The following procedures are done at the source decoder after the variable length decoding or arithmetic decoding:

- Motion compensation.
- Inverse quantization.
- Inverse DCT.
- Reconstruction of blocks.

Motion Compensation
The motion vector for each macroblock is obtained by adding predictors to the vector differences. In the case of one vector per macroblock (i.e., when the advanced prediction mode is not used), the candidate predictors are taken from three surrounding macroblocks, as described previously. Also, half pixel values are found using bilinear interpolation, as shown in Figure 3.22. In the case of H.263, the integer pixel positions are indicated by the Xs and the half pixel positions are indicated by the Os. Integer pixel positions are denoted with the letters A, B, C and D, and the half pixel positions are denoted with the

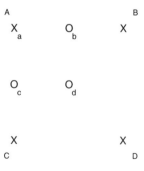

Figure 3.22 Half-pixel prediction is done using bilinear interpolation

letters a, b, c and d. It should be noted that pixel A is the same as pixel a. Then, the half pixel values are:

$$a = A \tag{3.43}$$

$$b = (A + B+1)/2 \tag{3.44}$$

$$c = (A + C+1)/2 \tag{3.45}$$

$$d = (A + B + C + D+2)/4 \tag{3.46}$$

Inverse Quantization

It should be pointed out that the term 'inverse quantization' does not imply that the quantization process is invertible. The quantization operation is clearly not invertible. This term simply implies the process of obtaining the reconstructed transform coefficient from the transmitted quantization level. Thus, if COF' is the reconstructed transform coefficient and LEVEL is the absolute value of the quantized version of the transform coefficient, the inverse quantization is done as follows for the example of H.263:

For INTRA blocks and INTER blocks, except for the dc coefficient:

$$|\text{COF}'| = 0 \quad \text{if LEVEL} = 0 \tag{3.47}$$

Otherwise:

$$|\text{COF}'| = 2 \times \text{QP} \times \text{LEVEL} + \text{QP} \qquad \text{if QP is odd,} \tag{3.48}$$

$$|\text{COF}'| = 2 \times \text{QP} \times \text{LEVEL} + \text{QP} - 1 \qquad \text{if QP is even.} \tag{3.49}$$

The sign of COF is then added to obtain $|\text{COF}'|$:

$$\text{COF}' = \text{Sign(COF)} \times |\text{COF}'| \tag{3.50}$$

For the dc coefficient of an INTRA block:

$$\text{COF}' = \text{LEVEL} \times 8 \tag{3.51}$$

Inverse DCT

After inverse quantization, the resulting 8 × 8 blocks are processed by a separable two-dimensional inverse DCT of size 8 × 8. The output from the inverse transform ranges from − 256 to 255 after clipping to be represented with 9 bits.

The Inverse DCT is given by the following equation:

$$f(x, y) = \frac{1}{4} \sum_{u=0}^{7} \sum_{v=0}^{7} C(u)C(v)F(u, v) \cos[\pi(2x + 1)u/16] \cos[\pi(2y + 1)v/16] \tag{3.52}$$

with $u, v, x, y = 0, 1, 2, \ldots, 7$, where:

$$C(u) = \begin{cases} 1/\sqrt{2} & \text{if } u = 0 \\ 1 & \text{otherwise,} \end{cases} \qquad (3.53)$$

$$C(v) = \begin{cases} 1/\sqrt{2} & \text{if } v = 0 \\ 1 & \text{otherwise,} \end{cases} \qquad (3.54)$$

x, y are the spatial coordinates in the original block domain and u, v are the coordinates in the transform domain.

Reconstruction of Blocks
After the inverse DCT, a reconstruction is formed for each luminance and chrominance block. For INTRA blocks, the reconstruction is equal to the result of the inverse transformation. For INTER blocks, the reconstruction is formed by summing the prediction and the result of the inverse transformation. The transform is performed on a pixel basis.

3.5 Wavelet-Based Video Compression

We next present the basics of video compression using the Discrete Wavelet Transform (DWT). The DWT has been very successful in video compression and DWT-based codecs have been proven to outperform the DCT-based ones [16, 17]. Also, the JPEG-2000 standard is wavelet-based [18]. These image codecs, besides offering good compression, also offer efficient *SNR scalability*. A *scalable* encoder produces a bitstream that can be partitioned into layers. One layer is the *base* layer and can be decoded by itself and provide a basic signal quality. One or more *enhancement* layers can be decoded along with the base layer to provide an improvement in signal quality. For image coding, the improvement will be in terms of image quality. Thus, the PSNR of the image will be increased (SNR scalability). For video coding, the improvement may also be in terms of spatial resolution (*spatial scalability*) or temporal resolution (*temporal scalability*). DWT-based image codecs produce a bitstream that can be partitioned into arbitrary layers. Thus, the bitstream can be 'chopped-off' at any point and the first part of it can be decoded. Furthermore, the image quality of this first part of the bitstream will typically be better than that of non-scalable DCT-based coding at the same bit rate.

The efficient compression and scalability of DWT-based image codecs compared to their DCT-based counterparts has inspired research on wavelet-based video compression. However, so far, no video compression standard is wavelet-based, except for some limited use of wavelet-based coding for intra frames in standards such as MPEG-4 [19]. We will next attempt to discuss some of the problems and challenges involved in applying the DWT to video coding. We will assume that the reader has some basic knowledge on wavelet based image coding. More information can be found in [16–18].

Figure 3.23 shows a block diagram for a DWT-based video encoder. It can be seen that such an encoder is very similar to a classic motion-compensated DCT-based encoder. Motion estimation and compensation are done in the spatial domain and the encoding of

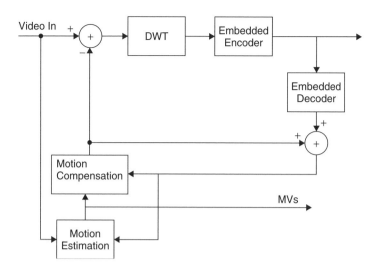

Figure 3.23   A DWT -based video encoder with motion estimation and compensation in the spatial domain

the prediction error is done using DWT instead of DCT using an encoder like SPIHT, EZW or JPEG-2000. Such a codec does not outperform DCT-based codecs. The main reason is the following: Motion estimation and compensation is done on a block basis. Thus, the prediction error will inevitably have discontinuities along block edges. In DCT-based codecs, the DCT is taken along blocks as well (typically of size 8×8) and the DCT block edges coincide with the discontinuities of the prediction error. Thus, there are no discontinuities that are due to the block-based motion compensation in the 8×8 DCT blocks. On the other hand, the DWT is taken on the whole image (in this case, the prediction error) instead of blocks. Thus, the encoder will have to spend bits to encode these discontinuities and this reduces the efficiency of the codec in Figure 3.23.

Figure 3.24 shows another DWT-based video encoder. Compared to the encoder in Figure 3.23, the encoder in Figure 3.24 performs motion estimation in the DWT domain instead of the spatial domain. Thus, the prediction error is already in the DWT domain and just needs to be encoded into a bitstream. This eliminates the efficiency problems that were due to the discontinuities that were present in the prediction error when performing block-based motion compensation in the spatial domain. However, another problem arises, which is due to the fact that the DWT is not shift-invariant. Assuming a one-dimensional signal and one level of decomposition, the DWT corresponds to filtering the signal using a high-pass and a low-pass discrete-time filter and sub-sampling the outputs by a factor of two. A system is shift-invariant if a shift in its input corresponds to a shift in its output. In DWT, if we shift the input by, say, one sample, we will not get the output of the original signal shifted by one sample due to the fact that the sub-sampling will discard different samples in this case. Thus, the DWT is not shift-invariant. This poses a problem when performing motion estimation and compensation in the DWT domain. Let us assume a video sequence consisting of two frames. The second frame is basically the first frame shifted by one pixel in the horizontal direction. If motion estimation and compensation

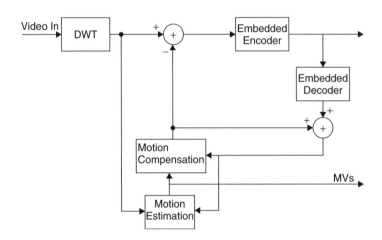

Figure 3.24 A DWT-based video encoder with motion estimation and compensation in the wavelet domain

are performed in the spatial domain, the second frame can be predicted perfectly from the first (except of course for the first pixel column of the image). However, if motion estimation and compensation are performed in the DWT domain, the second frame cannot be perfectly predicted from the first, exactly because the DWT is not shift-invariant. Thus, motion estimation and compensation are less efficient when performed in the DWT domain. Another way of looking at it is that the DWT spatial sub-bands have a lower resolution than the original video frames (due to the sub-sampling), thus the motion estimation and compensation is not as efficient.

A solution to the problem above is to use the *Overcomplete Discrete Wavelet Transform (ODWT)* instead of the DWT. The ODWT corresponds to the DWT without the sub-sampling. Thus, the ODWT representation of a signal has more samples than the DWT and the original signal, that's why it is called overcomplete. Since the DWT is by itself sufficient to represent a signal, it is always possible to derive the ODWT from the DWT. Thus, it is possible to convert the reference frame from the DWT to the ODWT and predict the DWT of the current frame from the ODWT of the previous (reference) frame. Clearly, the prediction error will be in the DWT domain. Thus, the use of the ODWT instead of the DWT does not increase the number of pixels to be encoded. Figure 3.25 illustrates the OWDT.

The use of the ODWT improves compression efficiency significantly. Since we are using image codecs such as SPIHT or JPEG-2000 for the encoding of the prediction error, we might think that we have efficient scalability, just like when applying these codecs to image compression. Unfortunately, this is not the case. Figure 3.26 illustrates why. If frame k is encoded using the full bitstream but is decoded using part of the bitstream, the decoder will have a different frame k than the encoder. Recall that the encoder has a decoder in it so that encoder and decoder have the same reference frame. This would be impossible in this case. Thus, the quality of frame $k+1$ and subsequent frames will deteriorate due to the fact that different reference frames will be used at the encoder and decoder. We will refer to this discrepancy between encoder and decoder as *drift*. Drift

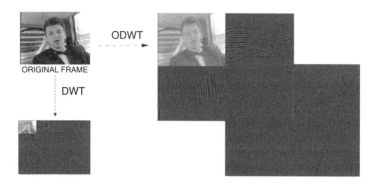

Figure 3.25   An illustration of Overcomplete Discrete Wavelet Transform (ODWT)

can be avoided by using only base layer information for prediction. Thus, encoder and decoder will always have the same reconstructed frame to use as reference. However, this results in a reduction of compression efficiency because the reference frames will be of lesser quality, thus, more information will have to be encoded with the prediction error, thus resulting in an increase of the number of bits to be transmitted. For these reasons, initial efforts towards applying wavelets to video compression were not very successful.

A way of avoiding drift problems is to utilize a 3D wavelet transform, the third dimension being time. Thus, a three dimensional signal is constructed by concatenating a group of frames (GOFs) and a three-dimensional DWT is applied. Then, the resulting signal in the DWT domain can be encoded temporal sub-band by temporal sub-band using, for example, SPIHT, or it can be encoded all at once using a technique like 3D-SPIHT [20]. Then, the resulting bitstream will have the same nice SNR scalability properties that DWT-based image coding has. However, if the DWT is applied in a straightforward fashion without any motion compensation, the encoding will not take full advantage of the temporal redundancy in the video signal.

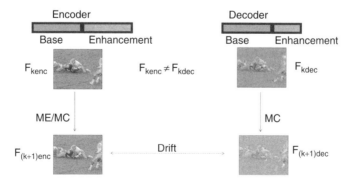

Figure 3.26   Illustration of the drift problems of the wavelet-based codecs of Figure 3.23 and Figure 3.24

The main theoretical development that promises efficient 3D wavelet-based video codecs with perfect invertibility is motion compensated temporal filtering (MCTF) using lifting, which was first introduced in [21, 22]. MCTF may be performed in two ways:

- Two-dimensional spatial filtering followed by temporal filtering (2D+t) [23–27].
- Temporal filtering followed by two-dimensional spatial filtering (t+2D) [28–31].

In the case of 2D+t motion compensated temporal filtering, the motion estimation and compensation is performed in the ODWT domain. Figure 3.27 illustrates 2D+t motion compensated temporal filtering while Figure 3.28 illustrates t+2D motion compensated temporal filtering.

The resulting wavelet coefficients can be encoded using different algorithms like 3D-SPIHT [20] or 3D-ESCOT [32]. Several enhancements have been made recently to the MCTF schemes presented either by introducing longer filters or by optimizing the operators involved in the temporal filter [23, 27, 28, 33]. Recently, a JPEG2000-compatible scalable video compression scheme has been proposed that uses the 3/1 filter for MCTF [34].

Though 3D schemes offer drift-free scalability with high compression efficiency, they introduce considerable delay, which makes them unsuitable for some real time video applications like tele-conferencing. In contrast to 2D methods, frames cannot be encoded one by one but processing is done in groups of frames. Thus, a certain number of frames must be available to the encoder to start encoding. The number of frames required depends on the filter length. Similarly, the group of frames must be available at the receiver before decoding can start. Thus, 3D video coding schemes offer better performance but also relax the causality of the system.

3.5.1 Motion-Compensated Temporal Wavelet Transform Using Lifting

Lifting allows for the incorporation of motion compensation in temporal wavelet transforms while still guaranteeing perfect reconstruction. Any wavelet filter can be implemented using lifting. Let us consider as an example the Haar wavelet transform:

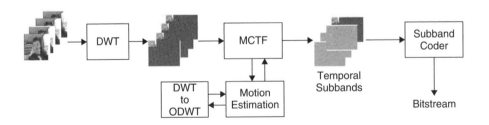

Figure 3.27   Illustration of 2D+t motion compensated temporal filtering

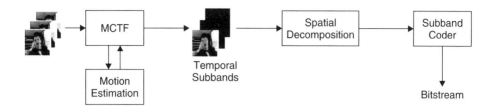

Figure 3.28 Illustration of t+2D motion compensated temporal filtering

$$h_k[x, y] = f_{2k+1}[x, y] - f_{2k}[x, y] \qquad (3.55)$$

$$l_k[x, y] = \frac{1}{2}\{f_{2k}[x, y] + f_{2k+1}[x, y]\} \qquad (3.56)$$

where $f_k[x, y]$ denotes frame k and $h_k[x, y]$ and $l_k[x, y]$ represent the high-pass and low-pass temporal sub-band frames. The lifting implementation of the above filter calculates the high-pass temporal sub-band first and then uses it in order to calculate the low-pass temporal sub-band:

$$h_k[x, y] = f_{2k+1}[x, y] - f_{2k}[x, y] \qquad (3.57)$$

$$l_k[x, y] = f_{2k}[x, y] + \frac{1}{2}h_k[x, y] \qquad (3.58)$$

Using lifting, the Haar filter along the motion trajectories with motion compensation can be implemented as:

$$h_k[x, y] = f_{2k+1}[x, y] - W_{2k \to 2k+1}(f_{2k}[x, y]) \qquad (3.59)$$

$$l_k[x, y] = f_{2k}[x, y] + \frac{1}{2}W_{2k+1 \to 2k}(h_k[x, y]) \qquad (3.60)$$

where $W_{i \to j}(f_i)$ denotes the motion-compensated mapping of frame f_i into frame f_j. Thus, the operator $W_{i \to j}(.)$ gives a per pixel mapping between two frames and this is applicable to any motion model.

In the case of the biorthogonal 5/3 wavelet transform, the analysis equations using motion compensated lifting are

$$h_k[x, y] = f_{2k+1}[x, y] - \frac{1}{2}\{W_{2k \to 2k+1}(f_{2k}[x, y]) + W_{2k+2 \to 2k+1}(f_{2k+2}[x, y])\} \qquad (3.61)$$

$$l_k[x, y] = f_{2k}[x, y] + \frac{1}{4}\{W_{2k-1 \to 2k}(h_{k-1}[x, y]) + W_{2k+1 \to 2k}(h_k[x, y])\} \qquad (3.62)$$

In the lifting operation, the prediction residues (temporal high pass sub-bands) are used to update the reference frame to obtain a temporal low sub-band. We will refer to this as the update step (equation (3.62)) in the following discussion.

If the motion is modeled poorly, the update step will cause ghosting artifacts to the low pass temporal sub-bands. The update step for longer filters depends on a larger number of future frames. If a video sequence is divided into a number of fixed sized GOFs that are processed independently, without using frames from other GOFs, high distortion will be introduced at the GOF boundaries for longer filters. When longer filters based on lifting are used with symmetric extension, the distortion will be in the range of 4–6 dB (PSNR) at the GOF boundaries irrespective of the motion content or model used [27, 28]. Hence, to reduce this variation at the boundaries, we need to use frames from past and future GOFs. Thus, it is observed that the introduced delay (in frames) is greater than the number of frames in the GOF. The encoding and decoding delay will be very high, as the encoder has to wait for future GOFs. In [28], the distortion at the boundaries for the 5/3 filter is reduced to some extent by using a sliding window approach. But this clearly introduces delay both at the encoder and at the decoder. We should note that, even when the delay is high, if the motion is not modeled properly, then the low pass temporal sub-bands will not be free from ghosting artifacts.

By skipping entirely the update step for 5/3 filter [31, 35], the analysis equations can be modified as:

$$h_k[x, y] = f_{2k+1}[x, y] - \frac{1}{2}\{W_{2k \to 2k+1}(f_{2k}[x, y]) + W_{2k+2 \to 2k+1}(f_{2k+2})[x, y]\} \quad (3.63)$$

$$l_k[x, y] = f_{2k}[x, y] \quad (3.64)$$

We refer to this filter set as the 3/1 filter.

Filters without update step will minimize the dependency on future frames thereby reducing the delay. Also, the low pass temporal sub-bands are free from ghosting artifacts introduced by the update step. Hence, by avoiding the update step, we get high quality temporal scalability with reduced delay. But at full frame rate resolution, the 3/1 filter suffers in compression efficiency compared to the 5/3 filter.

In 3D coding schemes, a high level of compression efficiency is achieved by applying a temporal filter to a group of frames. The number of frames in a buffer will increase with the length of the filter and the number of temporal decomposition levels. This introduces a delay both at the encoder and decoder. As mentioned earlier, the 3/1 filter offers less delay than the 5/3 filter at the expense of compression efficiency. A family of temporal filters with the following requirements has been proposed [26]:

- Any GOF length N can be used.
- Each GOF can be processed independently, without need for frames from neighboring GOFs.
- Motion compensated temporal filtering is used.
- Compression efficiency of the proposed filter set should be at least competitive with the 5/3 filter.

Thus, a filter set is proposed that is defined by the filter length N and the number of lifting steps involved (S). The filter length N here refers to the number frames being filtered. We refer to the filter set as (N, S) temporal filter. The number of frames N can vary from two to any number and need not be in some power of two. Unlike 5/3 and other longer filters, the proposed (N, S) filter can be processed independently (without reference to other GOFs) and thus finite fixed size GOFs can be created without introducing high distortions at the boundary. Any combination of (N, S) filters can be chosen to achieve the given delay requirements.

A detailed explanation of the (N,S) filter set is beyond the scope of this chapter. More information can be found in [26]. It should be pointed out that, unlike the 5/3 and 3/1 temporal filters, the (N, S) temporal filter does not correspond to a temporal wavelet transform.

Figure 3.29 shows an illustration of the (5, 2) filter with the update step. It can be seen that the GOF consists of five frames and there are two lifting steps (levels of temporal decomposition). As in all (N,S) filters, the first frame in the GOF is not updated. Thus, low pass temporal sub-bands L_0^1 and L_0^2 (where the superscript corresponds to the level of decomposition) are identical to the original frame F_0. However, sub-bands L_2^1 and L_1^2 are not identical to F_4 due to the update steps. The following temporal sub-bands will need to be encoded and transmitted: $L_0^2, H_0^1, H_0^2, H_1^1, L_1^2$. The low pass temporal sub-bands are analogous to intra frames in motion-compensated DCT-based video coding, whereas the high pass temporal sub-bands are analogous to inter frames. Of course, if update steps are used, the low pass temporal sub-bands are not identical to the original frames. A (N,S) temporal filter produces two low pass temporal sub-bands and N-2 high pass temporal sub-bands. Since low pass sub-bands have more information and require more bits to encode, we would prefer to have only one low pass sub-band per GOF. It is possible to concatenate another temporal filter to the (N,S) filter in order to obtain a GOF with only one low pass temporal sub-band [26].

Figure 3.30 illustrates the (5, 2) temporal filter for the case without an update step. Also, actual frames and temporal sub-bands are shown. Now, both the first and fifth frames of the GOF are not updated and are identical to the original video frames. By looking at the

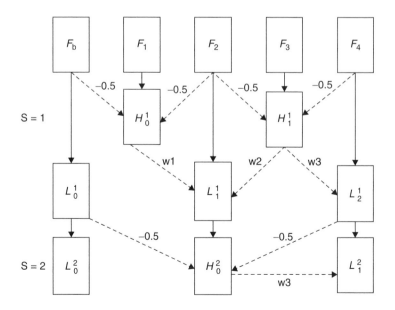

Figure 3.29 Illustration of the (5, 2) temporal filter (with update step)

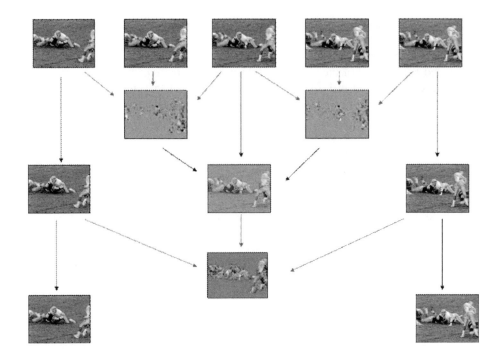

Figure 3.30   Illustration of the (5, 2) temporal filter (without update step)

temporal sub-bands of Figure 3.30, it becomes clear that low pass frames are analogous to intra frames and high pass frames are analogous to inter frames.

References

1. BT 601 Recommendation ITU-R BT.601-6, Studio encoding parameters of digital television for standard 4:3 and wide-screen 16:9 aspect ratios, Year 2007.
2. J. Keith, "Video demystified, 4th Edition (Demystifying Technology)," Elsevier Inc., 2005.
3. Schanda, J., "Colorimetry: Understanding the CIE System." John Wiley & Sons, Inc, US, 2007.
4. ISO/IEC JTC 1 and ITU-T, "Generic coding of moving pictures and associated audio information – Part 2: Video," ISO/IEC 13818-2 (MPEG-2) and ITU-T Rec. H.262, 1994 (and several subsequent amendments and corrigenda).
5. AVC ISO/IEC JTC 1 and ITU-T, "Advanced video coding of generic audiovisual services," ISO/IEC 14496-10 (MPEG-4 Part 10) and ITU-T Rec. H.264, 2005 (and several subsequent amendments and corrigenda).
6. ISO/IEC JTC 1, "Coding of moving pictures and associated audio for digital storage media at up to about 1,5 Mbit/s – Part 2: Video," ISO/IEC 11172 (MPEG-1), Nov. 1993.
7. C. Poynton, *A Technical Introduction to Digital Video*, John Wiley & Sons, US, 1996.
8. T.M. Cover and J. A. Thomas, *Elements of Information Theory*, John Wiley & Sons, Inc, US, 1991.
9. R. G. Gallager, *Information Theory and Reliable Communication*, John Wiley & Sons, Inc, US, 1968.
10. Y. Wang, J. Ostermann and Y.-Q Zhang, *Video Processing and Communications*, Prentice-Hall, 2002.
11. A. Papoulis and S. U. Pillai, *Probability, Random Variables and Stochastic Processes*, McGraw-Hill, 2002.
12. K. Sayood, *Introduction to Data Compression*, Morgan Kaufmann, 1996.
13. A. N. Netravali and B. G. Haskell, "Digital Pictures – Representation and Compression," Plenum Press, June 1989.

14. A. V. Oppenheim, R. W. Schafer and J. R. Buck, "Discrete-Time Signal Processing, 2nd Edition," Prentice Hall, December 1998.
15. ITU-T Rec. H.263, *Video Codec for low bit rate communication*, 1996.
16. A. Said and W. Pearlman, "A new, fast, and efficient image codec based on set partitioning in hierarchical trees", *IEEE Transactions on Circuits and Systems for Video Technology*, Vol. 6, pp. 243–50, June 1996.
17. J. M. Shapiro, "Embedded image coding using zerotrees of wavelet coefficients," *IEEE Transactions on Signal Processing*, Vol. 41, No. 12, pp. 3445–62, 1993.
18. C. Christopoulos, A. Skodras, and T. Ebrahimi, "The JPEG-2000 still image coding system: an overview," *IEEE Transactions on Consumer Electronics*, Vol. 46, No. 4, pp. 1103–27, 2000.
19. MPEG4 video group, "Core experiments on multifunctional and advanced layered coding aspects of MPEG-4 video", Doc. ISO/IEC JTC1/SC29/WG11 N2176, May 1998.
20. B.-J. Kim, Z. Xiong, and W. A. Pearlman, "Low bit-rate scalable video coding with 3-D set partitioning in hierarchical trees (3-D SPIHT)," *IEEE Transactions on Circuits and Systems for Video Technology*, Vol. 10, No. 8, pp. 1374–87, 2000.
21. B. Pesquet-Popescu and V. Bottreau, "Three-dimensional lifting schemes for motion compensated video compression," in *Proc. IEEE International Conference on Acoustics, Speech and Signal Processing*, Salt Lake City, UT, 2001, pp. 1793–6.
22. A. Secker and D. Taubman, "Motion-compensated highly scalable video compression using an adaptive 3-D wavelet transform based on lifting," in *Proc. IEEE Conference on Image Processing*, Thessaloniki, Greece, 2001, pp. 1029–32.
23. Y. Andreopoulos, A. Munteanu, J. Barbarien, M. van der Schaar, J., and P. Schelkens, "In-band motion compensated temporal filtering," *Signal Processing: Image Communication*, Vol. 19, No. 7, pp. 653–73, 2004.
24. X. Li, "Scalable video compression via overcomplete motion compensated wavelet coding," *Signal Processing: Image Communication*, Vol. 19, No. 7, pp. 637–51, 2004.
25. V. Seran and L. P. Kondi, "Quality variation control for three-dimensional wavelet-based video coders," *EURASIP Journal on Image and Video Processing*, Volume 2007, Article ID 83068, 8 pages, doi:10.1155/2007/83068.
26. V. Seran and L. P. Kondi, "New temporal filtering scheme to reduce delay in wavelet-based video coding," *IEEE Transactions on Image Processing*, Vol. 16, No. 12, pp. 2927–35, December 2007.
27. Y. Wang, S. Cui, and J. E. Fowler, "3D video coding using redundant-wavelet multihypothesis and motion-compensated temporal filtering," in *Proceedings of IEEE International Conference on Image Processing (ICIP '03)*, Vol. 2, pp. 755–8, Barcelona, Spain, September 2003.
28. A. Golwelkar and J. W. Woods, "Scalable video compression using longer motion compensated temporal filters," in *Visual Communications and Image Processing*, Vol. 5150 of *Proceedings of SPIE*, pp. 1406–16, Lugano, Switzerland, July 2003.
29. S. T. Hsiang and J. W. Woods, "Embedded video coding using motion compensated 3-D sub-band/wavelet filter bank," in *Proceedings of the Packet Video Workshop*, Sardinia, Italy, May 2000.
30. G. Pau, C. Tillier, B. Pesquet-Popescu, and H. Heijmans, "Motion compensation and scalability in lifting-based video coding," *Signal Processing: Image Communication*, Vol. 19, No. 7, pp. 577–600, 2004.
31. A. Secker and D. Taubman, "Lifting-based invertible motion adaptive transform (LIMAT) framework for highly scalable video compression," *IEEE Transactions on Image Processing*, Vol. 12, No. 12, pp. 1530–42, 2003.
32. J. Xu, S. Li and Y. Q. Zhang, "A wavelet codec using 3-D ESCOT," in *IEEE PCM*, December 2000.
33. N. Mehrseresht and D. Taubman, "An efficient content adaptive motion compensation 3-D-DWT with enhanced spatial and temporal scalability," in *Proc. IEEE International Conference on Image Processing*, 2004, Vol. 2, pp. 1329–32.
34. T. Andre, M. Cagnazzo, M. Antonini, and M. Barlaud, "JPEG2000-compatible scalable scheme for wavelet-based video coding," *EURASIP Journal on Image and Video Processing*, 2007, DOI: 10.1155/2007/30852.
35. M. van der Schaar and D. S. Turaga, "Unconstrained motion compensated temporal filtering (UMCTF) framework for wavelet video coding," in *Proceedings of the IEEE International Conference on Acoustics, Speech, and Signal Processing (ICASSP '03)*, Vol. 3, pp. 81–4, Hong Kong, April 2003.

4

4G Wireless Communications and Networking

4.1 IMT-Advanced and 4G

Today, wireless technologies and systems which are claimed to be "4G" represent a market positioning statement by different interest groups. Such claims must be substantiated by a set of technical rules in order to qualify as 4G. Currently, the ITU (International Telecommunications Union) has been working on a new international standard for 4G, called IMT-Advanced, which is regarded as an evolutionary version of IMT-2000, the international standard on 3G technologies and systems.

With the rapid development of telecommunications technologies and services, the number of mobile subscribers worldwide has increased from 215 million in 1997 to 946 million (15.5% of global population) in 2001, as shown in Figure 4.1. It is predicted that by the year 2010 there will be 1700 million terrestrial mobile subscribers worldwide. A substantial portion of these additional subscribers is expected to be from outside the countries that already had substantial numbers of mobile users by the year 2001. Figure 4.1 shows the user trends of mobile and wireline telecommunications services and applications.

4G technologies can be thought of as an evolution of the 3G technologies which are specified by IMT-2000. The framework for the future development of IMT-2000 and IMT-Advanced and their relationship to each other are depicted in Figure 4.2. Systems beyond IMT-2000 will encompass the capabilities of previous systems. Other communication relationships will also emerge, in addition to person-to-person, such as machine-to-machine, machine-to-person and person-to-machine.

One of the unique features of 4G networks is that they will accommodate heterogeneous radio access systems, which will be connected via flexible core networks. Thus, an individual user can be connected via a variety of different access systems to the desired networks and services. The interworking between these different access systems in terms of horizontal and vertical handover and seamless service provision with service negotiation including mobility, security and QoS management will be a key requirement, which may be handled in the core network or by suitable servers accessed via the core network. This

4G Wireless Video Communications Haohong Wang, Lisimachos P. Kondi, Ajay Luthra and Song Ci
© 2009 John Wiley & Sons, Ltd

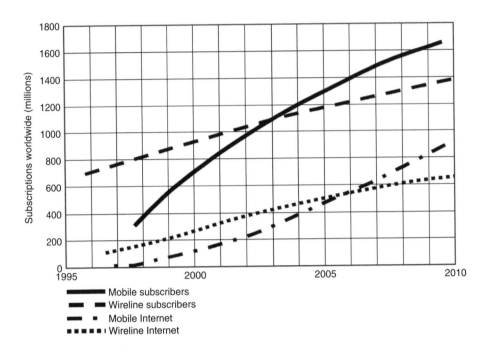

Figure 4.1 Global growth of mobile and wireline subscribers. Reproduced with kind permission from © ITU, 2008

optimally connected anywhere, anytime view could be realized by a network comprising a variety of interworking access systems connected to a common packet-based core network, as seen in Figure 4.3. Figure 4.4 illustrates a flexible and scalable environment which can be used for the allocation of system capacity in a deployment area, where one or more systems may be deployed according to need.

The timelines associated with these different factors are depicted in Figure 4.5. When discussing the time phases for IMT-advanced, it is important to specify the time at which the standards are completed, when spectrum must be available, and when deployment may start. Currently, IMT-Advanced is still at the call-for-proposal stage.

IMT-Advanced has been developed to provide true end-to-end IP services to mobile users at "anytime anywhere". Although the standardization process is still ongoing, the major design goals of 4G are quite certain, which are: 1) 4G will be all IP networks, meaning that circuit switching will be eliminated in the next-generation cellular networks. 2) 4G will have a very high data rate. It is expected that 4G networks will be capable of providing 100 Mbps data rate under high mobility, which is much faster than 3G. 3) 4G will provide Quality of Service (QoS) and security to the end users, which has been lacking in 3G. 4) IP-based multimedia services such as Voice over IP (VoIP) and video streaming are expected to be the major traffic types in 4G. In this chapter, we will discuss three major contenders for 4G technologies: 3GPP Long Term Evolution (LTE), WiMax IEEE 802.16m and 3GPP2 Ultra Mobile Broadband (UMB).

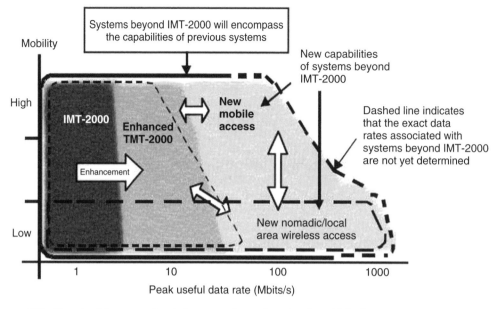

Figure 4.2 Illustration of capabilities of IMT-2000 and IMT-Advanced. Reproduced with kind permission from © ITU, 2008

4.2 LTE

LTE, which stands for 3rd Generation Partnership Project (3GPP) Long Term Evolution, is one of the next major steps in mobile radio communications designed to ensure competitiveness in a longer time frame, i.e. for the next 10 years and beyond. The increasing usage of mobile data and newly-emerged applications such as Multimedia Online Gaming (MMOG), mobile TV and streaming services has motivated the 3GPP to work on this standard. The aim of LTE is to improve the Universal Mobile Telecommunications System (UMTS) mobile phone standard and provide an enhanced user experience for next generation mobile broadband. LTE is planned to be introduced in 3GPP Release 8.

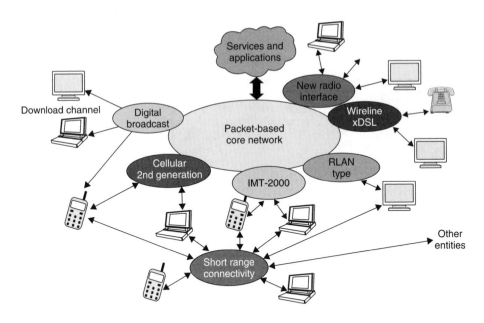

Figure 4.3 Future IMT-Advanced network, including a variety of potential interworking access systems. Reproduced with kind permission from © ITU, 2008

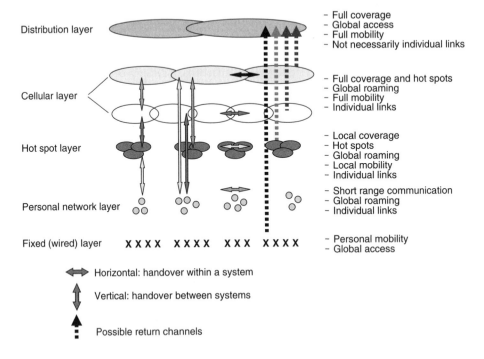

Figure 4.4 Illustration of complementary access systems. Reproduced with kind permission from © ITU, 2008

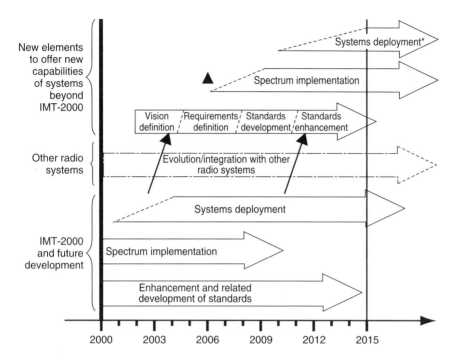

The sloped dotted lines indicate that the exact starting point of the particular subject can not yet be fixed

▲ :Possible spectrum identification at WRC-07

* :Possible wide deployment around the year 2015 in some countries

Figure 4.5 Phases and expected timelines for IMT-Advanced. Reproduced with kind permission from © ITU, 2008

4.2.1 Introduction

The high-level LTE requirements for 3GPP radio-access technology include reduced latency, higher user data rates, improved system capacity and coverage and reduced costs for operators. Therefore, an evolution of the radio interface as well as the radio network architecture should be considered. It was also recommended that the Evolved UTRAN (E-UTRAN) should bring significant improvement so as to justify the standardization effort and should avoid unnecessary options. The main advantages of LTE are high throughput, low latency, plug and play, FDD and TDD in the same platform, superior end-user experience and simple architecture resulting in low Operating Expenditures (OPEX). Furthermore, LTE also supports seamless connection to existing networks, such as GSM, CDMA and HSPA.

The feasibility study on the UTRA and UTRAN Long Term Evolution was started in December 2004, with the objective being "to develop a framework for the evolution of the 3GPP radio-access technology towards a high-data-rate, low-latency and packet-optimized radio-access technology" [1, 2]. The study focused on supporting services provided from the PS-domain, concerning the radio interface physical layer for both downlink and uplink, the radio interface layers 2 and 3, the UTRAN architecture and RF issues. Furthermore, the Next Generation Mobile Networks (NGMN) initiative provided a set of recommendations for the creation of networks suitable for the competitive delivery of mobile broadband services, with the goal of "to provide a coherent vision for technology evolution beyond 3G for the competitive delivery of broadband wireless services" [1, 2]. The long-term objective of NGMN is to "establish clear performance targets, fundamental recommendations and deployment scenarios for a future wide area mobile broadband network" [1, 2].

The goals of LTE include improving spectral efficiency, lowering costs, improving services, making use of new spectrum and reframed spectrum opportunities and better integration with other open standards. The architecture that results from this work is called Evolved Packet System (EPS) and comprises Evolved UTRAN (E-UTRAN) on the access side and Evolved Packet Core (EPC) on the core side. EPC is also known as System Architecture Evolution (SAE) and E-UTRAN is also known as LTE.

As a result, the detailed requirements laid down for LTE can be found in the Table 4.1 [3]. Generally, LTE meets the key requirements of next generation networks, including downlink peak rates of at least 100 Mbit/s, uplink peak rates of at least 50 Mbit/s and Radio Access Network (RAN) round-trip times of less than 10 ms. Moreover, LTE supports flexible carrier bandwidths, from 1.4 MHz up to 20 MHz, as well as both Frequency Division Duplex (FDD) and Time Division Duplex (TDD).

4.2.2 Protocol Architecture

4.2.2.1 E-UTRAN Overview Architecture

As shown in Figure 4.6, the E-UTRAN architecture consists of eNBs in order to provide the E-UTRA user plane (RLC/MAC/PHY) and control plane (RRC) protocol terminations towards the UE. The eNBs interface to the aGW via the S1, and are inter-connected via the X2.

4.2.2.2 User Plane and Control Plane

The E-UTRAN protocol stack is shown in Figure 4.7, where RLC and MAC sub-layers perform the function of scheduling, ARQ and HARQ. The PDCP sub-layer performs functions such as header compression, integrity protection (FFS: For Further Study) and ciphering for the user plane.

The E-UTRAN control plane protocol is shown in Figure 4.8, where RLC and MAC sub-layers perform the same functions as for the user plane. The RRC sub-layer performs functions such as broadcasting, paging, RB control, RRC connection management, etc. The PDCP sub-layer performs functions such as integrity protection and ciphering for the control player. The NAS performs for the user plane SAE bearer management, authentication, idle mode mobility handling, security control, etc.

Table 4.1 The detailed requirements for 3GPP LTE

Peak Data Rate	DL	100 Mb/s (5bps /Hz) in 20 MHz DL spectrum
	UL	50 Mb/s (2.5 bps/Hz) in 20 MHz UL spectrum
User Throughput (per MHz)	DL	Average user throughput is 3–4 times of Rel-6 HSDPA
	UL	Average user throughput is 2–3 times of Rel-6 HSUPA
Latency	C-plane	Less than 100 ms from camped to active, less than 50 ms from dormant to active
	U-plane	Less than 5 ms
	C-plane Capacity	Support at least 200 users in active per cell for 5 MHz spectrum
Spectrum Efficiency (bits/sec/Hz/site)	DL	3-4 times of Rel-6 HSDPA
	UL	2-3 times of Rel-6 HSUPA
Spectrum Flexibility	Spectrum allocation of [1.25, 2.5, 5, 10, 15, 20] MHz	
Mobility	(0–15 km/h)	Optimized for low mobile speed
	(15–120 km/h)	High performance for higher mobile speed
	120–350 km/h, or even up to 500 km/h	Mobility across the cellular network shall be maintained
Further Enhanced MBMS	Re-use unicast physical layer component	
	Simultaneous, tightly integrated and efficient provisioning of	
	Dedicated voice and MBMS services to the user	
	Unpaired MBMS operation in unpaired spectrum arrangements	
Deployment Scenarios	Standalone	
	Integrating with existing UTRAN and/or GERAN	
Co-existence and interworking with 3GPP RAT	Real-time service handover	Less than 300 ms interruption time between UTRAN and E-UTRAN/GERAN
	Non real-time service handover	Less than 500 ms interruption time between UTRAN and E-UTRAN/GERAN
RRM	Enhanced support for end to end QoS	
	Efficient support for transmission of higher layers	
	Support of load sharing and policy management cross different RATs	

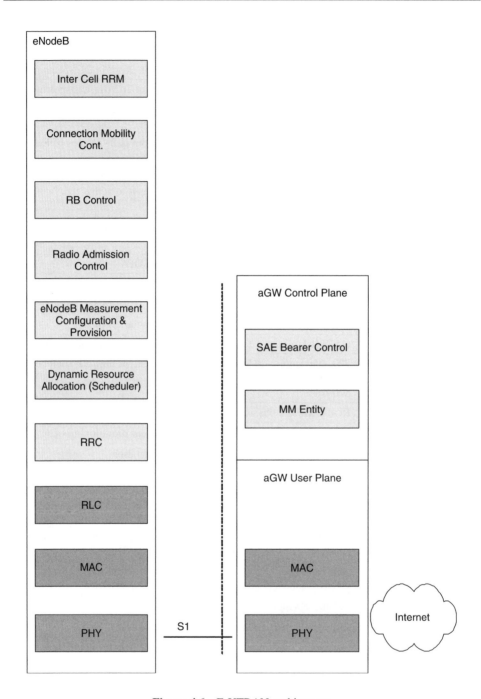

Figure 4.6   E-UTRAN architecture

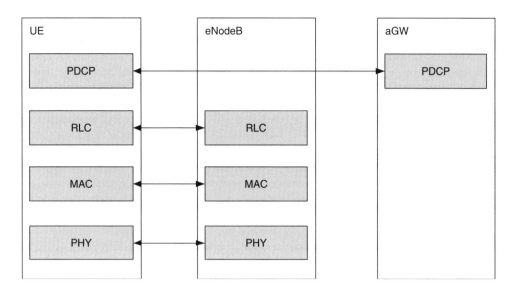

Figure 4.7 User plane protocol stack

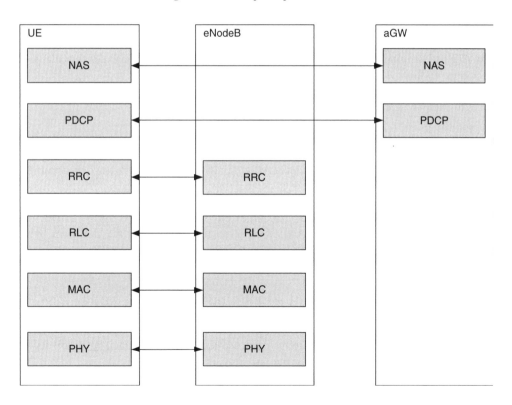

Figure 4.8 Control plane protocol stack

4.2.2.3 LTE Physical Layer

In the 30th conference of 3GPP in December 2005, TSG RAN decided to use downlink OFDMA and uplink SC-FDMA for the physical layer [4] which means that OFDMA won the LTE competition against CDMA. This outcome comes from, on the one hand, technological considerations – the adoption of OFDMA and SC-FDMA can reduce peak-to-average power ratio (PAPR) at the receiver's end, leading to a smaller end user terminal with lower cost. On the other hand, it avoids the restrictions and monopolies of the core CDMA technology [5, 6]. The following are some of the key technologies used in LTE physical layer [7].

LTE uses OFDM for the downlink, which meets the LTE requirement of 100 Mbits/s data rate, the spectral efficiency and enables cost-efficient solutions for very wide carriers with high peak rates. OFDM is a well-established technology, widely used in standards such as IEEE 802.11a/b/g, 802.16, HIPERLAN-2, DVB and DAB. By configuring the quantities of sub-carriers, it can achieve flexible bandwidth configurations ranging from 1.25 Hz to 20 MHz. In the time domain a radio frame is 10 ms long and consists of 10 sub-frames of 1 ms each. Every sub-frame consists of two slots where each slot is 0.5 ms. The sub-carrier spacing in the frequency domain is 15 kHz. Twelve of these sub-carriers together (per slot) is called a resource block, so one resource block is 180 kHz. Six resource blocks fit in a carrier of 1.4 MHz, and 100 resource blocks fit in a carrier of 20 MHz. In addition, the Cyclic Prefix (CP) of 4.7 μs can ensure the handling of time delay while not increasing processing time. Another longer CP (16.7 μs) can be used to increase cell coverage or multi-cell broadcasting services. By using OFDM, a new dimension is added to Adaptive Modulation and Coding (AMC), i.e. the adaptive frequency variable, leading to more flexible and efficient resource scheduling. Inheriting from the HSDPA/HSUPA concept, LTE adopts link adaptation and fast re-transmission in order to increase gain, avoiding macro-diversity which requires the support of network architecture. The supported modulation formats on the downlink data channels are QPSK, 16QAM and 64QAM. For MIMO operation, a distinction is made between single user MIMO, for enhancing one user's data throughput, and multi-user MIMO for enhancing the cell throughput, usually with the antenna configuration of 2 × 2, namely, two transmitting antennae are set in the eNodeB, while two receiving antennae are set in UE. For higher speed downlink, four antennae are used in the eNodeB.

In the uplink, LTE uses a pre-coded version of OFDM called Single Carrier Frequency Division Multiple Access (SC-FDMA). This is to compensate for the drawback with normal OFDM, which has a very high Peak to Average Power Ratio (PAPR). High PAPR requires expensive and inefficient power amplifiers with high requirements of linearity, which increases the terminal cost and drains the battery very fast. SC-FDMA solves this problem by combining resource blocks in such a way that reduces the need for linearity, and so power consumption, in the power amplifier. A low PAPR also improves coverage and the cell-edge performance. Within each TTI, the eNodeB allocates a unique frequency for transmitting data. Data for different users are separated from each other by using frequency space or time slot, ensuring the orthogonality among uplink carriers within the cell, avoiding frequency interference. Slow power control can resist the path loss and shading effect. Thanks to the orthogonality of uplink transmission, fast power control is no longer needed in order to deal with the near-far effect. In the meantime, with the help of CP, multi-path interference can be wiped out. The enhanced AMC mechanism applies to uplink as well. The supported modulation schemes on the uplink data channels are QPSK, 16QAM and 64QAM. If virtual MIMO/Spatial division multiple access (SDMA) is introduced the data rate in the uplink direction can be increased depending on the number of antennae at the base station. With this technology more than one mobile can reuse the same resources. Similarly, the uplink channel coding uses Turbo code. The basic MIMO configuration for uplink single user is also 2×2. Two transmitting antennae are installed in the UE, and another two for receiving are installed in the eNodeB [8].

4.2.3   LTE Layer 2

The LTE layer 2 is split into three sub-layers, i.e. Medium Access Control (MAC), Radio Link Control (RLC) and Packet Data Convergence Protocol (PDCP). The PDCP/RLC/MAC architecture for downlink and uplink are depicted in Figure 4.9 and Figure 4.10, respectively.

Service Access Points (SAP) between the physical layer and the MAC sub-layer provide the transport channels. The SAPs between the MAC sub-layer and the RLC sub-layer provide the logical channels, and the SAPs between the RLC sub-layer and the PDCP sub-layer provide the radio bearers. Several logical channels can be multiplexed onto the same transport channel. The multiplexing of radio bearers with the same QoS onto the same priority queue is FFS. In the uplink, only one transport block is generated per Transmission Time Interval (TTI) in the case of non-MIMO. In the downlink, the number of transport blocks is FFS.

The MAC sub-layer provides the following services and functions: multiplexing or demultiplexing of RLC PDUs belonging to one or different radio bearers into/from transport blocks (TB) delivered to and from the physical layer on transport channels, mapping between logical channels and transport channels, traffic volume measurement reporting, error correction through HARQ, priority handling between logical channels of one UE, priority handling between UEs by means of dynamic scheduling, transport format selection, mapping of Access Classes to Access Service Classes (FFS for RACH), padding (FFS) and in-sequence delivery of RLC PDUs if RLC cannot handle the out of sequence delivery caused by HARQ (FFS).

For the RLC sub-layer, the main services and functions are to transfer of upper layer PDUs supporting AM, UM or TM data transfer (FFS), error correction through ARQ, segmentation according to the size of the TB, re-segmentation when necessary, concatenation of SDUs for the same radio bearer is FFS, in-sequence delivery of upper layer PDUs, duplicate detection, protocol error detection and recovery, flow control, SDU discard (FFS) and reset, etc.

The main services and functions of the PDCP sub-layer include: header compression and decompression, transfer of user data, ciphering of user plane data and control plane data (NAS Signalling), integrity protection of control plane data (NAS signalling) and integrity protection of user plane data are FFS.

LTE RRC

The RRC sub-layer fulfills a variety of services and functions, including the broadcast of system information related to the non-access stratum (NAS) and access stratum (AS), paging, the establishment, maintenance and release of an RRC connection between the UE

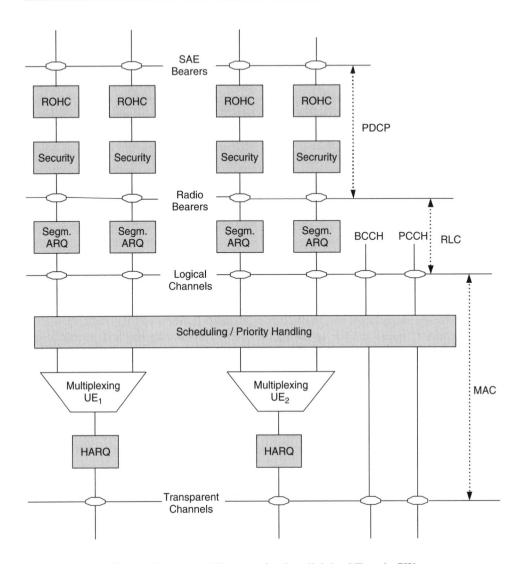

Figure 4.9   Layer 2 Structure for downlink in eNB and aGW

and E-UTRAN, mobility functions, notification for multicast/broadcast services (FFS), the establishment, maintenance and release of radio bearers for multicast/broadcast services, QoS management functions (FFS if spread across multiple layers), UE measurement reporting and control of the reporting, MBMS control (FFS) and NAS direct message transfer to/from NAS from/to UE.

NAS Control Protocol
The services and functions of Non-Access Stratum (NAS) control protocol include SAE bearer control, paging origination, configuration and control of PDCP and security.

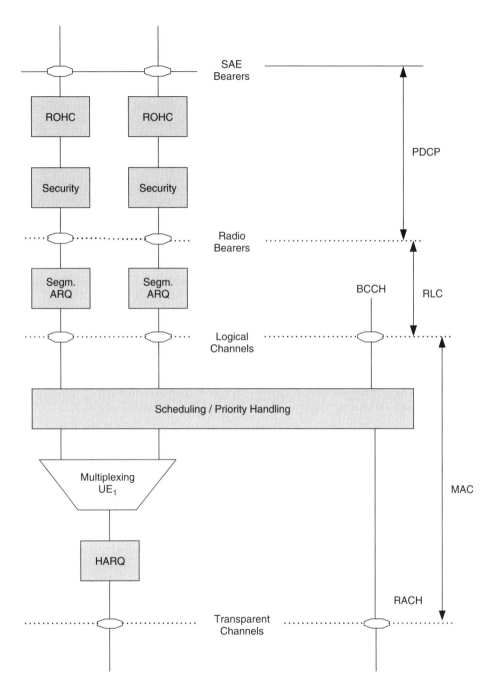

Figure 4.10   Layer 2 Structure for uplink in UE

4.2.4   The Evolution of Architecture

Figure 4.11 depicts the network architecture of UMTS R6 [9], in which NodeB provides a wireless access point for end users, while controlling and managing network traffic in the meantime. The Radio Network Controller (RNC) controls and manages eNodeB, including wireless resources, local users, and wireless access status, as well as optimization for transmission. GPRS provides service to the Serving GPRS Support Node (SGSN), responsible for the control and management of data flow of packet-switching networks, while also taking care of the sending and receiving packet data between NodeB and Gateway GPRS Support Node (GGSN). GGSN connects to the core network, serving as the gateway between the local networks and the external packet-switching networks, the so-called GPRS router. The IP protocol on backbone networks is used to connect with SGSN and GGSN [10].

The network architecture of LTE, shown in Figure 4.12, uses IP for the lower level transmission, thus forming a mesh network. With this architecture, UE mobility within the entire network can be achieved, ensuring seamless handover. Every eNode connects to aGW through mesh networks, with one eNodeB capable of connecting with multiple aGW and vice versa [4, 11].

Finally, Figure 4.13 shows the evolved system architecture, relying possibly on different access technologies, while the major entities and reference points are defined in Table 4.2.

4.2.5   LTE Standardization

The LTE project was initiated by the TSG RAN work group, consisting four sub-groups assigned to work on different areas. The groups of RAN1, RAN2 and RAN3 work on

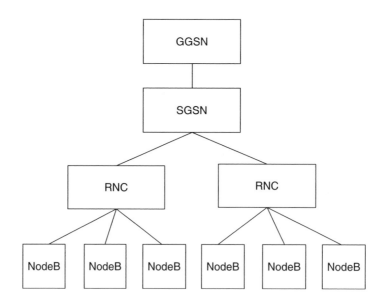

Figure 4.11   Network architecture for 3GPP R6

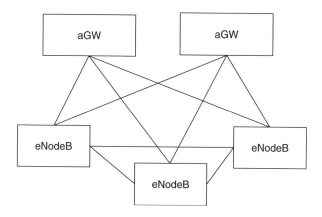

Figure 4.12   Network architecture for 3GPP LTE

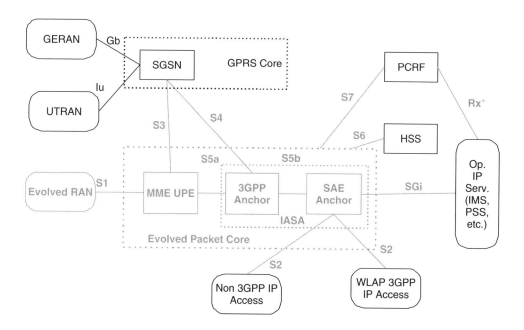

Figure 4.13   Logical high level architecture for the evolved system

feasibility research, concerning the physical layer of air interface, protocol stacks and the interfaces and architecture of wireless access networks. These three groups provide their work reports to TSG RAN, who standardizes them.

According to the 3GPP work flow, there are two stages in the standardization work. The first stage is the Study Item (SI) stage, from December 2004 to June 2006, engaged in feasibility research. A variety of technical research reports form the output of this stage. From June 2006 to June 2007 is the Work Item (WI) stage, during which work on

Table 4.2 Functions of major entities and reference points

3GPP Anchor	The functional entity that anchors the user plane for mobility between the 2G/3G access system and the LTE access system
SAE Anchor	The functional entity that anchors the user plane for mobility between 3GPP access systems and non-3GPP access systems
S1	Access to Evolved RAN radio resources for the transport of user plane and control plane traffic
S2a	Providing the user plane with related control and mobility support between a trusted non 3GPP IP access and the SAE Anchor
S2b	Providing the user plane with related control and mobility support between ePDG and the SAE Anchor
S3	Enabling user and bearer information exchange for inter 3GPP access system mobility in idle and/or active state. It is based on Gn reference point as defined between SGSNs
S4	Providing the user plane with related control and mobility support between GPRS Core and the 3GPP Anchor and is based on Gn reference point as defined between SGSN and GGSN
S5a	Providing the user plane with related control and mobility support between MME/UPE and 3GPP anchor
S5b	Providing the user plane with related control and mobility support between 3GPP anchor and SAE anchor
S6	Enabling transfer of subscription and authentication data for authenticating/authorizing user access to the evolved system
S7	Providing transfer of (QoS) policy and charging rules from PCRF to Policy and Charging Enforcement Point (PCEP)
Sgi	Reference point between the Inter AS Anchor and the packet data network

the formulation and regulation for technical standard of system evolution led to concrete technical standards. After this, it is planned to commercialize in 2009 or 2010.

4.3 WIMAX-IEEE 802.16m

Similar to LTE, WiMAX is also competing for a place in the IMT-Advanced 4G standard. As a result, the 802.16m working group has been tasked by the IEEE to enhance the systems. IEEE 802.16m is an amendment to IEEE 802.16-2004 and IEEE 802.16e-2005. WiMAX is backwards compatible, i.e. 802.16e and 802.16m mobile devices can be served by the same base station on the same carrier. However, in contrast to 802.16e which uses only one carrier, 802.16m can use two or even more carriers in order to increase increase overall data transfer rates. A theoretical data rate requirement for 802.16m is a target of 100 Mbps in mobile and 1 Gbps in stationary. IEEE 802.16m systems can operate in RF frequencies of less than 6 GHz as well as licensed spectra allocated to the mobile and fixed broadband services. In this section, we will provide an overview of the current status of IEEE 802.16m.

4.3.1 Network Architecture

As shown in Figure 4.14, the network architecture includes three functional entities: Mobile Station (MS), Access Service Network (ASN) and Connectivity Service Network (CSN).

An ASN is comprised of one or more Base Station(s) and one or more ASN Gateway(s), which may be shared by more than one CSN. The ANS provides radio access to an IEEE 802.16e/m MS. Specifically, the ASN provides the following functions [12]: 1) IEEE 802.16e/m Layer-1 (L1)/Layer-2 (L2) connectivity among IEEE 802.16e/m MSs; 2) Transfer of AAA (authentication, authorization and accounting) messages to IEEE 802.16e/m MS's Home Network Service Provider (H-NSP); 3) Network discovery and selection of the MS's preferred NSP; 4) Relay functionality for establishing Layer-3 (L3) connectivity with an MS; 5) Radio resource management; 6) ASN anchored mobility; 7) CSN anchored mobility; 8) Paging; and 9) ASN-CSN tunneling.

A CSN may be comprised of network elements such as routers, AAA proxy/servers, user databases and Interworking gateway MSs. The CSN provides IP connectivity services to the IEEE 802.16e/m MS(s). Specifically, the CSN provides the following functions: 1) MS IP address and endpoint parameter allocation for user sessions; 2) AAA proxy or server; 3) Policy and admission control based on user subscription profiles; 4) ASN-CSN tunneling support; 5) IEEE 802.16e/m subscriber billing and inter-operator settlement; 6) Inter-CSN tunneling for roaming; and 7) Inter-ASN mobility.

An IEEE802.16m MS usually has four states: 1) Initialization state, in which an MS decodes BCH information and selects one target BS; 2) Access state, in which the MS performs network entry to the selected BS; 3) Connected state, in which the MS maintains at least one connection as established during Access State, while MS and BS may establish additional transport connections, consisting of three modes: sleep mode, active mode and

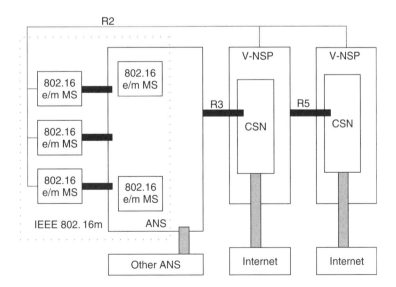

Figure 4.14   IEEE 802.16m overall network architecture

scanning mode; and 4) Idle state which consists of two separated modes, paging available mode and paging unavailable mode. The MS may perform power saving by switching between these two modes.

4.3.2   System Reference Model

As shown in Figure 4.15, the reference model for IEEE 802.16m is very similar to that of IEEE 802.16e [13, 14]. The difference is that IEEE 802.16m performs soft classification of the MAC common part sub-layer into radio resource control and management functions and medium access control functions (i.e., no SAP is required between the two classes of functions).

4.3.3   Protocol Structure

4.3.3.1   MAC Layer

MAC of 802.16m BS/MS: The 802.16m MAC consists of two sub-layers: Convergence sub-layer (CS) and Common Part sub-layer (CPS).

The MAC common part sub-layer can be further divided into Radio Resource Control and Management (RRCM) functions and Medium Access Control (MAC) functions. As shown in Figure 4.16, the RRCM sub-layer provides radio resource-related functions such

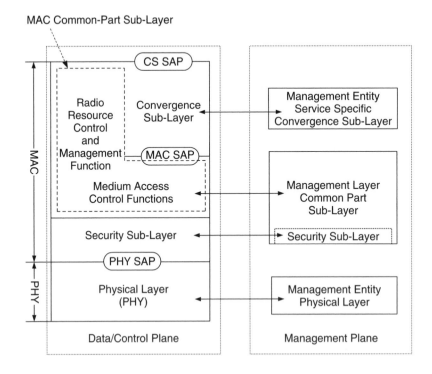

Figure 4.15   IEEE 802.16m system reference model

as: 1) Radio Resource Management; 2) Mobility Management; 3) Network-entry Management; 4) Location Management; 5) Idle Mode Management; 6) Security Management; 7) System Configuration Management; 8) MBS; 9) Connection Management; 10) Relay functions; 11) Self Organization; and 12) Multi-Carrier. The main function of each block in Figure 4.16 is listed in Table 4.3.

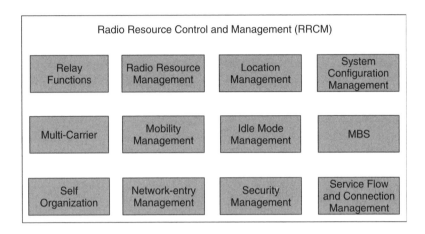

Figure 4.16 Functions of IEEE 802.16m Radio Resource Control and Management

Table 4.3 Block functions of BS/MS RRCM in Figure 4.16

Radio Resource Management block	Adjustment of radio network parameters based on traffic load function of load control, admission control and interference control
Mobility Management block	Functions related to Intra-RAT/ Inter-RAT handover
Network-entry Management block	Location based service (LBS)
Idle Mode Management block	Location update operation during idle mode
Security Management block	Key management for secure communication
System Configuration Management block	System configuration parameters, and system parameters and system configuration information for transmission to the MS
MBS (Multicast and Broadcasting Service) block	Management messages and data associated with broadcasting and/or multicasting service
Service Flow and Connection Management block	MS identifier allocation and connection identifiers during access/handover/ service flow creation procedures
Relay Functions block	Multihop relay mechanisms
Self Organization block	Self configuration and self optimization mechanisms
Multi-carrier (MC) block	A common MAC entity to control a PHY spanning over multiple frequency channels

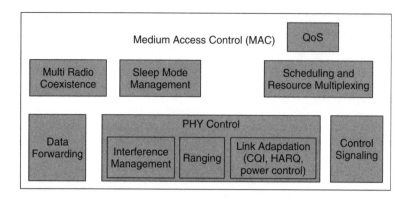

Figure 4.17   Functions of IEEE 802.16m Medium Access Control (MAC)

As shown in Figure 4.17, the Medium Access Control (MAC) includes function blocks which are related to the physical layer and link layer such as: 1) PHY Control; 2) Control Signaling; 3) Sleep Mode Management; 4) QoS; 5) Scheduling and Resource Multiplexing; 6) ARQ; 7) Fragmentation/Packing; 8) MAC PDU formation; 9) Multi-Radio Coexistence; 10) Data forwarding; 11) Interference Management; and 12) Inter-BS coordination. The main function of each block in Figure 4.17 is listed in Table 4.4.

Figures 4.16 and 4.17 form the control plane of 802.16m BS/MS MAC. Figure 4.18 shows the data plane which includes ARQ, fragmentation/packing, MAC PDU formation.

MAC of 802.16m RS: In IEEE 802.16m, Relay Stations (RSs) may be deployed to provide improved coverage and/or capacity. A 802.16m BS that is capable of supporting a 802.16j RS will communicate with the 802.16j RS in the "legacy zone". The 802.16m BS is not required to support 802.16j protocols [15] in the "802.16m zone". The design of 802.16m relay protocols should be based on the design of 802.16j wherever possible, although 802.16m relay protocols used in the "802.16m zone" may be different from 802.16j protocols used in the "legacy zone".

The 802.16m RS MAC is divided into two sub-layers: 1) Radio Resource Control and Management (RRCM) sublayer; and 2) Medium Access Control (MAC) sublayer. The function blocks of 802.16m RS are defined as in Figure 4.19. Note that most of the function blocks of 802.16m RS are similar to those of 802.16m BS/MS, and the functional blocks and the definitions listed here do not imply that these functional blocks will be supported in all RS implementations.

As shown in Figure 4.19, the 802.16m RS RRCM sublayer includes the functional blocks: 1) Mobility Management; 2) Network-entry Management; 3) Location Management; 4) Security Management; 5) MBS; 6) Path Management functions; 7) Self Organization; and 8) Multi-Carrier. The main function of each block in Figure 4.19 is listed in Table 4.5.

As shown in Figure 4.20, the 802.16m RS Medium Access Control (MAC) sublayer includes the function blocks related to the physical layer and link controls: 1) PHY Control; 2) Control Signaling; 3) Sleep Mode Management; 4) QoS; 5) Scheduling and Resource Multiplexing; 6) ARQ; 7) Fragmentation/Packing; 8) MAC PDU Formation;

Table 4.4   Block functions of BS/MS MAC in Figure 4.17

PHY Control block	PHY signaling such as ranging, measurement/feedback (CQI), and HARQ ACK/NACK
Control Signaling block	Resource allocation messages
Sleep Mode Management block	Sleep mode operation
QoS block	QoS management based on QoS parameters input from Connection Management function for each connection
Scheduling and Resource Multiplexing block	Packet scheduling and multiplexing based on properties of connections
ARQ block	Handles MAC ARQ function.
Fragmentation/Packing block	MSDU fragmentation or packing of MSDUs based on scheduling results from Scheduler block
MAC PDU formation block	Construction of MAC protocol data unit (PDU)
Multi-Radio Coexistence block	Concurrent operations of IEEE 802.16m and non-IEEE 802.16m radios collocated on the same mobile station
The Data Forwarding block	Forwarding functions when RSs are present on the path between BS and MS
Interference Management block	Management of the inter-cell/sector interferences
Mobility Management block	Supports functions related to Intra-RAT/ Inter-RAT handover
Inter-BS coordination block	Coordination of the actions of multiple BSs by exchanging information for e.g., interference management

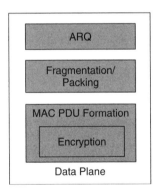

Figure 4.18   Data plane of IEEE 802.16m BS/MS MAC

9) Data Forwarding; and 10) Interference Management. Each function block is defined as follows [12]: the main function of each block in Figure 4.20 is listed in Table 4.6.

Figures 4.19 and 4.20 form the control plane of 802.16m BS/MS MAC. Figure 4.21 shows the data plane which includes ARQ, Fragmentation/Packing and MAC PDU Formation.

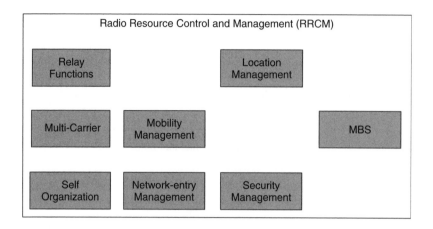

Figure 4.19 Functions of IEEE 802.16m RS Radio Resource Control and Management

Table 4.5 Block functions of IEEE 802.16m RS RRCM in Figure 4.19

Mobility Management block	MS handover operations in cooperation with the BS
Network-entry Management block	RS/MS initialization procedures and performing RS network entry procedure to the BS
Location Management block	Management of supporting location based service (LBS)
Security Management block	The key management for the RS
MBS (Multicast and Broadcasting Service) block	Coordination of with the BS to schedule the transmission of MBS data
Path Management Functions block	Procedures to maintain relay paths
Self Organization block	Supports RS self configuration and RS self-optimization mechanisms coordinated by BS
Multi-carrier (MC) block	A common MAC entity to control a PHY spanning over multiple frequency channels at the RS

MAC Addressing: Each MS has a global address and logical addresses that identify the MS and connections during operation. The global address of an MS is the 48-bit globally-unique IEEE 802 MAC address which uniquely identifies the MS. The MAC logical addresses are assigned to the MS by management messages from a BS. Logical addresses are used for resource allocation and management of the MS. During network operation, a "Station Identifier" is also assigned by the BS to the MS to uniquely identify the MS for the BS. Each MS registered in the network has an assigned "Station Identifier". Some specific "Station Identifiers" are reserved, for example, for broadcast, multicast and ranging. For the identification of connections attached to an MS, each MS connection is assigned a "Flow Identifier" that uniquely identifies the connection within the MS. "Flow Identifiers" identify management connections and active transport service flows.

HARQ in MAC: To improve robustness and performance, HARQ is supported in downlink and uplink packet (re)transmissions in both BS and MS.

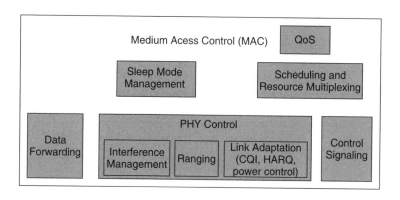

Figure 4.20 Functions of IEEE 802.16m RS medium access control

Table 4.6 Block functions of IEEE 802.16m RS MAC in Figure 4.20

PHY Control block	PHY signaling such as ranging, measurement/feedback (CQI), and HARQ ACK/NACK at the RS
Control Signaling block	RS resource allocation messages such as MAP as well as specific control signaling messages
Sleep Mode Management block	Sleep mode operation of its MSs in coordination with the BS
QoS block	Rate control based on QoS parameters
Scheduling and Resource Multi-plexing block	Scheduling of the transmission of MPDUs
ARQ block	MAC ARQ function between BS, RS and MS
Fragmentation/Packing block	Fragmentation or packing of MSDUs based on scheduling results from Scheduler block
MAC PDU formation block	Construction of MAC protocol data units (PDUs)
Data Forwarding block	Forwarding functions on the path between BS and RS/MS
Interference Management block	Functions at the RS to manage the inter-cell/sector and inter-RS interference among RS and BS

The following HARQ-related parameters are defined: 1) Maximum retransmission delay; 2) Maximum number of retransmissions; 3) Maximum number of HARQ processes; and 4) ACK/NACK delay. For the choice of HARQ schemes in both downlink and uplink, the following three options are being considered: 1) synchronous; 2) asynchronous; and 3) a combination of the previous schemes.

For the resource allocation of HARQ retransmissions, in synchronous HARQ, resources allocated for retransmissions in the downlink can be fixed or adaptive according to control signaling; in asynchronous HARQ, an adaptive HARQ scheme is used in the downlink. In adaptive asynchronous HARQ, the resource allocation and transmission format for the HARQ retransmissions may be different from the initial transmission. In the case of

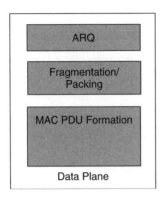

Figure 4.21   Data plane of IEEE 802.16m RS MAC

retransmissions, control signaling is required in order to indicate the resource allocation and transmission format along with other necessary HARQ parameters.

Handover: The 802.16m MS handover in 802.16m considers the following cases: 1) handover from legacy serving BS to legacy targeting BS; 2) handover from 16 m serving BS to legacy targeting BS; 3) handover from legacy serving BS to 16 m targeting BS; and 4) handover from 16 m serving BS to 16 m targeting BS.

4.3.3.2 PHY Layer

Duplex modes: IEEE 802.16m supports TDD and FDD duplex modes, including H-FDD MS operation, in accordance with the IEEE 802.16m system requirements [16]. Unless otherwise specified, the frame structure attributes and baseband processing are common to all duplex modes.

Multiple Access Schemes: Both downlink and uplink use OFDMA as the multiple access scheme in IEEE 802.16m. Table 4.7 provides the OFDMA parameters.

Frame Structure: In practice, the IEEE 802.16m frame structure with frame lengths of up to 20 milliseconds is not flexible. The downside of such long frames is slow network access, since each device only has one transmission opportunity per frame. The IEEE 802.16m uses a new frame structure, consisting of super-frames (20 ms), which is further divided into frames (5 ms), and then each frame is divided into eight sub-frames (0.617 ms). Each super-frame begins with a DL sub-frame that contains a superframe header. Sub-frames are classified to two types depending on the size of cyclic prefix: 1) the type-1 sub-frame which consists of six OFDMA symbols; and 2) the type-2 sub-frame that consists of seven OFDMA symbols. The basic frame structure is applied to FDD and TDD duplexing schemes, including H-FDD MS operation. The number of switchings between DL and UL in each radio frame in TDD systems is either two or four.

The frame structure in the IEEE 802.16m is designed for supporting the following functions: 1) multi-carrier operation; 2) legacy frames; 3) legacy frames with a wider-bandwidth channel; 4) relay function; and 5) channel coexistence.

Downlink/Uplink Physical Structure: The 5 ms radio frame is divided into eight sub-frames. Each of the sub-frames can be allocated for downlink transmission or uplink

Table 4.7 OFDMA parameters of IEEE 802.16m

Nominal Channel Bandwidth (MHz)		5	7	8.75	10	20
Over-sampling Factor		28/25	8/7	8/7	28/25	28/25
Sampling Frequency (MHz)		5.6	8	10	11.2	22.4
FFT Size		512	1024	1024	1024	2048
Sub-Carrier Spacing (kHz)		10.937500	7.812500	9.765625	10.937500	10.937500
Useful Symbol Time T_u (μs)		91.429	128	102.4	91.429	91.429
Cyclic Prefix (CP) $T_g = 1/8T_u$	Symbol Time T_s (μs)	102.857	144	115.2	102.857	102.857
	Number of OFDM symbols per Frame	48	34	43	48	48
	Idle time (μs)	62.86	104	46.40	62.86	62.86
Cyclic Prefix (CP) $T_g = 1/16T_u$	Symbol Time T_s (μs)	97.143	136	108.8	97.143	97.143
	Number of OFDM symbols per Frame	51	36	45	51	51
	Idle time (μs)	45.71	104	104	45.71	45.71

transmission. Each sub-frame is then divided into a number of frequency partitions, where each partition consists of a set of physical resource units across the total number of OFDMA symbols available in the sub-frame. Each frequency partition can include contiguous (localized) and/or non-contiguous (distributed) physical resource units. Each frequency partition can be used for different purposes such as fractional frequency reuse (FFR) or multicast and broadcast services (MBS).

For sub-carrier allocation of OFDMA symbols in downlink/uplink transmissions, similar to 802.16e, 802.16m also defines two types of resource units: physical resource unit (PRU) and logical resource unit (LRU). A PRU is the basic physical unit for resource allocation that is comprised of $P_{subcarrier}$ consecutive sub-carriers by N_{symbol} consecutive OFDMA symbols. A LRU is the basic logical unit for distributed and localized resource allocations. A LRU is further classified into two types: logical distributed resource unit (LDRU) and logical localized resource unit (LLRU). The LDRU contains a group of sub-carriers which are spread across the distributed resource allocations within a frequency partition. Therefore, the LDRU can be used to achieve frequency diversity gain. The size of the LDRU is same as the size of PRU, i.e., $P_{subcarrier}$ sub-carriers by N_{symbol} OFDMA symbols. The minimum unit for forming the LDRU is equal to one sub-carrier. The LLRU contains a group of sub-carriers which are contiguous across the localized resource allocations. Therefore, the LLRU can be used to achieve frequency-selective scheduling gain. The size of the LLRU is also equal to the size of the PRU, i.e., $P_{subcarrier}$ sub-carriers by N_{symbol} OFDMA symbols.

Pilot Structure: Similar to IEEE 802.16e, pilot structure is also used to perform channel estimation, measurements (channel quality indicators (CQI) such as Signal-to-Interference-Noise Ratio (SINR) and interference mitigation/cancellation) as well as frequency offset estimation and time offset estimation.

For downlink transmission, IEEE 802.16m supports both common and dedicated pilot structures. The common pilot sub-carriers can be used by all MSs. Pilot sub-carriers that can be used only by a group of MSs is a special case of common pilots and are termed shared pilots. Dedicated pilots can be used with both localized and diversity allocations. The dedicated pilots, associated with a specific resource allocation, can be only used by the MSs allocated to the specific resource allocation. Therefore, they can be precoded or beamformed in the same way as the data sub-carriers of the resource allocation. The pilot structure is defined for up to four transmission (Tx) streams, and there are unified and non-unified pilot pattern designs for common and dedicated pilots. Equal pilot density per Tx stream is usually adopted, while unequal pilot density per OFDMA symbol of the downlink subframe might be used. Furthermore, each PRU of a data burst assigned to one MS has the equal number of pilots.

For uplink transmission, the pilot structure is also defined for up to four Tx streams with orthogonal patterns. Pilot patterns enabling active interference supression algorithms should be employed, which have the following characteristics: 1) pilot locations fixed within each DRU and LRU; and 2) pilot sequences with low cross correlation.

DL Control Structure: DL control channels are needed to convey information essential for system operation. In order to reduce overhead and network entry latency as well as improve the robustness of the DL control channel, information is transmitted hierarchically over different time scales from the superframe level to the sub-frame level. Generally speaking, control information related to system parameters and system configuration is transmitted at the super-frame level, while control and signaling related to traffic transmission and reception are transmitted at the frame/sub-frame level.

The information carried by the DL control channels includes: 1) synchronization information; 2) essential system parameters and system configuration information which include deployment-wide common information, downlink sector-specific information, and uplink sector-specific information; 3) extended system parameters and system configuration information; 4) control and signaling for DL notifications; and 5) control and signaling for traffic.

The transmission of DL Control Information can be performed by using the following types of control channels:

- *Synchronization Channel (SCH)*. The SCH is a DL physical channel which provides a reference signal for time, frequency, and frame synchronization, RSSI estimation, channel estimation and BS identification.
- *Broadcast Channel (BCH)*. The BCH carries essential system parameters and system configuration information. The BCH is divided into two parts: Primary Broadcast Channel (PBCH) and Secondary Broadcast Channel (SBCH).
- *Unicast Service Control Channels*. These channels carry Unicast service control information including both user-specific control information and non-user-specific control information.
- *Multicast Service Control Channels*. These channels carry Multicast service control information/content.

DL MIMO Transmission Scheme: IEEE 802.16m supports both single-user MIMO (SU-MIMO) and multiple-user MIMO (MU-MIMO) schemes. In SU-MIMO, only one user is scheduled in a Resource Unit (RU). In MU-MIMO, multiple users can be scheduled in a RU. Single-user MIMO schemes are used to improve per-link performance. Multi-user MIMO schemes are used to enable a resource allocation to transfer data to two or more MSs. IEEE 802.16m uses Multi-user MIMO to boost system throughput.

SU-MIMO supports both open-loop single-user MIMO and closed-loop single-user MIMO. For open-loop single-user MIMO, both transmit diversity and spatial multiplexing schemes are supported. Transmit diversity modes include:

- 2Tx rate-1: STBC/SFBC, and rank-1 precoder;
- 4Tx rate-1: STBC/SFBC with precoder, and rank-1 precoder;
- 8Tx rate-1: STBC/SFBC with precoder, and rank-1 precoder.

Spatial multiplexing modes include:

- 2Tx rate-2: rate 2 SM;
- 4Tx rate-2: rate 2 SM with precoding;
- 8Tx rate-2: rate 2 SM with precoding;
- 4Tx rate-3: rate 3 SM with precoding;
- 8Tx rate-3: rate 3 SM with precoding;
- 4Tx rate-4: rate 4 SM;
- 8Tx rate-4: rate 4 SM with precoding.

For open-loop single-user MIMO, CQI and rank feedback may still be transmitted in order to assist the base station with rank adaptation, transmission mode switching and rate adaptation. Note that CQI and rank feedback may or may not be frequency dependent. For closed-loop single-user MIMO, codebook based precoding is supported for both TDD and FDD systems. To assist the base station scheduling, resource allocation, and rate adaptation, the following information may be sent by a mobile station:

- Rank (Wideband or sub-band);
- Sub-band selection;
- CQI (Wideband or sub-band, per layer);
- PMI (Wideband or sub-band for serving cell and/or neighboring cell);
- Doppler estimation.

For closed-loop single-user MIMO, sounding based precoding is also supported for TDD systems.

MU-MIMO supports multi-layer transmission with one stream per user. MU-MIMO includes the MIMO configuration of 2Tx antennae to support up to two users and 4Tx or 8Tx antennae to support up to four users. Both CQI feedback and CSI are supported in MU-MIMO. In CQI feedback, both wideband CQI and sub-band CQI may be transmitted by a MS. In CSI feedback, codebook-based feedback is supported in both FDD and TDD, and sounding-based feedback is supported in TDD.

The architecture of downlink MIMO on the transmitter side is shown in Figure 4.22. Similar to IEEE 802.16e, the concepts of "layer" and "stream" are also defined in IEEE

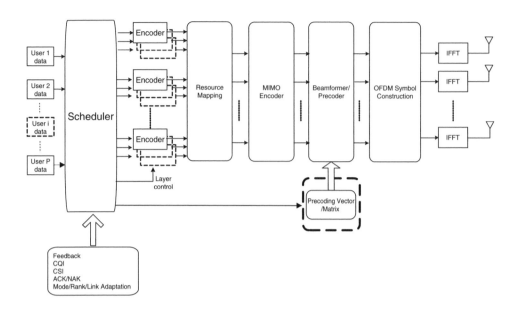

Figure 4.22   IEEE 802.16m Downlink MIMO architecture

802.16m. A "layer" is defined as a coding/modulation path fed to the MIMO encoder as an input, and a "stream" is defined as each output of the MIMO encoder that is passed to the beamformer/precoder. If vertical encoding is utilized, there is only one encoder/modulator block (one "layer"). If horizontal encoding is utilized, there are multiple encoders/modulators (multiple "layers"). The functions of the blocks in Figure 4.22 are listed in Table 4.8.

The BS employs a minimum of two transmit antennae, and the MS employs a minimum of two receive antennae. The antenna configurations are $(N_T, N_R) = (2, 2), (4,2), (4,4),$ $(8,2),$ and $(8,4),$ where N_T denotes the number of BS transmit antennas and N_R denotes the number of MS receive antennae. For layer-to-stream mapping, The number of streams, M, for SU-MIMO is $M \leq \min(N_T, N_R)$, where M is no more than four (eight streams are for further study). MU-MIMO can have up to two streams with two Tx antennae, and up to four streams for four Tx antennae and eight Tx antennae. For SU-MIMO, vertical encoding (SCW) is employed. For MU-MIMO, MCW (or horizontal) encoding is employed at the base-station while only one layer is transmitted to a mobile station.

To support the time-varying radio environment for IEEE 802.16m systems, both MIMO mode and rank adaptation are supported based on the feedback information such as the system load, the channel information, MS speed and average CINR. Switching between SU-MIMO and MU-MIMO is also supported. Some advanced MIMO implementation schemes such as Multi-cell MIMO and MIMO for Multi-cast Broadcast Services are also supported in IEEE 802.16m.

UL Control Structure: Several types of UL control channels are defined in IEEE 802.16m including: 1) UL Fast Feedback Channel; 2) UL HARQ Feedback Channel; 3) UL Sounding Channel; 4) UL Ranging Channel; and 5) Bandwidth Request Channel.

Table 4.8 Block functions of the downlink MIMO architecture in Figure 4.23

Encoder block	Contains the channel encoder, interleaver, rate-matcher, and modulator for each layer.
Resource Mapping block	Maps the modulation symbols to the corresponding time-frequency resources in the allocated resource units (RUs).
MIMO Encoder block	Maps $L(\geq 1)$ layers onto $M(\geq L)$streams, which are fed to the precoding block.
Precoding block	Maps streams to antennae by generating the antenna-specific data symbols according to the selected MIMO mode.
OFDM Symbol Construction block	Maps antenna-specific data to the OFDM symbol.
Feedback block	Contains feedback information such as CQI and CSI from the MS.
Scheduler block	Schedules users to resource blocks and decide their MCS level, MIMO parameters.

UL control channels carry multiple types of UL control information including: 1) Channel quality feedback; 2) MIMO feedback; 3) HARQ feedback; 4) Synchronization; 5) Bandwidth request; and 6) E-MBS feedback.

4.3.4   Other Functions Supported by IEEE 802.16m for Further Study

Other functions supported by 802.16m for further study include: 1) Security; 2)Inter-Radio Access Technology; 3) Location Based Services; 4) Enhanced Multicast Broadcast Service; 5) Multi-hop Relay; 6) Solutions for Co-deployment and Co-existence; 7) Self-organization; and 8) Multi-carrier Operation.

4.4   3GPP2 UMB

UMB stands for Ultra Mobile Broadband, defined by the Technical Specification Group C of the Third Generation 3 Partnership Project 2 (3GPP2). In general, UMB is an evolution of 3G CDMA 1xEv-DO systems. In order to enhance the backward compatibility, 3GPP2 has made UMB to be connected with 3GPP LTE systems, making LTE the single path for 4G wireless network. In this section, we will review the basic features of UMB.

As shown in Figure 4.23, a UMB network usually includes the following key elements, which are defined as follows.

eBS: Evolved Base Station, which provides the over-the-air (OTA) signaling and user-data transport that is used by the AT for connectivity to the radio access network. In addition, an eBS provides the following functions such as: 1) providing a layer 2 attachment point for the AT; 2) acting as a layer 1 attachment point for both forward and reverse links; 3) encryption/decryption of packets at the radio-link protocol (RLP) level for OTA transmission/reception; 4) scheduling for OTA transmission; and 5) header compression. The eBS also provides the following important functions, which are: 1) forward-link serving eBS (FLSE)-Serving eBS for the forward-link physical layer; 2) reverse-link serving

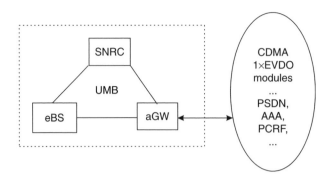

Figure 4.23   3GPP2 Ultra Mobile Broadband system

eBS (RLSE)-Serving eBS for the reverse-link physical layer; 3) signaling radio network controller (SRNC)-Session anchor eBS that stores the session information of ATs and serves as a permanent route to the AT; and 4) data attachment point (DAP)-Receiving eBS for all the data packets from the common AGW. Additionally, an eBS can look into user IP packets and can optimize OTA scheduling or perform other value-added functions.

AGW: Access Gateway, which provides user IP connectivity to the Internet. In other words, an AGW is the first-hop router for a mobile terminal. The AGW performs layer 3 services and above, including hot-lining, policy enforcement and more.

SRNC: Signaling Radio Network Controller, which maintains radio-access-specific information for the AT in the converged access network (CAN). The SRNC is responsible for maintaining the session reference (session storage point for negotiated air-interface context), supporting idle-state management, and providing paging-control functions when the AT is idle. The SRNC is also responsible for access authentication of the AT. The SRNC function may be hosted by an eBS or may be located in a separate location on the wired network.

AAA: Authentication, Authorization and Accounting Function. This functional entity provides functions including authentication, authorization and accounting for the ATs' use of network resources.

HA: Home Agent. The HA is used to provide a mobility solution to the AT in a 3GPP2 packet-data network. However, in an evolved network, the HA may also be used for mobility among networks using different technologies.

PDSN: Packet Data Serving Node, which is the node that provides IP connectivity to the end user in the existing EV-DO or CDMA2000 1X packet- data networks.

PCRF: Policy and Charging Rules Function, which provides rules to the AGW. The purpose of the PCRF rules are to: 1) detect a packet belonging to a service data flow; 2) provide policy control for a service data flow; and 3) provide applicable charging parameters for a service data flow.

4.4.1 Architecture Reference Model

As is shown in Figure 4.24, the architecture reference model of UMB is a flattened network architecture, which is much different from the traditional hierarchical network

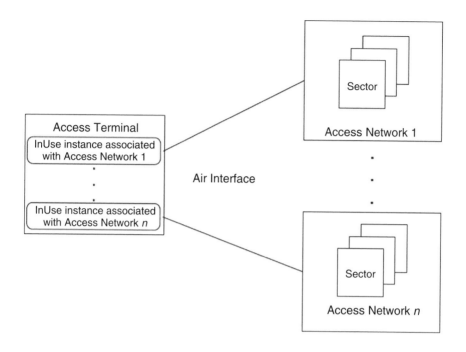

Figure 4.24   UMB architecture reference model

architecture [17]. In the flattened network architecture, each access terminal (AT) can associate with multiple access networks through multiple instantiations of the AT protocol stack (i.e. InUse instances of the protocol stack, which are called routes) via either directly radio link or Inter-Route Tunneling Protocol. This feature will enable UMB to enhance the throughput as well as provide seamless mobility to AT.

4.4.2   Layering Architecture and Protocols

In this section, we will review the UMB layering architecture and the protocols of each layer. In the current UMB standard, there are eight functional entities defined in the layering architecture: application layer, radio link layer, MAC layer, physical layer, security functions, route control plane, session control plane and connection control plane.

Physical Layer: The Physical Layer defines the physical layer protocols, which provide the channel structure, frequency, power output, modulation, encoding, and antenna specifications for both Forward and Reverse Channels. More detail about the Physical Layer protocols can be found in [18].

MAC Layer: The Medium Access Control (MAC) Layer defines the procedures used to receive and to transmit over the Physical Layer. The MAC Layer protocols are shown in Figure 4.26.

The Packet Consolidation Protocol provides transmission prioritization and packet encapsulation for upper layer packets. Superframe Preamble MAC Protocol provides the necessary procedures to be followed by an access network transmit and access terminal

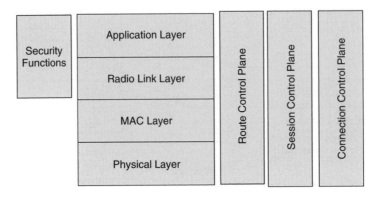

Figure 4.25   UMB layering architecture

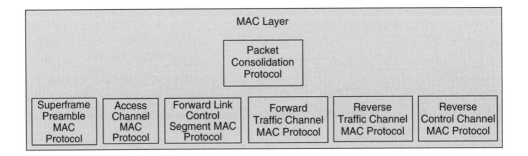

Figure 4.26   UMB MAC layer and protocols

in order to receive the superframe preamble. Access Channel MAC Protocol defines the procedures followed by the access terminal in order to transmit to as well as by an access network in order to receive from the Access Channel. The Forward Link Control Segment MAC Protocol provides the procedures followed by an access network in order to transmit to as well as by the access terminal in order to receive from the Forward Control Channel. The Forward Traffic Channel MAC Protocol provides the procedures followed by an access network in order to transmit to as well as by the access terminal in order to receive from the Forward Data Channel. The Reverse Control Channel MAC Protocol provides the procedures followed by the access terminal in order to transmit to as well as by an access network in order to receive from the Reverse Control Channel. The Reverse Traffic Channel MAC Protocol provides the procedures followed by the access terminal in order to transmit to as well as by an access network in order to receive from the Reverse Data Channel.

Radio Link Layer: The protocols in the Radio Link Layer provide key services to support various applications including reliable and in-sequence delivery of application layer packets, multiplexing application layer packets and Quality of Service (QoS) negotiation. The QoS Management Protocol provides the procedures of negotiation for flow and filter specifications in order to support appropriate Quality of Service (QoS) for application layer

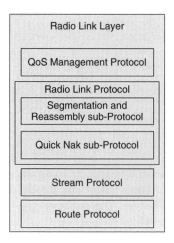

Figure 4.27 Radio link layer and protocols

packets, including negotiation of packet filters and Quality of Service (QoS) for IP packets and mapping of reservations to data streams. The Radio Link Protocol (RLP) provides the functions of fragmentation and reassembly, retransmission and duplicate detection for upper layer packets. The basic RLP includes the Segmentation and Reassembly (SAR) Sub-Protocol, which provides segmentation and reassembly, retransmission and duplicate detection of higher layer packets, and the Quick Nak Sub-Protocol, which provides an indication of higher layer packet losses. RLP uses the Signaling Protocol to transmit and receive messages.

The Stream Protocol provides ways to identify the stream on which the upper layer fragments are being carried. Up to 31 streams can be multiplexed in the Basic Stream Protocol, and each of them can be specified with different QoS (Quality of Service) requirements. The Route Protocol provides procedures to route Stream Protocol packets through the serving route between an access terminal and an access network. The basic protocol functions are: 1) adding Route Protocol Header to identify the protocol stack associated with this route during route initialization; and 2) determining whether Route Protocol packets for transmissions should be delivered to the Packet Consolidation Protocol of this route or the Inter-Route Tunneling Protocol of this or another route. Usually, a route consists of an InUse protocol stack associated with an access network.

Application Layer: The Application Layer provides multiple key applications as shown in Figure 4.28. It provides the Signaling Protocol for transporting air interface protocol messages. It also provides the Inter-Route Tunneling Protocol for transporting packets to or from other routes. Other Application Layer protocols include the EAP Support Protocol for authentication services, the Internet Protocol (IP) for user data delivery, the RoHC Support Protocol for compressing packet headers, and protocols for transporting packets from other air interfaces. These Application Layer protocols are defined in [19]. The Signaling Protocol provides message transmission services for signaling message delivery.

The protocols in each layer of the protocol stack use the Basic Signaling Protocol (BSP) to exchange messages. BSP provides one or two octet headers to define the type of protocol

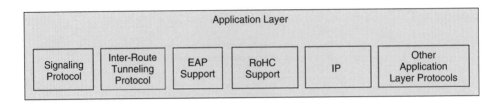

Figure 4.28   Application layer and protocols

used by each message and uses the header to route the message to the corresponding protocol. BSP is a message-routing protocol and routes messages to protocols specified by the Type 7 field in the BSP header. The protocol type is assigned to InUse as well as InConfiguration in the case of each protocol [17].

The Inter-Route Tunneling Protocol provides the transportation of packets from other routes, which performs tunneling of data associated with different routes. The Inter-Route Tunneling Protocol Header indicates the route taken by the payload. The Inter-Route Tunneling Protocol allows one route to carry payload that was originally taking another route as well as its own payload. A route consists of an InUse protocol stack connected with an access network. At the sender, the Inter-Route Tunneling Protocol receives packets from another route or from the same route, as shown in Figure 4.28. The Inter-Route Tunneling Protocol adds an Inter-Route Tunneling Protocol Header to the tunneled packet in order to identify the destination route and sends this packet via the Radio Link Protocol.

Other Application Layer Protocols such as IP, EAP Support Protocol, RoHC Support Protocol and others may create payload to be carried over the UMB air interface.

Connection Control Plane: The Connection Control Plane provides air link connection establishment and maintenance services, which are defined in [20]. The protocols in the Connection Control Plane are control protocols and do not carry the data of other protocols. The protocols in this plane use the Signaling Protocol to deliver messages, except that the Overhead Messages Protocol can send information blocks directly through the MAC Layer. The Connection Control Plane controls the air-link state by managing the states of MAC Layer protocols as well as by changing the operating parameters of MAC Layer protocols.

The Air Link Management Protocol provides the following functions: 1) providing the Connection Control Plane with a general state machine and state-transition rules followed by an access terminal and an access network; 2) activating and deactivating Connection

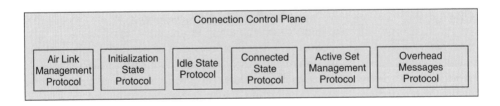

Figure 4.29   Connection control plan and protocols

Control Plane protocols applicable to each protocol state; 3) responding to supervision failure indications from other protocols; and 4) providing mechanisms to enable the access network to redirect the access terminal to another network.

The Initialization State Protocol defines the procedures and messages required for an access terminal to associate with a serving network. This protocol poses two requirements on an access terminal: 1) getting channel band information by using the Air Link Management Protocol; and 2) preventing the access terminal from connecting to an access network with a protocol stack which is not supported by the access network as defined through InitialProtocolSetIdentifiers.

The Idle State Protocol provides the procedures and messages used by both the access terminal and the access network, when the access terminal has acquired a network without an open connection. This protocol operates in one of the following states: inactive state, sleep state, monitoring state and access state. The Connected State Protocol provides the procedures and messages needed by both the access terminal and the access network while a connection is open. This protocol operates in one of the following five states: inactive state, BindATI state, open state, semi-connected state and closed state.

The Active Set Management Protocol provides the procedures to maintain the air link between an access terminal and an access network for transmission, reception and supervision of messages such as the SystemInfo block, the QuickChannelInfo block, the ExtendedChannelInfo message and the SectorParameters message. The SystemInfo and QuickChannelInfo blocks can be broadcast by the access network directly via the Superframe Preamble MAC Protocol. The ExtendedChannelInfo and SectorParameters messages are broadcast by using the Signaling Protocol. This protocol operates in one of the two states: active and inactive. The Overhead Messages Protocol defines the procedures and messages used by the access terminal and the access network in order to maintain the radio link when the access terminal moves among the coverage areas of different sectors. It operates in one of three states: inactive, idle, and connected.

Session Control Plane: The Session Control Plane provides services for session negotiation and configuration. A session is defined as a shared state maintained between the access terminal and the access network. During a session, both the access terminal and the access network can open and close a connection many times. In other words, sessions are closed rarely, except on occasions such as when an access terminal leaves the current coverage area or the access terminal is unavailable. The Session Control Plane is defined in [21]. The Session Control Protocol provides the means to allow for the negotiation of the protocol configuration parameters during a session. Moreover, this protocol can determine that a session is still alive as well as close a session.

Route Control Plane: The Route Control Plane provides procedures for the creation, maintenance and deletion of routes, which is defined in [22]. The Route Control Protocol performs mainly the following functions: 1) managing the creation and deletion of routes; 2) maintaining the mapping between route identifier (RouteID) and Access Network Identifier (ANID); 3) maintaining the identifier of DataAttachmentPoint (DAP) route and SessionAnchor route; and 4) handling UATI and PagingID assignment. The protocol can be operated in one of three states: WaitingToOpen, Open and WaitingToClose.

Security Functions: The security functions include key services for key exchange, ciphering and message integrity protection, which are defined in [23].

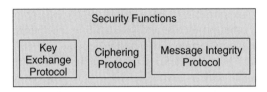

Figure 4.30 Security functions and protocols

The Key Exchange Protocol specifies the procedures for security key generation used by both access networks and access terminals for message integrity protection and ciphering based on a Pair-wise Master Key, which is established by higher layer protocols. The basic functions for the protocol are: 1) authenticating that both access terminal and access network having the same Pair-wise Master Key; 2) calculating security keys from the Pair-wise Master Key; and 3) blocking a man-in-the-middle attack, where a rogue entity causes both an access terminal and an access network to agree upon a weaker security protocol. The Message Integrity Protocol defines the procedures needed by both access network and access terminal for protecting the integrity of signaling messages through applying the AES CMAC function. The Ciphering Protocol specifies the procedures used by an access network and an access terminal for securing the traffic, where the AES procedures are adopted to encrypt and decrypt the Radio Link Protocol packets.

Broadcast-Multicast Service (BCMCS) Upper Layer Protocols: The broadcast packet data system provides a streaming mechanism in order to deliver higher layer packets over an access network to multiple access terminals. The Forward Broadcast and Multicast Services Channel transports packets from a content server, which can also carry Forward Link signaling messages generated by the Broadcast Protocol Suite as well as payload from other routes. The Forward Broadcast and Multicast Services Channel has a Forward Link, but it does not have a Reverse Link. Forward Link messages are transmitted directly on the Forward Broadcast and Multicast Services Channel or tunneled via the Inter-Route Tunneling Protocol of a unicast route. Reverse Link messages are tunneled through the Inter-Route Tunneling Protocol of a unicast route. Furthermore, the Forward Broadcast and Multicast Services Channel includes Broadcast Physical Channels and Broadcast Logical Channels.

Broadcast-Multicast flows (also called BCMCS flows) and signaling messages are associated with Broadcast Logical Channels and are transmitted over Broadcast Physical Channels. The Broadcast Physical Channels consist of several sub-channels called interlace-multiplex pairs, which may be different across sectors. The Basic Broadcast Protocol Suite specifies a Broadcast MAC Protocol and a Broadcast Physical Layer Protocol which describe the structure of Broadcast Physical Channels.

A Broadcast Logical Channel (also called logical channel) refers to a set of interlace-multiplex pairs of the Broadcast Physical Channel associated with a sector over which broadcast content is transmitted. Each logical channel can carry multiple BCMCS flows. An interlace-multiplex pair associated with a sector can be assigned to at most one logical channel. A logical channel is identified by the parameter pair (sector, BCIndex), where a sector is determined by the parameter pair (SectorId, BCMCS Channel). The BCMCS Channel indicates the frequency assignment of a single

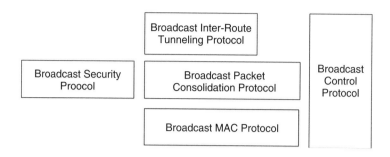

Figure 4.31   Broadcast-Multicast service upper layer protocols

channel. BCIndex refers to the index value of the first PHY frame, starting from zero, among all PHY frames of the set of interlace-multiplex pairs associated with the logical channel. A Broadcast-Multicast Service Flow identifier (BCMCSFlowID) identifies a Broadcast-Multicast flow (also called a BCMCS flow). The content of a given BCMCS flow may change with time (for example, BCMCS flow is a single multimedia flow). The content of a BCMCS flow cannot be further divided across multiple logical channels.

As was discussed earlier, the protocols in the BCMCS Upper Layer Protocol Suite provide functions that offer broadcast-multicast service. The Broadcast MAC Protocol defines the procedures for data transmission via the Forward Broadcast and Multicast Services Channel. The Broadcast MAC Protocol also provides Forward Error Correction (FEC) and multiplexing in order to reduce the radio link packet error rate. The Broadcast Packet Consolidation Protocol provides a framing service to higher layer packets and multiplexes, higher layer packets and signaling messages. Broadcast Inter-Route Tunneling Protocol provides a tunneling service to packets generated by the unicast routes on the Broadcast Physical Channel. The Broadcast Security Protocol provides a ciphering service to the packet payload of the Broadcast Packet Consolidation Protocol. The Broadcast Control Protocol defines the control procedures used to control various aspects of the operation of the broadcast packet data system, such as BCMCS flow registration and the Broadcast Parameters message. The Broadcast Physical Layer provides the channel structure for the Forward Broadcast and Multicast Services Channel.

Acknowledgements

The authors appreciate the kind permission from ITU, 3GPP, and 3GPP2 to reproduce the related materials in this chapter. The 3GPP2 figures and related material are reproduced by the written permission of the Organizational Partners of the Third Generation Partnership Project 2 (http://www.3gpp2.org).

References

1. "3GPP TD RP-040461: Proposed Study Item on Evolved UTRA and UTRAN".
2. "3GPP, TR25.896, Feasibility Study for Enhanced Uplink for UTRA FDD".
3. "3GPP TR 25.913, Requirements for evolved Universal Terrestrial Radio Access (UTRA) and Universal Terrestrial Radio Access Network (UTRAN)".

4. "3GPP TR 25.814, Physical Layer Aspects for Evolved Universal Terrestrial Radio Acces (UTRA)", 2006.

5. "3GPP R1-061118, E-UTRA physical layer framework for evaluation, Vodafone, Cingular, DoCoMo, Orange, Telecom Italia, T-Mobile, Ericsson, Qualcomm, Motorola, Nokia, Nortel, Samsung, Siemens".

6. "3GPP RP-060169, Concept evaluation for evolved UTRA/UTRAN, Cingular Wireless, CMCC, NTT DoCoMo, O2, Orange, Telecom Italia, Telefonica, T-Mobile, Vodafone".

7. "3GPP, TR25.848, Physical Layer Aspects of UTRA High Speed Downlink Packet Access".

8. "3GPP, TR 25.942 V3.3.0 (2002-06), RF System Scenarios, June 2002".

9. H. E. *et al*, "Technical Solutions for the 3G Long-Term Evolution[J]", *IEEE Commun. Mag.*., pp. 38–45, Mar. 2006.

10. "ETSI TR 101 112 (V3.1.0): Universal Mobile Telecommunications System (UMTS); Selection procedures for the choice of radio transmission technologies of the UMTS (UMTS 30.03 version 3.1.0)".

11. "3GPP, TR 23.882, 3GPP System Architecture Evolution".

12. The Draft IEEE 802.16m System Description Document (SDD), IEEE 802.16 Broadband Wireless Access Working Group, Jul. 2008.

13. IEEE Std. 802.16-2004: IEEE Standard for Local and metropolitan area networks Part 16: Air Interface for Fixed Broadband Wireless Access Systems, Jun. 2004.

14. IEEE std. 802.16e-2005: IEEE Standard for Local and metropolitan area networks Part 16: Air Interface for Fixed and Mobile Broadband Wireless Access Systems, Amendment 2: Physical and Medium Access Control Layers for Combined Fixed and Mobile Operation in Licensed Bands, and IEEE Std. 802.16-2004/Cor1-2005, Corrigendum 1, Dec. 2005.

15. IEEE Std 802.16j-2008: Draft Amendment to IEEE Standard for Local and Metropolitan Area Networks – Part 16: Air Interface for Fixed and Mobile Broadband Wireless Access Systems – Multihop Relay Specification, Jun. 2008.

16. IEEE 802.16m System Requirements Document (SRD), IEEE 802.16m-07/002r5, Jun. 2008.

17. C.S0084-000-0, Overview for Ultra Mobile Broadband (UMB) Air Interface Specification, Aug. 2007.

18. C.S0084-002-0, MAC Layer for Ultra Mobile Broadband (UMB) Air Interface Specification, Aug. 2007.

19. C.S0084-004-0, Application Layer for Ultra Mobile Broadband (UMB) Air Interface Specification., Aug. 2007.

20. C.S0084-006-0, Connection Control Plane for Ultra Mobile Broadband (UMB) Air Interface Specification, Aug. 2007.

21. C.S0084-007-0, Session Control Plane for Ultra Mobile Broadband (UMB) Air Interface Specification, Aug. 2007.

22. C.S0084-008-0, Route Control Plane for Ultra Mobile Broadband (UMB) Air Interface Specification, Aug. 2007.

23. C.S0084-005-0, Security Functions for Ultra Mobile Broadband (UMB) Air Interface Specification, Aug. 2007.

5

Advanced Video Coding (AVC)/ H.264 Standard

5.1 Digital Video Compression Standards

It has long been recognized that to promote interoperability and achieve economics of scale there is a strong need to have international standards. Two organizations – ITU-T and ISO/IEC – have been most involved in setting those standards. In ITU-T, the Video Coding Experts Group (VCEG) under Study Group 16 has focused on developing video coding standards. In ISO/IEC the Moving Picture Experts Group (MPEG), formally known as Working Group 11 (WG-11) under SC-29 has concentrated on developing international coding standards for compressing digital video and audio. In ITU-T the main focal applications have been Video Conferencing and Video Telephony and in MPEG the main focal applications have been Digital Television transmission and storage. However, more recently, those lines have been blurring. ITU-T and ISO/IEC jointly developed two video coding standards: 1) MPEG-2/H.262 [1]; and 2) MPEG-4 Part 10/H.264 [2] also known as Advanced Video Coding (AVC). As the latter standard is known as both MPEG-4 AVC and H.264, in this text it is referred to as AVC/H.264 or sometimes simply as AVC. This standard is expected to play a critical role in the distribution of digital video over 4G networks. Full detailed description of this standard will require a complete book dedicated only to this subject. Due to limited space, only a high level description of this standard is provided to whet the appetite and give reader a good basic understanding of the coding tools used in the standard that provide higher coding efficiency over previous standards.

The movement towards creating international standards started in the mid 1980s with ITU-T approving H.120 standard in 1988. After that, as shown in Figure 5.1, several international standards have been developed and are maintained by ITU-T and ISO/IEC. This figure provides only a rough approximate order of the development time frame of these standards. The time line and time periods shown in the figure correspond approximately to the time when the bulk of the technical work was done. Amendments, extensions and corrections to these standards continued long after the periods shown.

4G Wireless Video Communications   Haohong Wang, Lisimachos P. Kondi, Ajay Luthra and Song Ci
© 2009 John Wiley & Sons, Ltd

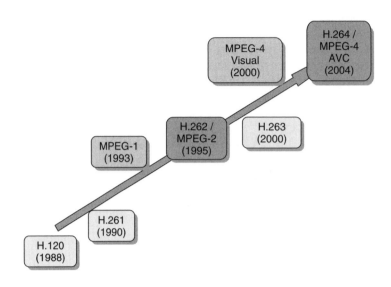

Figure 5.1   International digital video coding standards

In the 1980s and 1990s it was generally believed that one would need different basic codec structures and algorithms depending upon the compressed bit rate. In video conferencing it was believed that one would need a separate standard for compression for bit rates that were a multiple of 64 (m × 64) up to 384 kbps and for bit rates higher than 384 kbps. It was not the case. Similarly, it was also widely believed that one would need separate compression standards for compressing Standard Definition Television (SDTV) and High Definition Television (HDTV). MPEG-2 was in the beginning targeted towards SDTV and the name MPEG-3 was reserved for compressing HDTV. As the understanding of compression algorithms matured, it was discovered that there was no such need and that the basic algorithm developed for SDTV was also equally appropriate for HDTV. The development of MEPG-3 was dropped and the MPEG-2 specification was extended to cover the resolutions and bit rates corresponding to HDTV.

The MPEG-2 standard was developed jointly by ISO/IEC's WG-11 under SC29 and ITU-T and is also known as the ITU-T H.262 standard. After the completion of this standard, WG-11's and ITU-T's video coding groups moved in two separate directions. WG-11 started to work on MPEG-4 Part 2 [3], which was focused not only on improving coding efficiency over MPEG-2 but also on the interactive applications. This was the first standard that developed object oriented compression methodology. However, there have been few applications developed for the object oriented representation of video. Therefore, the most used profiles for MPEG-4 Part 2 so far are Simple Profile (SP) and Advanced Simple Profile (ASP). SP was focused on lowering the complexity of the compression algorithm over MPEG-2 and ASP was focused on increasing coding efficiency over MPEG-2. ASP was successful in improving coding efficiency by a factor of about 1.4 over MPEG-2. ITU-T's Video Coding Experts Group (VCEG) focused on developing more efficient video coding standards, known as H.263 and H.263++, for video phone and conferencing applications [4]. After these standards were developed, VCEG began to work towards improving compression efficiency under the informal project name of H.26L. The resulting H.26L codec showed promising coding efficiency and performed very well in the Call for Proposal (CfP) sent out by MPEG for the next generation codec beyond MPEG-4 Part 2 and MPEG-2. As a result, it was agreed by WG-11 and SG-16 that MPEG and VCEG should join forces with the aim of developing jointly the next generation of advanced video standards. The Joint Video Team (JVT), consisting of experts from MPEG and VCEG, was formed in December 2001 to develop the next generation of video coding standards. As H.26L performed well in MPEG's testing of various algorithms, it was decided to keep the basic syntax of H.26L and to extend it so as to expand the focus from video phone and video conferencing applications to also include digital TV and digital video creation in studio applications and to add new coding tools in order to improve its performance for those applications. One of the prime goals was to improve the coding efficiency by a factor of 2 over MPEG-2.

Figure 5.2 provides some examples of the coding efficiency comparisons of MPEG-2, MPEG-4 ASP and MPEG-4 AVC/H.264 for standard MPEG test sequences 'Mobile & Calendar' and 'Bus'. In Figure 5.2, CIF stands for the Common Intermediate Format video with a picture (frame) resolution of 352×288 pixels, while HHR represents the Half-Horizontal Resolution video with a picture (frame) resolution of 352×480 pixels. It is important to note that the results will vary from codec to codec and sequence to sequence.

In this figure the bit rates are such that they provide 32 dB Peak Signal to Noise (PSNR) and are normalized for the MPEG-2 bit rate to be 100% for each sequence. As shown in equation (5.1), PSNR is obtained by taking the ratio of the peak signal power to the compression noise power, where the compression noise power is obtained by subtracting the decompressed video from the original video and obtaining the average power. PSNR

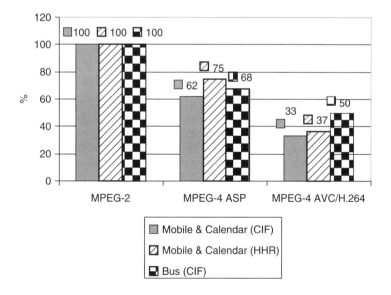

Figure 5.2   Percentage bit rate required for the same PSNR ($\sim$32 dB)

in dB is given by:

$$PSNR = 10\log_{10}\left[\frac{255^2}{\dfrac{1}{Total\ pixels}\displaystyle\sum_{pixels}(original\ pixel\ value - decoded\ pixel\ value)^2}\right]$$

(5.1)

Over the last 15 years, the coding efficiency of the coding algorithms and standards has improved by a factor of about 4. Interestingly, the basic core structure has remained the same – motion compensated hybrid transform-DPCM. The transform has also remained the same – Discrete Cosine Transform (DCT) or DCT-like integer transform. Several rounds of testing have been done at the international level and this hybrid coding structure with DCT or DCT-like transform always came out ahead in terms of coding performance. On the algorithm side, the primary reason for the improvement in coding efficiency has been the ability to perform more accurate predictions of the pixels from their spatially and temporally neighboring pixels. This is due to a significant increase in the complexity of the spatial and temporal prediction algorithms over the years. As implementation and integration technology progressed, more and more complex algorithms for performing predictions could be added to the standards. This has resulted in lower prediction error in the pixels being compressed thus requiring a lesser number of bits and therefore providing better coding efficiency. In addition, these new algorithms also made AVC/H.264 the first standard to bridge the gap between video conferencing and digital television applications. Until the development of this standard, there were two categories of standards. One, such as MPEG-2, that was more efficient and used for entertainment video (TV) and another, such as H.263 or H.263++, was more efficient and used for video conferencing and telephony. AVC/H.264 was the first standard that was equally applicable and highly efficient for both categories and the entire range of bit rates – from 32 kbps to more than 250 Mbps. That is five orders of magnitude of bit rate range!

5.2   AVC/H.264 Coding Algorithm

As is also the case with the previous standards, the AVC/H.264 standard standardizes decoder and bit stream syntax. This specifies indirectly the coding tools which can be used by an encoder to generate a bitstream that can be decoded by a compliant decoder. Figure 5.3 shows basic coding structure of the corresponding encoder.

Digital video is a three – two spatial and one temporal – dimensional signal. Therefore, the prediction of a certain pixel value can be done based on the neighboring pixels in the same picture (spatial prediction) or the in other pictures in the past or the future (temporal prediction). Once the predicted value (at point D) is obtained, the difference of the true value and the predicted value is calculated. This difference is also called the prediction error. A DCT-like (sometimes also called Integer DCT) transform of the difference signal is taken and the transform coefficients are quantized. In some coding modes, described later, a scaling matrix can also be applied to each coefficient in order to shape the coefficient values so that higher frequency coefficients are quantized more than the lower ones. The quantized coefficients are then scanned out as a one dimensional signal and passed through two possible lossless entropy coding algorithms – Context

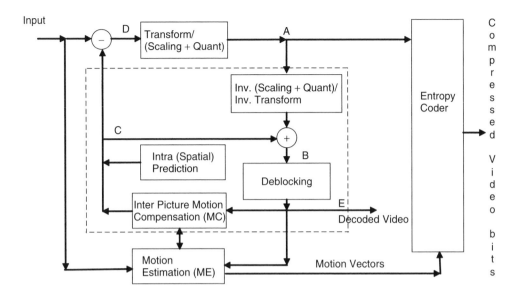

Figure 5.3   AVC encoder structure

Adaptive Variable Length Coder (CAVLC) or Context Adaptive Binary Arithmetic Coder (CABAC).

In this figure, one more layer of temporal prediction – Motion Estimation (ME) and Motion Compensation (MC) – and intra (Spatial) prediction are shown. As described in sections 3.4.1, 3.4.2 and 3.4.3, the area within the dotted box corresponds to the local decoder loop within the encoder and the signal at point E is the same as the output of a decoder. Motion is estimated between current blocks and the *decoded* blocks in previously compressed pictures. As explained in section 3.4.1, this avoids drifting between an encoder and a decoder. In many implementations ME is done in two stages. In the first stage motion is estimated as a pre-encoding step (which can be done ahead of the rest of the encoding process) by using original pictures and MVs are then refined to the final values by using the decoded reference pictures to obtain the prediction error values at point D in Figure 5.3. There is an additional functional block called Deblocking in that loop. At a high level, the functional blocks of the encoder are as shown in Figure 5.4 and are described in more detail below.

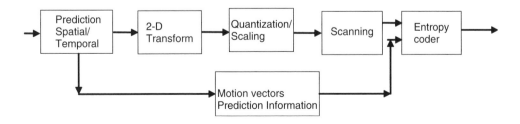

Figure 5.4   Functional blocks of an AVC encoder

5.2.1 Temporal Prediction

Temporal prediction consists of two steps – 1) Motion Estimation (ME); and 2) Motion Compensation (MC). Motion Estimation is the single most computationally intensive step in an encoder. As the name suggests, in this step an estimate of the motion of the pixels in a picture is obtained. This allows one to estimate where the pixels belonging to a certain object in a picture were in the past and will be in the future. The Motion Compensation step uses the motion vectors obtained in the ME step and provides the estimate of the current pixel value. If the estimate is exact, the difference between the pixel values in other pictures and that in the current picture will be zero. However, there are also some changes that happen between the pixel value in the current picture and other pictures. Those could be due to the rotation of an object, occlusion, change in the intensity and many other such reasons. Therefore, the difference between the predicted value and the true value of a pixel is not always zero. The difference is then transformed and quantized. The decoder also needs to know the motion vectors used to create the difference. The encoder compresses the motion vectors (MVs) used for prediction and sends them along with the coefficient values.

5.2.1.1 Motion Estimation

It is not only very hard and costly to estimate the motion of every pixel, but it also takes a large number of bits to send the motion vectors (MVs) to the decoder. Therefore, motion is estimated for a group of pixels together. The group of pixels was selected to be 16×16 (16 pixels horizontally and 16 pixels vertically) in earlier standards such as H.261 and MPEG-2. This group of pixels is called the macroblock (MB). In MPEG-2 for interlaced video, the size of the macroblock was 16×8 for the macroblocks that were coded in the field mode. The size of the MB was reached as a compromise between implementation complexity and compression efficiency. In the 1990s it was very costly to implement ME and MC blocks smaller than 16×16 in an encoder and a decoder. When AVC/H.264 was developed the Very Large Scale Integration (VLSI) technology had progressed significantly and it was reasonable to implement motion estimation for an area smaller than 16×16. In this standard, motion can be estimated for groups of pixels of sizes 16×16, 16×8, 8×16, 8×8, 8×4, 4×8 and 4×4, as shown in Figure 5.5. A 16×16 MB can be further partitioned into two 16×8, or two 8×16, or four 8×8 size MBs. An 8×8 MB can be further partitioned into two 8×4, or two 4×8, or four 4×4 sub-macroblocks. In MPEG-2 the term macroblock was used to refer to the group of pixels (16×16 or 16×8 for field MB) to which MV was assigned for motion prediction. This size was different than the size of the group of pixels for which transform (8×8) was taken and that group was referred as a 'block'. In AVC/H.264 that line of separation

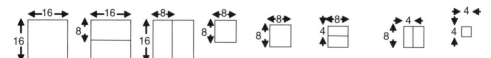

Figure 5.5(a)   MB sizes

Figure 5.5(b)   Sub-macroblock partitions of 8×8 MB

between the MV block and transform size block has blurred. Therefore, the terms MB, sub-MB, 'block' and 'partition' are used interchangeably in this text unless there is a need to be explicit. The context of the terms makes it clear whether a block corresponding to MV is referred to or a block corresponding to transform is referred to.

An encoder is given the flexibility to select the size that is the most appropriate for a given region of a picture and the cost of ME. As shown in Figure 5.6, in the areas closer to the moving edges, encoders can break a MB in smaller sizes.

In this way a better motion estimation can be obtained around moving edges which will result in a smaller prediction error. Note that breaking an MB into smaller parts increases

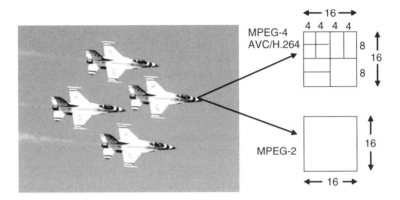

Figure 5.6(a) Picture to be compressed

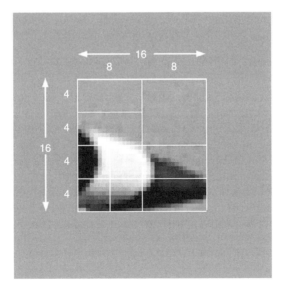

Figure 5.6(b) Zoomed portion at the nose of a plane in Figure 5.6(a) showing need for variable block size for the motion estimation

the number of bits required to send the motion information to the decoder as there will be more motion vectors associated with a greater number of sub-macroblocks. Therefore, an encoder needs to make smart decisions and do a trade-off between the higher cost (number of bits) of sending more motion vectors versus the lower cost (number of bits) associated with sending smaller prediction errors.

As with earlier standards, the AVC/H.264 standard does not describe how motion is estimated for a given macroblock/sub-macroblock. It is left up to the encoders to develop their own algorithms. There are many different algorithms developed in order to estimate the motion [5, 6]. They provide different implementation cost versus coding efficiency trade offs. They can be classified into two main categories:

1. Time domain.
2. Frequency domain.

The majority of motion estimation algorithms fall within the time domain category. In the time domain block matching based ME approaches are common where the current block is matched against the blocks in past or future pictures. The match with the smallest difference – typically the Sum of Absolute Difference (SAD) – is taken to be the location of that sub-macroblock in that picture and is used as reference for prediction. The full search method, where every sub-macroblock in the reference picture is compared against the current sub-macroblock in the current picture, provides the minimum SAD value. However, it sometimes becomes very computationally intensive especially for an HD picture. Therefore, many fast motion estimation techniques have been developed. They provide various cost-performance compromises and are used depending upon the implementation architecture.

A Rate-Distortion (RD) optimization based approach for ME is also used by many encoders. In this approach the cost of representing motion vectors and the prediction error are modeled as the following Lagrangian cost function:

$$C = D + \lambda \times R \tag{5.2}$$

where D is the measure of the distortion in the picture and R is the total number of bits for representing motion information. Typically, D is based on the sum of the absolute difference of the luminance components corresponding to various possible motion vectors. In this approach a motion vector is selected that minimizes the cost function C. Unfortunately, as there is no close form value available for λ, its value is decided experimentally. This type of RD optimization requires significant computational power that may not be available in some implementations. Thus, encoders need to make implementation specific compromises on RD optimization or use other motion estimation methods. For example, many encoders do not use sub-8 × 8 sizes while encoding HD or SD size pictures as they are more useful at SIF/CIF/QVGA or smaller picture sizes and some encoders may choose to use only 16 × 16 and 8 × 8 partitions.

5.2.1.2   P and B MBs

A motion predicted MB can be of one of the two types – Predicted (P) or Bi-predicted (B). In a P MB there is only one MV associated with that MB or its partitions. However, unlike MPEG-2, that motion vector may be derived from reference pictures that can either be in the past or in the future in a captured video sequence relative to the current picture being compressed. In B MB, up to two motion vectors may be used by the MB or sub-MBs. Unlike MPEG-2, both the reference pictures corresponding to those motion vectors can either be in the past or future or one in the past and one in the future relative to the current picture being compressed.

5.2.1.3 Multiple References

In AVC/H.264, more than one picture can be used as references in order to estimate the motion of various different parts of a picture, as shown in Figure 5.7. Therefore, if motion is such that there is an occlusion and parts of an object in the current picture are hidden in the picture that is in the immediate past or future, one can look for a better match in other pictures that are further away and may contain a better match. This is also useful in the situation where the motion is periodic and a better match for the current macroblock is not necessarily in the immediate neighboring picture. Many times, a video consists of pictures with sudden changes in brightness due to flashes. Those flashes interfere with motion estimation. In that case, a better estimate of motion can be obtained by skipping over the flashes and using pictures further away in time. Therefore, in AVC, the use of multiple reference frames allows for an encoder to optimize and adapt the ME process to the content being compressed.

5.2.1.4 Motion Estimation Accuracy

A picture is a sampled image in which the motion is continuous. Therefore, a better estimate of motion can be obtained by interpolating between the pixels, especially for the sharp moving edges. Due to implementation costs, the interpolation in MPEG-2 was

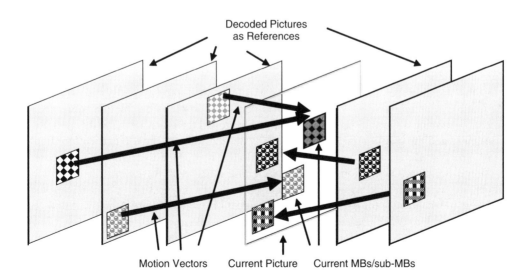

Figure 5.7 Multiple reference pictures

limited to a factor of 2 and motion was estimated with a $1/2$ pixel accuracy. In AVC/H.264 (and also in MPEG-4 Part 2) the interpolation was expanded to be up to a factor of 4 and the motion can be estimated with 1/4th pixel accuracy. For the luma samples, the 1/4th pixel interpolation is done in two steps. In the first step, pixel values at half way between the pixels in the captured picture are calculated by using a 6-tap filter $(1, -5, 20, 20, -5, 1)/32$. The pixel values at 1/4th the pixel locations are then calculated by using linear interpolation. For chroma samples, the prediction values are obtained by bilinear interpolation. For 4:2:0, since the sampling grid of chroma has half the resolution of the luma sampling grid, the displacements used for chroma have a 1/8th sample position accuracy [12].

5.2.1.5 Weighted Prediction

Unlike prior standards, the reference pixels can also be weighted by a weighting factor before obtaining the prediction error. This can improve coding efficiency significantly for scenes containing fades, and can be used flexibly for other purposes as well. In a fade, the intensity of the video picture dims to black or comes out of black to a regular intensity. Owing to changes in intensity between pictures, a proper weighting in order to compensate for the variation in the intensity provides a better estimate of the current pixel values and hence results in smaller prediction error. The standard does not specify an algorithm for finding the weighting and it is left up to the encoders to implement any desired algorithm. The standard only specifies the syntax corresponding to the weighting values and how to send them to a decoder, if used.

5.2.1.6 Frame and Field MV

A video sequence may consist of a progressively or interlaced scanned picture. In progressively scanned pictures the lines in a video frame are captured contiguously. Interlaced scanned frame consists of two fields. In interlaced scanned pictures the alternate lines are captured at one given time (see Figure 3.1). In a 525 line standard the spacing between the two fields is 1/60th of second. As two fields are captured at different time, parts of the scene that are in motion appear at different locations in a frame, even though they belong to the same object. AVC/H.264 gives the encoder the option (there are Profile related restrictions that are discussed below) of breaking a MB in a frame into two fields and estimating the motion separately for each field. This is beneficial in obtaining motion more accurately for moving objects, especially for the pixels around the edges or when the motion is non-linear and the displacement of the pixels is not the same in the top and bottom field.

AVC/H.264 allows for two degrees of freedom with regard to frame or field based motion estimation. One option encoders may chose is to break the entire frame into two fields and treat each field as an independent picture for compression. This decision can be made independently for each picture. This process is called Picture Adaptive Frame or Field (PicAFF or PAFF) based compression. Another choice which encoders have is to compress the frames in the frame mode and make the frame and field decision adaptively and independently for a portion of a frame. The smallest size of the frame for which a frame or field decision can be made is 16×32, which is 2 MB high. This process is called MacroBlock Adaptive Frame Field (MBAFF) compression. Notice that, unlike the

MPEG-2 standard, in MBAFF the field/frame decision is made at MB-pair level and not at each MB level. The reason for selecting (vertical) MB-pair as the smallest quanta for that decision was that once a 16×32 size area is divided into two fields, it provides two macroblocks, each of 16×16 size and now all the coding tools, e.g. 16×16 Spatial Prediction, 16×16 or 16×8 or 8×16 ME, etc., used in the standard for a frame MB can now also be used for a field MB. However, note that the same numbers of lines in a field span twice as large a space as the same numbers of lines in a frame. This not only impacts upon the decision to choose the frame or field MB mode but also other mode decisions made while compressing a MB. Ideally, it would provide the best compression by compressing an MB in all three PicAFF, MBAFF and frame only modes and picking the one that provides the best compression. However, that becomes a very computationally intensive and expensive process. Therefore, the decision of which of these three modes to use is made fairly early in the compression process. Those decisions are largely based on the motion in the frame. Frames with large motion (e.g. frames with panning) are generally compressed using PicAFF. The MBAFF compression tool can provide better coding efficiency for frames with mixed, moving and static, zones. In PicAFF, as each field is compressed as independent pictures, the MBs in the second field of a frame can use MBs in the first field as references. However, in MBAFF the decoding order is as shown in Figure 5.8 and the current MB can not use the MBs in the same frame that occurs in the future because they are not available (i.e. compressed and decompressed) when the current MB is being compressed.

Although the decoding process for the interlaced scanned video is not significantly harder or more complex than a progressively scanned video, compressing an interlaced scanned video at the encoding end is significantly more complex. That complexity gives rise to relatively less coding efficiency in a typical encoder available today for an interlaced scanned video than that for a progressively scanned video.

5.2.1.7   MV Compression

The motion vectors used to estimate the motion and prediction of the current MB need to be conveyed to the decoder so that it can also use the same motion

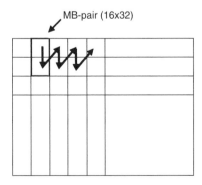

Figure 5.8   Decoding order for MB AFF frame

vectors to calculate the pixel values from the prediction errors and motion vector information. Therefore, the motion vector information is also compressed before sending it to a decoder. Techniques used to compress MV can be divided into two categories.

In the first category, motion vector corresponding to a sub-MB is compressed by a DPCM-like process. A reference MV is computed based on the neighboring blocks. As shown in Figure 5.9(a), for 8×16 partitions the motion vector of neighboring block to the left or up in the diagonal right position is used as reference. For 16×8 partitions the motion vector of neighboring blocks up or to the left of the partition is used as reference.

For other partitions the median MV of the upper, upper right and left neighboring blocks is taken as reference. An example is shown in Figure 5.9(b) where neighboring blocks N_1, N_2 and N_3 are the neighbors used. The details of exactly which neighbors to use for various cases (such as MBAFF, non-availability of neighbors, etc.) are provided in the standard. The difference of the MV corresponding to the current MB and the reference MV is compressed using lossless variable length coding.

In the second category, the motion vector of a MB is not sent explicitly to a decoder and is derived by the decoder based on the motion vectors of neighboring pixels in temporal and spatial dimensions. An encoder can use one of the two possible modes: Direct and Skip modes. There are two types of Direct modes: Temporal and Spatial. In Temporal Direct mode the MV is derived by scaling the MV of the co-located pixels, as shown in Figure 5.9(c).

In Spatial Direct mode the MV of the current block is derived by examining the motion vectors of a collocated MB without the scaling process, and using the motion vectors of neighboring blocks for generating prediction motion vectors.

Direct modes are very helpful when there is high correlation among the MVs of the blocks within a region. This allows for a significant reduction in bit rate as the MVs of the current blocks are not sent explicitly in the bit streams.

Like Direct mode, there is another mode where the MVs are not sent explicitly and are derived by the decoders based on the MVs of the neighboring blocks. It is called the Skip mode. It exploits the fact that sometimes not only the motion is highly correlated but also the prediction is so good that there are no prediction errors after quantization. In this mode neither the MV are sent explicitly nor the coefficient values.

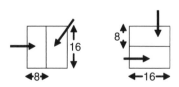

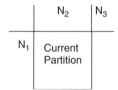

Figure 5.9(a) Prediction directions for motion vectors for 8×16 and 16×8 partitions

Figure 5.9(b) MV prediction based on median MV of neighbors' MVs

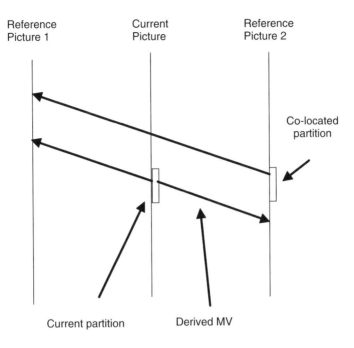

Figure 5.9(c)   MV prediction in direct mode

5.2.2 *Spatial Prediction*

AVC.H.264 allows for extensive spatial prediction modes. These modes are used when temporal prediction is not used. MB with no temporal prediction is also called Intra (I) MB. The spatial prediction of I MB is called intra prediction. In AVC/H.264 such prediction is done in pixel-domain. By comparison, in MPEG-2 only simple intra predictions of the DC coefficients are performed.

An encoder may use one of the three possible spatial prediction modes: 16×16 luma (for chroma, corresponding chroma block size is used), 8×8 luma and 4×4 luma (there are profile related restrictions, discussed below, on when these modes can be used). In 16×16 mode, the pixel values of a 16×16 MB can be predicted in one of four different modes: Vertical, Horizontal, DC and Plane. In Horizontal prediction, the pixels in an MB are predicted from the pixels in the left side, $P(-1,j)$ of the MB, as shown in Figure 5.10.

The -1 value in the notation $P(-1,j)$ denotes the column of pixels left of the 16×16 macroblock and j, from 0 to 15, is the pixel number along the y-axis. In Vertical prediction, the pixels in an MB are predicted from the pixels on the top of the MB, pixels $P(i, -1)$, shown in Figure 5.10, where -1 signifies the row above the top row of the 16×16 macroblock and i, ranging from 0 to 15, is the pixel number along the x-axis. In the DC prediction mode, the pixels are predicted from the DC (average) values of the neighboring pixels. In the Plane mode it is assumed that intensity of the pixels increases or decreases linearly in the MB. The prediction mode that gives the least prediction error is chosen.

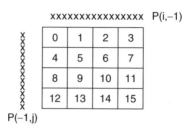

Figure 5.10   Spatial prediction for 16×16 mode

In 8×8 luma prediction mode, one of the eight different directions, as shown in Figure 5.11, and DC prediction can be used. Including DC, there are nine possible intra prediction modes. The directional modes are as shown in the figure. Mode 0 is the vertical direction, 1 is horizontal and so on. DC prediction is given mode number 2. Similarly, in 4×4 luma prediction mode, one of the eight different directions, shown in Figure 5.11, and DC prediction can be used for predicting pixels of a 4×4 block. The spatial prediction mode, selected for the prediction, is further compressed and sent in the bit stream.

Chroma intra prediction operates using full macroblock prediction. Because of differences in the size of the chroma arrays for the macroblock in different chroma formats (i.e., 8×8 chroma in 4:2:0 MBs, 8×16 chroma in 4:2:2 MBs and 16×16 chroma in 4:4:4 MBs), chroma prediction is defined for three possible block sizes. The prediction type for the chroma is selected independently of the prediction type for the luma.

In large flat zones the 16×16 mode is more efficient to use. The 4×4 mode is used for a region with high detail and fast changing intensity. The 8×8 mode is used for regions that fall in between the two extremes. This flexibility allows AVC/H.264 to achieve a very high coding efficiency for the Intra blocks.

5.2.3   The Transform

As shown in Figure 5.3, prediction errors are transformed before quantization. In the initial stage of the development of the standard, only 4×4 sized transform was used.

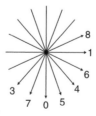

Figure 5.11   Possible spatial (intra) prediction directions

As the focus of the standard moved towards large picture sizes, such as HD and 4k ×
2k, and applications such as video creation in studios, an additional flexibility of using
8 × 8 transform was allowed for encoders.

5.2.3.1   4 × 4 Integer DCT and Inverse Integer DCT Transform

As mentioned above, the standard provides the specification of a decoder and not that of
an encoder. Therefore, it standardizes inverse transform used in the decoder. To match that
transform, an encoder needs to use a matching forward transform. The 4 × 4 transform,
corresponding to the inverse transform specified in the standard, is given by:

$$
C = \begin{bmatrix} 1 & 1 & 1 & 1 \\ 2 & 1 & -1 & -2 \\ 1 & -1 & -1 & 1 \\ 1 & -2 & 2 & -1 \end{bmatrix}
$$

Note that all the coefficients of this transform are integers. This is not only relatively
simple to implement but also allows an encoder to implement the transform exactly
without causing precision related errors. In prior standards, like H.261 or MPEG-2, true
DCT was used. Some coefficients of the DCT are irrational numbers and cannot be
implemented without introducing precision errors. It poses two problems: 1) they cannot
be implemented without truncation errors; and, more importantly, 2) the decoder part
in an encoder (the dotted area in Figure 5.3) will likely not be same as the decoder in
a receiver designed by some other company. As encoders and decoders will generally
implement those DCT coefficients with differing precision, there will be a mismatch
between the two. It can potentially cause some drifting of encoded and decoded pixel
values. To avoid that drifting, a small amount of dithering (called mismatch control) was
introduced in the MPEG-2 standard. Although this potential drifting does not happen
in real signals and is not a problem, it was decided to remove this potential source of
error in AVC/H.264. In addition to the simplicity of the design and implementation, this
turns out to be another benefit of the integer forward and inverse DCT transform in
AVC/H.264.

One way to analyze the forward transform is as follows [7, 8]. A 4 × 4 DCT transform
is given by:

$$
Y = \begin{bmatrix} a & a & a & a \\ b & c & -c & -b \\ a & -a & -a & a \\ c & -b & b & -c \end{bmatrix} \begin{bmatrix} p_{11} & p_{12} & p_{13} & p_{14} \\ p_{21} & p_{22} & p_{23} & p_{24} \\ p_{31} & p_{32} & p_{33} & p_{34} \\ p_{41} & p_{42} & p_{43} & p_{44} \end{bmatrix} \begin{bmatrix} a & b & a & c \\ a & c & -a & -b \\ a & -c & -a & b \\ a & -b & a & -c \end{bmatrix}
$$

where, $a = \dfrac{1}{2}$, $b = \sqrt{\dfrac{1}{2}} \cos\left(\dfrac{\pi}{8}\right) = 0.653281482438\ldots$, and $c = \sqrt{\dfrac{1}{2}} \cos\left(\dfrac{3\pi}{8}\right) = $
$0.382683432365\ldots$ and p_{ij} is $(i, j)^{th}$ pixel value. This can be manipulated into the

following form:

$$
Y = \begin{bmatrix} a & 0 & 0 & 0 \\ 0 & b & 0 & 0 \\ 0 & 0 & a & 0 \\ 0 & 0 & 0 & b \end{bmatrix} \begin{bmatrix} 1 & 1 & 1 & 1 \\ 1 & d & -d & -1 \\ 1 & -1 & -1 & 1 \\ d & -1 & 1 & -d \end{bmatrix} \begin{bmatrix} p_{11} & p_{12} & p_{13} & p_{14} \\ p_{21} & p_{22} & p_{23} & p_{24} \\ p_{31} & p_{32} & p_{33} & p_{34} \\ p_{41} & p_{42} & p_{43} & p_{44} \end{bmatrix} \begin{bmatrix} 1 & 1 & 1 & d \\ 1 & d & -1 & -1 \\ 1 & -d & -1 & 1 \\ 1 & -1 & 1 & -d \end{bmatrix} \begin{bmatrix} a & 0 & 0 & 0 \\ 0 & b & 0 & 0 \\ 0 & 0 & a & 0 \\ 0 & 0 & 0 & b \end{bmatrix}
$$

$$
= \left[\begin{bmatrix} 1 & 1 & 1 & 1 \\ 1 & d & -d & -1 \\ 1 & -1 & -1 & 1 \\ d & -1 & 1 & -d \end{bmatrix} \begin{bmatrix} p_{11} & p_{12} & p_{13} & p_{14} \\ p_{21} & p_{22} & p_{23} & p_{24} \\ p_{31} & p_{32} & p_{33} & p_{34} \\ p_{41} & p_{42} & p_{43} & p_{44} \end{bmatrix} \begin{bmatrix} 1 & 1 & 1 & d \\ 1 & d & -1 & -1 \\ 1 & -d & -1 & 1 \\ 1 & -1 & 1 & -d \end{bmatrix} \right] \otimes \begin{bmatrix} a^2 & ab & a^2 & ab \\ ab & b^2 & ab & b^2 \\ a^2 & ab & a^2 & ab \\ ab & b^2 & ab & b^2 \end{bmatrix}
$$

$$
= C \cdot P \cdot C^T \otimes A
$$

where, $d = \dfrac{c}{b} = 0.41421356237\ldots$ and $\otimes$ denotes element by element multiplication for the matrices to the left and the right of the symbol. The matrix A can now be absorbed into the quantization matrix which also performs element by element operation. So, the transform is now given by the C matrix above. However, it still contains d which is an irrational number. Therefore, the value of d was truncated to 0.5. With this change the approximated 'DCT' transform matrix becomes:

$$
C = \begin{bmatrix} 1 & 1 & 1 & 1 \\ 1 & \frac{1}{2} & -\frac{1}{2} & -1 \\ 1 & -1 & -1 & 1 \\ \frac{1}{2} & -1 & 1 & -\frac{1}{2} \end{bmatrix}
$$

However, the division by 2 causes loss of accuracy. Therefore, the second and fourth rows are multiplied by 2 and appropriate changes are done down the encoding path. Thus, the final form of the transform becomes:

$$
C = \begin{bmatrix} 1 & 1 & 1 & 1 \\ 2 & 1 & -1 & -2 \\ 1 & -1 & -1 & 1 \\ 1 & -2 & 2 & -1 \end{bmatrix}
$$

Transform can now not only be implemented accurately but also easily. As it can be implemented without the use of a multiplier and using only additions and shifts, it is sometimes also stated to be a multiplier free transform. Note that owing to the approximation made above, the transform is no longer the true cosine transform. However, it does not hurt the coding efficiency in any significant way.

5.2.3.2 8 × 8 Transform

In a similar style to the 4 × 4 transform, integer 8 × 8 inverse DCT transform is specified in the standard and the corresponding forward 8 × 8 integer DCT transform is given by:

$$
C = \begin{bmatrix}
8 & 8 & 8 & 8 & 8 & 8 & 8 & 8 \\
12 & 10 & 6 & 3 & -3 & -6 & -10 & 12 \\
8 & 4 & -4 & -8 & -8 & -4 & 4 & 8 \\
10 & -3 & -12 & -6 & 6 & 12 & 3 & -10 \\
8 & -8 & -8 & 8 & 8 & -8 & -8 & 8 \\
6 & -12 & 3 & 10 & -10 & -3 & 12 & -6 \\
4 & -8 & 8 & -4 & -4 & 8 & -8 & 4 \\
3 & -6 & 10 & -12 & 12 & -10 & 6 & -3
\end{bmatrix}
$$

5.2.3.3   Hadamard Transform for DC

When the 16×16 Intra prediction mode is used with the 4×4 transform, the DC coefficients of the sixteen 4×4 luma blocks in the macroblock are further transformed by using the following Hadamard transform:

$$
H_{4\times4} = \begin{bmatrix}
1 & 1 & 1 & 1 \\
1 & 1 & -1 & -1 \\
1 & -1 & -1 & 1 \\
1 & -1 & 1 & -1
\end{bmatrix}
$$

The DC coefficients of the 4×4 chroma block samples in a macroblock are also further transformed by Hadamard transform. In 4:2:0 video format, there are only four DC coefficients. They are transformed by using the following 2×2 Hadamard transform:

$$
H_{2\times2} = \begin{bmatrix}
1 & 1 \\
1 & -1
\end{bmatrix}
$$

For 4:4:4 video format, $H_{4\times4}$ above is used to further transform 4×4 chroma DC coefficients.

5.2.4   Quantization and Scaling

When 4×4 transform is used, coefficients cannot be weighted explicitly before quantization. Coefficients in a macroblock can be quantized using one of the 52 possible values for 8 bit video. For higher bit depth video, the number of possible values increases by 6 for each additional bit. The quantization step is controlled by the quantization parameter. Quantization step sizes are not linearly related to the quantization parameter. The step size doubles for every six increments of the quantization parameter.

When 8×8 transform is used, the encoder is given the additional flexibility of using its own weighting matrix during the quantization step. The weighting matrix can be adapted from picture to picture.

5.2.5   Scanning

The two dimensional transformed coefficients are scanned out in a particular order so as to produce a one dimensional string of coefficients, which is then input to the variable

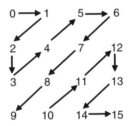

Figure 5.12   Zig-zag scanning for progressive video

length encoder, as shown in Figure 5.3. Two different styles of scanning are used – one for progressively scanned and another for interlaced scanned video where MBs are compressed in the field mode. In progressively scanned video or for frame MBs, a traditional zig-zag scanning order, as shown in Figure 5.12, is used. The zig-zag scanning is used as it provides longer run of zeros, in a typical compressed video, which can be compressed more efficiently by the VLCs used in the standard.

However, for interlaced video, the statistics of the coefficients is different for the field MBs than it is for the MBs in a progressively scanned video. Therefore, for field MBs the scanning order is as shown in Figure 5.13. These scanning orders are tuned more optimally for a field MB.

Similarly, zig-zag and alternate scans for 8 × 8 transformed blocks are also specified in the standard.

5.2.6   Variable Length Lossless Codecs

The AVC/H.264 standard includes three different types of lossless encoders – Exp-Golomb, CAVLC (Context Adaptive VLC) and CABAC (Context Adaptive Binary Arithmetic Coding). Exp-Golomb codes are used only for those high level syntax elements that are not voluminous in the number of bits, e.g. chroma format, bit depth, number of reference frames, etc. Quantized transform coefficients are coded using either CAVLC or CABAC. As described below, some receivers are required to implement only CAVLC while others are required to implement both. Both of these coding techniques recognize the fact that digital video is a non-stationary signal and there are large

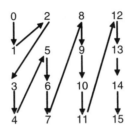

Figure 5.13   Alternate scan for field MBs in interlaced video

variations in the probability of various syntax elements from one sequence to another. They both adopt the philosophy of adapting the coder to the characteristics of the digital video being compressed.

5.2.6.1 Exp-Golomb Code

This code is simple to implement. The bit stream is generated for a given code value, as shown in Table 5.1. In this table x_i is either 0 or 1. These codes consist of prefix part: 1, 01, 001, etc. and the suffix bits are as shown in the table. For n bits in the suffix, there are 2^n codes for a given prefix part.

Various syntax elements are mapped into the code values as specified in the standard. Table 5.2 shows in explicit form the bit stream generated corresponding to those values.

For the syntax elements with signed value, like change in the quantization values, mapping of the signed value to the code value is specified in the standard and is shown in Table 5.3.

Table 5.1 Exp-Golomb code

Bits	Code Values
1	0
0 1 x_0	1–2
0 0 1 x_1 x_0	3–6
0 0 0 1 x_2 x_1 x_0	7–14
0 0 0 0 1 x_3 x_2 x_1 x_0	15–30
0 0 0 0 0 1 x_4 x_3 x_2 x_1 x_0	31–62
$\cdots$	$\cdots$

Table 5.2 Exp-Golomb code examples

Bits	Code Values
1	0
0 1 0	1
0 1 1	2
0 0 1 0 0	3
0 0 1 0 1	4
0 0 1 1 0	5
0 0 1 1 1	6
0 0 0 1 0 0 0	7
0 0 0 1 0 0 1	8
0 0 0 1 0 1 0	9
$\cdots$	$\cdots$

Table 5.3 Mapping of signed syntax values to
code values

Code Values	Signed Syntax Element Value
0	0
1	1
2	−1
3	2
4	−2
5	3
6	−3
. . .	. . .

5.2.6.2 CAVLC (Context Adaptive VLC)

The core philosophy of CAVLC is similar to that in the previous standards where the more frequently occurring data elements are coded with shorter length codes and the length of run of zeros is also taken into account while developing the code tables. A major basic change in comparison to the VLCs used in previous standards is that it exploits the fact that in the tail-end there are small valued (±1) non-zero coefficients. In addition, it also adapts the table used for coding based on past information such that one can use more optimal statistics that are adapted to the video. These coefficients are coded and conveyed by sending: total non-zero coefficients; trailing ones, sign of trailing ones, total zeros, the number of consecutive zero valued coefficients, run of zeros and non-zero coefficient values. To give a high level view let us consider an example. Let the coefficients after zig-zag scan be 10,0,0,6,1,0,0,−1,0,0,0,0,0,0,0,0. In this case, there are four non-zero coefficients, two trailing ones, four zeros between 10 and −1, the run of zeros between −1 and 1 is 2 and the run of zeros between 10 and 6 is also 2.

The statistics of the coefficient values have a narrower distribution in the end than in the beginning. Therefore, the coefficients are coded in reverse order. In the example above, 6 is the first coded coefficient value. A default coding table is used to code the first coefficient. The table used for the next coefficient is then selected based on the context as adapted by previously coded levels in the reverse scan. To adapt to a wide variety of input statistics several structured VLC tables are used.

5.2.6.3 CABAC

AVC/H.264 also allows the use of an Arithmetic Coder. Arithmetic coding has been used extensively for image and data compression [9, 10]. It is used for the first time here in a video coding standard. The Arithmetic Coder has an increased complexity over the traditional one dimensional run-length Huffman style encoder but it is capable of providing higher coding efficiency, especially for symbols that have fractional bit entropy. There are three basic steps for CABAC – 1) binarization; 2) context modeling and 3) binary arithmetic coding [11]. As the syntax elements, such as coding modes, coefficients, motion vectors, MB and Sub-MB types, etc., are not binary, the first step for CABAC is

Table 5.4   Example of binarization

Slice Type	Sub-MB Type	Bin string
	P 8 × 8	1
P Slice	P 8 × 4	0 0
	P 4 × 8	0 1 1
	P 4 × 4	0 1 0

to map those non-binary values into unique binary code words called 'bin strings'. As an example, the mapping of sub-MB types in a P slice is shown in Table 5.4.

The next step is context modeling. In this step, the probability models for various symbols are developed. As the probability model makes a significant impact on the coding efficiency of the coder, it is very important that a right model is developed and is updated during encoding and that the probability statistics of various symbols are adapted to the past video information. This adaptation allows the encoder to tailor the arithmetic encoding process to the video being compressed rather than having non-optimal fixed statistics based encoding.

The final stage of CABAC is a Binary Arithmetic Coder. Arithmetic encoding is based on recursive interval sub-division depending upon the probability of the symbols. The step of converting syntax elements to binary values allows one to simplify the implementation of the arithmetic coding engine. Low complexity binary arithmetic coding engines have been used in image compression algorithms and standards such as JPEG. In AVC, a multiplication free binary arithmetic coding scheme was developed in order to achieve a reasonable trade-off between complexity and coding efficiency.

CABAC is reported to achieve about 10 to 15% better coding efficiency in the range of video quality (PSNR) in a digital TV application [11–13]. This improvement can be lower or higher depending upon the video content.

5.2.7   Deblocking Filter

When a video is highly compressed, the compression artifacts start to become visible in the decoded pictures. Three of the most commonly seen artifacts are blocking, mosquito and ringing noises. In blocking noise, the blocks corresponding to the motion estimation and transformation start to become visible. In mosquito noise, small size noise starts to appear around moving edges. This is primarily because the motion of the region covered by a macroblock consists of pixels with mixed motion resulting in large prediction error and also the presence of edges produces higher frequency components that become quantized. In ringing noise, the sharp edges or lines are accompanied by their faint replicas. Although owing to flexible MB size in AVC/H.264 these noises are reduced significantly, they do start to appear as the compression factor increases. To minimize the impact of these noises, deblocking filters have been used. There are two schools of thought – deblocking as a post processing step or in-loop deblocking. In the first, deblocking filters are applied after the pictures are decoded. As in MPEG-2 and MPEG-4 Part 2, the design of those filters is not standardized and it is left up to the decoder manufacturers to decide how to design and implement these and whether to apply deblocking filters or not. This approach has the benefit of keeping the cost of decoder implementation low and leaves some room for innovation at the decoding and displaying end. A disadvantage in this approach is that it is hard to control the quality of the displayed video from the encoding/sending side. In addition, the pictures that are used as references contain these noises and they accumulate as more and more pictures use past pictures as references. Furthermore, these noises are a function of content as well as quantization. It becomes harder to implement a deblocking filter that takes those factors into account in the post processing step. To avoid those disadvantages, AVC/H.264 specifies an in-loop deblocking filter, as shown in Figure 5.3. As the name suggests, the deblocking is now done within the decoding loop.

The in-loop deblocking filter in AVC reduces the blockiness introduced in a picture. The filtered pictures are used to predict the motion for other pictures. The deblocking filter is a content and quantization adaptive filter that adjusts its strength depending upon the MB mode (Intra or Inter), the quantization parameter, motion vector, frame or field coding decision and the pixel values.

The deblocking filtering is designed so that the real edges in a scene are not smoothed out and the blocking artifacts are reduced. Several filters are used and their lengths and strengths are adjusted depending upon the coding mode (like Intra or Inter prediction, frame or field mode), size of the motion vector, neighboring pixels and quantization parameters. When the quantization size is small, the effect of the filter is reduced, and when the quantization size is very small, the filter is shut off. For the frame intra predicted blocks the filters are strongest and are four pixels long. In the field mode the vertical filters are only two pixels long as the field lines are twice as far apart as the frame lines. This reduces blurring due to filtering. All compliant decoders are required to implement the deblocking filter as specified in the standard.

An encoder is also allowed the option of not using the deblocking filter and signaling to a decoder in the syntax whether a deblocking filter is used or not. This allows encoders to choose between compression artifacts or deblocking artifacts. For very high bit rate applications, where the blocking noise is not significant, it may not be desirable to add blurring distortion due to deblocking filter. Encoders are also allowed the option of reducing the level of filtering by not using default parameters and optimizing them to a given application.

5.2.8   Hierarchy in the Coded Video

The basic coding structure of AVC/H.264 is similar to that of earlier standards and is commonly referred to as a motion-compensated-transform coding structure. The coding of video is performed picture by picture. Each picture to be coded is first partitioned into a number of slices (it is also possible to have one slice per picture). As in earlier standards, a slice consists of a sequence of integer numbers of coded MBs or MB pairs. To maximize coding efficiency, only one slice per picture may be used. In an error prone environment, multiple slices per picture are used so that the impact of an error in the bit stream is confined only to a small portion of a picture corresponding to that slice.

The hierarchy of video data organization is as follows:

Video sequence { picture { slices [MBs (sub-MBs (blocks (pixels)))] }}.

In AVC/H.264, slices are coded individually and are the coding units, while pictures are specified as the access units and each access unit consists of coded slices with associated data.

5.2.8.1   Basic Picture Types (I, P, B, B_R)

There are four different basic picture types that can be used by encoders: I, P, B and B_R (the standard does not formally use these names but they are most commonly used in the applications using the standard). I-pictures (Intra pictures) contain intra coded slices and consist of macroblocks that do not use any temporal references. As only spatial prediction is allowed, these are good places in a bitstream to perform random access, channel change, etc. They also stop the propagation of errors made in the decoding of pictures in the past. AVC also defines a special type of Intra picture called the Instantaneous Decoder Refresh (IDR) picture. As described in Chapter 10, it is a regular Intra picture with the constraint that pictures appearing after it in the bistream cannot use the pictures appearing before it as references. IDR pictures may be used as random access points as the decoding process does not depend upon the history before the IDR.

The P-pictures (Predicted pictures) consist of MBs or sub-MBs that can use up to one motion vector for prediction. Various parts of the P-pictures may use different pictures as references to estimate the motion. Unlike in MPEG-2, those reference pictures can be either in the past or in the future of the current frame in the video stream. P-pictures can also contain intra MBs. The B-pictures (Bi-predicted pictures) consist of MBs or sub-MBs that use up to two motion vectors for prediction. B pictures can contain MBs with one or zero (intra) motion vectors. Various parts of B-pictures also may use different pictures as references. Unlike in MPEG-2, both of the reference pictures, corresponding to the two MVs, can either be in the past or future or one can be in the past and the other in the future of the current picture in the video stream. B_R pictures are the B pictures that are used as references for temporal prediction by other pictures in the past or in the future. Note that in MPEG-2 one cannot use B-pictures as references.

5.2.8.2   SP and SI Pictures

SP and SI pictures are special types of pictures that are used for switching the bitstream from one rate to another. In some internet streaming environments with no Quality of Service there is no guaranteed bandwidth. Therefore, available channel capacity varies significantly and there is a need to change the compressed bit rate with time. To facilitate switching the bit rate quickly, SP and SI pictures can be used [14]. Consider a video compressed at two different bit rates, as shown in Figure 5.14. Let us assume that switching from Stream 1 to Stream 2 is desired. As shown in Figure 5.14, there are several switching points, like S_1 and S_2, created in the streams. In a normal bitstream, after switching, the pictures appearing in the future do not have the correct past references as the pictures received in the past belong to different bit rate. This causes distortion in the decoded video. To avoid this problem, S_1 and S_2 are created so that one can send another picture S_{12} which will allow one to create S_2 exactly but by using past pictures of Stream 1. Therefore, the decoding of Stream 2 picture after this point can occur without any

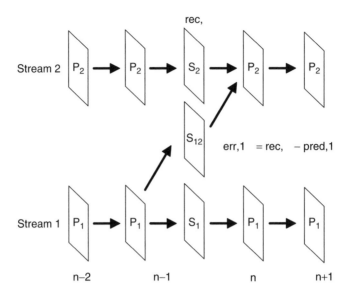

Figure 5.14   SP and SI pictures

distortion. If S_{12} is created by using temporal predictions it is called the SP picture. If only the spatial prediction is used then it is called the SI picture.

There are several issues related to SP and SI pictures. These pictures work well only when the switching occurs between the streams corresponding to the same video content. In addition, they are more complex and cause reduction in coding efficiency. Therefore, they are not widely used.

5.2.9   Buffers

The number of bits per picture varies significantly depending upon the content type and the picture type. Therefore, the compressed pictures are produced at an irregular and unpredictable rate. However, a picture typically needs to be decoded and displayed in a fixed picture time. Therefore, there is a mismatch between the number of bits that arrive at a decoder and the number of bits consumed by the decoder in one picture time. Therefore, the incoming bits are stored in a buffer called the Coded Picture Buffer (CPB) which holds the bits corresponding to the pictures yet to be decoded. To avoid the overflow and underflow of bits at the input of a decoder, it is important to ensure that an encoder knows how much input buffer a decoder has, to hold the incoming compressed bitstreams. To provide consistency and interoperability and not require a two way communication between an encoder and a decoder to convey that information, which is not practical in many applications, the standard specifies the minimum CPB buffer each decoder must have. It is an encoder's responsibility to generate the bitstream so that the CPB does not overflow and underflow.

In order to make sure that the encoders do not generate a bitstream that is not compliant with AVC/H.264, a hypothetical reference decoder (HRD) model is provided. This model contains input CPB, an instantaneous decoding process and an output decoded picture buffer (DPB). It also includes the timing models – rate and time when the bytes arrive at the input of those buffers and when the bytes are removed from those buffers. An encoder must create the bitstreams so that the CPB and DPB buffers of HRD do not overflow or underflow.

There are two types of conformance that can be claimed by a decoder – output order conformance and output timing conformance. To check the conformance of a decoder, test bitstreams conforming to the claimed Profile and Level are delivered by a hypothetical stream scheduler to both HRD and the decoder under test. For an output order conformant decoder, the values of all decoded pixels and the order of the output pictures must be the same as those of HRD. For an output timing conformant decoder, in addition, the output timing of the pictures must also be the same as that of HRD.

5.2.10 Encapsulation/Packetization

To be able to adapt the coded bit stream easily for diverse applications such as broadcasting over Cable, Satellite and Terrestrial networks, streaming over IP networks, video telephony or conferencing over wireless or ISDN channels, the video syntax is divided into two layers – the Video Coding Layer (VCL), and non-VCL layer. The VCL consists of bits associated with compressed video. Non-VCL information consists of sequence and picture parameter sets, filler data, Supplemental Enhancement Information (SEI), display parameters, picture timing, etc. VCL and non-VCL bits are encapsulated into NAL Units (NALU). Originally NAL stood for Network Abstraction Layer but as the standard progressed, its purpose was modified so that only the acronym NAL was kept in the standard. The format of a NALU is as shown in Figure 5.15.

The first byte of each NALU is a header byte and the rest is the data. The first bit of the header is a 0 bit. The next two bits indicate whether the content of NALU consists of a sequence or picture parameter set or a slice of a reference picture. The next five bits indicate the NALU type corresponding to the type of data being carried in that NALU. There are 32 types of NALUs allowed. These are classified in two categories: VCL NAL Units and non-VCL NAL Units. NALU types 1 through 5 are VCL NALUs and contain data corresponding to the VCL. NALUs with NALU type indicator value higher than 5 are non-VCL NALUs and carry information like SEI, Sequence and Picture Parameter set, Access Unit Delimiter etc. For example, NALU type 7 carries the Sequence Parameter Set and type 8 carries the Picture Parameter Set. Non-VCL NALU may or may not be present in bitstream and may be sent separately by any means of external communication.

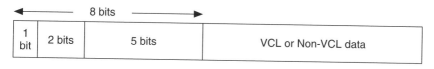

Figure 5.15 NAL unit format

5.2.11   Profiles

The AVC/H.264 standard was developed for a variety of applications ranging from video phone to entertainment TV to mobile TV to video conferencing to content creation in a studio. Those applications have different requirements. Requiring all of the decoders to implement all of the tools would make them unnecessarily complex and costly. Therefore, the standard divides the coding tools into different categories, called Profiles, based on various collections of applications. Each profile contains a sub-set of all of the coding tools specified in the standard. A decoder compliant with a certain profile must implement *all* of the tools specified in that profile. An encoder generating bitstream to be decoded by a decoder compliant to a certain profile is not allowed to use any tools that are not specified in that profile. However, it may choose to use a sub-set of tools specified in that profile.

Figures 5.16, 5.17 and 5.18 depict the various profiles specified in the standard. For consumer applications today, 4:2:0 format is used. Four profiles defined with those applications in mind are:

- Baseline
- Extended
- Main
- High.

At the time this text was written, JVT was in the process of creating fifth profile called Constrained Baseline Profile.

5.2.11.1   Baseline Profile

This was designed with mobile, cell phones and video conferencing applications in mind. In these applications the power consumption plays a very important role. Therefore, the complexity of encoders and decoders is desired to be low. B or B$_R$ pictures and CABAC

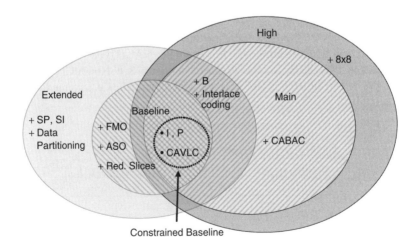

Figure 5.16   Profiles for 4:2:0 format

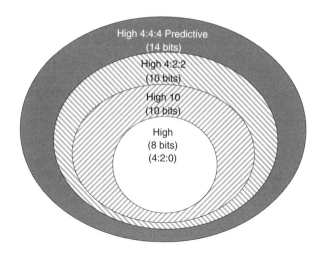

Figure 5.17 Hierarchy of high profiles

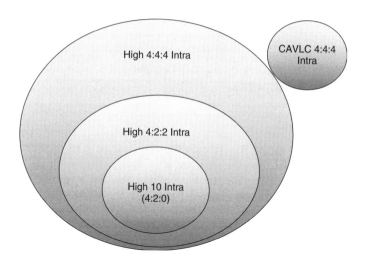

Figure 5.18 Intra-only profiles

coding tools are not allowed. Also, as the pictures in those applications are expected to be CIF or QVGA sizes, interlaced coding tools are also not included. Note that in this standard the motion in P pictures can be estimated based on the references in the future. To use the references that are in the future, an encoder has to wait for those references to arrive before starting the encoding. Therefore, P pictures do not necessarily imply that the bitstream is generated with low delay. Similarly, both the motion vectors in B pictures may use the pictures in the past as references. Therefore, the presence of B pictures in a bitsteam does not necessarily imply longer delays. In addition to the low power consumption, many of these applications operate in an environment that has a significant bit error rate. To allow for robust communication in that environment, three error resilience tools are also added – Flexible Macroblock Ordering (FMO), Arbitrary Slice Order (ASO) and Redundant Slices. In FMO, the macroblocks are not sent in the order in which they are scanned and displayed. Their order is selected by an encoder depending upon burst error length. As the MBs are not in display order, errors within them are scattered over a wider picture area and become relatively less noticeable. In ASO, the slices can be sent in an arbitrary order and in Redundant Slices, as the name suggests, the same slices can be sent more than once so that the receiver can pick the ones with less error. Note that Baseline profile is not a sub-set of Main and High profiles as these profiles do not support the error resilience tools mentioned above. Therefore, a Baseline profile compliant bitstream cannot be decoded by Main and High profile decoders unless the error resilience tools are not used.

Constrained Baseline Profile

In many environments (for example, in Portable Media Players), the bit error rates are not high and there is a strong desire to simplify the encoder and decoder implementations as well as to be compatible with Main and High profiles so that devices like Set Top Boxes and TVs, that implement those popularly used profiles in Digital Television, can also decode the bit stream. In order to achieve this goal, the AVC standard allows for the setting of a flag, called 'constraint_set1_flag', in the bitstream syntax to 0 or 1. When this flag is set to 1, it indicates that those error resilience tools are not used. Currently, there is no specific name assigned to this setting. However, at the time this text was written, JVT was in the process of formally adopting the name Constrained Baseline Profile corresponding to this setting.

5.2.11.2 Extended Profile

This profile is a superset of the Baseline profile and extends it to streaming video applications, as perceived at the time the standard was designed. In addition to including all of the tools allowed in the Baseline profile, it allows for the usage of SP and SI so that if there is a sudden change in the channel bandwidth the bitstream can be switched to the one using a lower bit rate. In order to increase coding efficiency and allow for the use of higher resolution (SD and HD) pictures, interlaced coding tools and B pictures are allowed. To keep the complexity of a decoder lower, it does not allow CABAC.

5.2.11.3 Main Profile

This profile was designed so as to provide high coding efficiency with Digital TV as one of the main applications. Therefore, it includes B-pictures, CABAC and Interlaced Coding tools. Note that it supports both CABAC and CAVLC. This allows for compatibility with Constrained Baseline profile. As, the error rates after FEC (Forward Error Correction) are not expected to be high (typically, less than 10^{-6}), the error resilience tools (FMO, ASO and RS) are not included in this profile.

5.2.11.4 High Profile

This profile provides the highest coding efficiency and highest possible flexibility with regard to the coding tools that can be used on the encoder side for a 4:2:0 color format video. It is a super set of the Main profile. In addition to all the tools used in the Main profile, it includes 8×8 transform, 8×8 Intra Prediction and downloadable Quantization/Weighting tables. On average, this profile is reported to provide about 10% higher coding efficiency in comparison to the Main Profile for 720p formats. This also allows one to fine tune the look and visual nature of video, e.g. preservation of film grain or picture by picture control of the distortion in order to provide more pleasing or desirable visual quality.

5.2.11.5 High10 Profile

This profile allows for the use of 10 bit video for 4:2:0 formats. This was developed in order to provide higher quality video for Electronic Cinema, contribution grade video, studios and possibly next generation High Definition DVD applications. As these applications do not typically use interlaced scanning, the 4:2:0 format is considered to be adequate for the lower end of those applications.

5.2.11.6 High 4:2:2 Profile

This profile is a super set of the High10 profile and allows for the use of the 4:2:2 color format for compression. As this format is very commonly used to capture, store and process video, the High 4:2:2 profile allows for compression without requiring the steps of converting video to the 4:2:0 format and then back to the 4:2:2 format. This avoids the artifacts introduced in converting 4:2:2 to 4:2:0 and back to the 4:2:2 format.

5.2.11.7 High 4:4:4 Predictive Profile

This profile is a super set of the High10 and High 4:2:2 profiles and allows for the highest color resolution and picture quality. Chroma has the same resolution as Luma. It allows for up to 14 bits per luma and chroma pixel values.

5.2.11.8 Intra Only Profiles

In studios and other environments where a great deal of video editing is done, it is desirable to have the capability to access each picture independently. To support those and Digital Cinema applications, a set of profiles is provided that allows no temporal prediction and where each picture is compressed in intra modes. As shown in Figure 5.18, these profiles cover the 4:2:0, 4:2:2 and 4:4:4 video chroma sampling formats. As is also shown in the figure, except for the CAVLC 4:4:4 Intra Profile, a decoder compliant with the higher chroma resolution profile is required to be able to decode the bitstream compliant with a profile with lower chroma resolution at the same level.

5.2.12 Levels

It is not practical to require every decoder to be able to decode a bitstream corresponding to all of the possible resolutions ranging from QCIF (176×144) at 15 frames/sec to Digital Cinema ($4k \times 2k$) resolution with 72 frames/sec. Therefore, the standard also specifies Levels for each Profile. A Level specifies the parameters that impact upon the processing power needed to decode the bitstream in real time and also the maximum memory required to decode a bitstream compliant with that level. Unlike MPEG-2, resolution is not specified in terms of maximum horizontal and vertical sizes. Realizing that one of the more fundamental parameters in the design of a decoder is pixels per picture, MPEG-4 Part 2 specified the Level constraints in terms of the number of pixels in a picture rather than the horizontal and vertical sizes of a picture. The same approach was followed in the AVC. Similarly, instead of frame rates, pixel rates are specified. Currently, the standard specifies 16 Levels. The Level of a decoder is another axis of the compliance plane. The compliance of a decoder is indicated by the Profile and the Level.

5.2.12.1   Maximum Bit Rates, Picture Sizes and Frame Rates

Table 5.5 shows the maximum bit rate (for VCL), the maximum number of 16×16 macroblocks (MBs) in a picture and the maximum MB rates for the levels defined in the standard. When NAL is included then the max bit rate increases by 20 %. The last column also provides the typical picture resolution of the picture and frame rate used in products compliant to that Level. Other picture sizes at different frame rates can also be used as long as the maximum number of MBs per picture is less than the one specified in the third column and the maximum frame rate is such that the pixel rate for the given picture size is less than the maximum MBs/sec specified in the fourth column. In addition to the constraints in Table 5.5, the aspect ratio of a picture was also constrained to be such that the horizontal and the vertical sizes cannot be more than *square root* (8 × *maximum frame size*). This is done to avoid placing the undue burden on the decoders of adding the capability to configure the memory arbitrarily. Furthermore, to limit the maximum frame rate corresponding to small picture sizes at a given level, the maximum frame rate is further restricted to be ≤172 fps.

As Table 5.5 shows, this standard is applied from the low rate of 10s of kbps to 100s of Mbps. Unlike previous standards, this standard provides high coding efficiency over 5 orders of magnitude of bit rate range!

Level 3 is also known as the SD level, Level 4 as the HD level and Level 4.2 is also known as the 1080 60p standard in digital TV applications. Currently Level 3 and Level 4 are widely deployed and various activities are under way to specify the use of Level 4.2 (1080 60p) for digital TV (see Chapter 10).

5.2.12.2   Maximum CPB, DPB and Reference Frames

To limit the maximum memory required in a decoder, the maximum CPB and DPB buffer sizes are also specified. Table 5.6 shows the maximum allowed CPB and DPB sizes at various levels. A decoder is required to have an input buffer memory at least as big as specified in this table. An encoder cannot produce a bit stream so as to require more than the specified buffer size. This standard specifies greater coded picture buffer sizes than previous standards. This was done in order to allow for maximum flexibility for encoders in optimizing video quality and decoding delay and thus achieving the best delay versus video quality trade-off required. By using a larger CPB buffer size an encoder can compress video with larger variation in bits per frame so as to accommodate larger

Table 5.5 Levels and associated bit rate, picture size and frame rate constraints for 4:2:0 format profiles

Level Number	Max. Compressed Bit Rate	Max. MBs per picture	Max. MB Rate (MB/sec)	Typical picture size and Rate
1	64 kbps	99	1485	SQCIF (128 × 96) × 30fps
				QCIF (176 × 144) × 15fps
1b	128 kbps	99	1485	QCIF (176 × 144) × 15fps
1.1	192 kbps	396	3000	QCIF × 30fps
1.2	384 kbps	396	6000	CIF (352 × 288) × 15fps
1.3	768 kbps	396	11 880	CIF × 30fps
2	2 Mbps	396	11 880	CIF × 30fps
2.1	4 Mbps	792	19 800	525 HHR (352 × 480) × 30fps
				625 HHR (352 × 576) × 25fps
2.2	4 Mbps	1620	20 250	525 SD (720 × 480) × 15fps
				625 SD (720 × 576) × 12.5fps
3	10 Mbps	1620	40 500	525 SD × 30fps
				625 SD × 25fps
				VGA (640 × 480) × 30fps
3.1	14 Mbps	3600	108 000	720p HD (1280 × 720) × 30fps
3.2	20 Mbps	5120	216 000	720p HD × 60fps
4	20 Mbps	8192	245 760	720p HD × 60fps
				1080 HD (1920 × 1088) × 30fps
				2k × 1k (2048 × 1024) × 30fps
4.1	50 Mbps	8192	245 760	
4.2	50 Mbps	8704	522 240	1080 HD × 60fps
				2k × 1k × 60fps
5	135 Mbps	22 080	589 824	2k × 1k × 72fps
5.1	240 Mbps	36 864	983 040	2k × 1k × 120fps
				4k × 2k (4096 × 2048) × 30fps

variations in the complexity of video quality. On the other hand, higher CPB buffer size causes higher decoding delay. The third column in the table provides examples of the maximum delays experienced by bitstreams that use the maximum CPB buffer sizes and are compressed at the maximum bit rates allowed by the levels.

The fourth column provides the maximum DPB size specified in the standard at various levels. This limits the maximum number of reference frames that can be used for those picture sizes. The fifth column shows the maximum number of reference frames corresponding to the DPB size limit specified in the standard. Note that for a given level one can use a smaller picture size and as a result use a greater number of reference frames as

Table 5.6 CPB and DPB sizes for 4:2:0 format profiles

Level Number	Max CPB size	Max delay at max bit rate	Max DPB size (Bytes)	Max number of reference frames for typical pictures sizes
1	175 kbits	2.7 sec	152 064	SQCIF: 8
				QCIF: 4
1b	350 kbits	2.7 sec	152 064	QCIF: 4
1.1	500 kbits	2.6 sec	345 600	QCIF: 9
1.2	1 Mbits	2.6 sec	912 384	CIF: 6
1.3	2 Mbits	2.6 sec	912 384	CIF: 6
2	2 Mbits	1 sec	912 384	CIF: 6
2.1	4 Mbits	1 sec	1 824 768	525 HHR: 7
				625 HHR: 6
2.2	4 Mbits	1 sec	3 110 400	525 SD: 6
				625 SD: 5
3	10 Mbits	1 sec	3 110 400	525 SD: 6
				625 SD: 5
				VGA: 6
3.1	14 Mbits	1 sec	6 912 000	720p HD: 5
3.2	20 Mbits	1 sec	7 864 320	720p HD: 5
4	25 Mbits	1.25 sec	12 582 912	720p HD: 9
				1080 HD: 4
				2k × 1k: 4
4.1	62.5 Mbits	1.25 sec	12 582 912	720p HD: 9
				1080 HD: 4
				2k × 1k: 4
4.2	62.5 Mbits	1.25 sec	13 369 344	1080 HD: 4
				2k × 1k: 4
5	135 Mbits	1 sec	42 393 600	2k × 1k: 13
5.1	240 Mbits	1 sec	70 778 880	2k × 1k:16
				4k × 2k: 5

more frames can in that case fit into the maximum allowed DPB size. To avoid the number of reference frames growing out of hand, a maximum limit of 16 frames, irrespective of the picture size, is also specified. Similarly, to cap the frame rate for small picture size for levels that allow for a high pixel rate, a maximum frame rate of 172 frames/sec is also specified.

A decoder compliant with a certain level is also required to be compliant with lower levels.

5.2.13 Parameter Sets

In addition to the compressed bits related to a video sequence, extra information is needed for the smooth operation of a digital video compression system. Furthermore, it is helpful to decoders if some information about some of the parameters used while compressing digital video is provided at a higher level in the syntax. Sequence Parameter Sets (SPS) and Picture Parameter Sets (PPS) provide that information.

5.2.13.1 Sequence Parameter Sets (SPS)

SPS provides information that applies to the entire coded video sequence. It includes information such as Profile and Level that is used to generate the sequence, the bit depth of the luma and chroma, the picture width and height and Video Usability Information (VUI) among other parameters.

Video Usability Information (VUI)
To be able to display a decoded video properly, it is helpful to have information such as color primaries, picture and sample aspect ratios and opto-electronic transfer characteristics, etc. These parameters are conveyed as part of the VUI. This information is sent as part of the Sequence Parameter Sets at the Sequence layer.

5.2.13.2 Picture Parameter Sets (PPS)

PPS provides information that applies to a coded picture. It includes information such as which of the two entropy coders is used, the scaling matrix, whether the default deblocking filter setting is used or not and whether 8×8 transform is used or not, etc.

5.2.14 Supplemental Enhancement Information (SEI)

The normative part of the standard is focused on the bitstream syntax and coding/decoding tools. It does not deal with other aspects of a digital video system that are required for the issues related to display and other information useful in a system. To help in providing that information the SEI messaging syntax is also developed in the standard. A system may or may not use SEI messages. Some key commonly used SEI messages are:

- Buffering period
- Picture timing
- Recovery point
- Pan scan rectangle
- User data

Buffering period SEI specifies parameters such as the delay related to the buffering and the decoding of the bitstream. Picture timing SEI provides information as to whether a frame is an interlace or progressive frame and if it is an interlaced frame then what is the order of fields (top and then bottom or bottom then top) and how should fields be repeated when 24 frames per sec movie material is displayed on TV with a 60 fields per sec refresh rate. Recovery point SEI provides information about how long it will take to fully recover a decoded picture without any distortion when one enters the stream at a given access point. Pan scan rectangle provides information on what region of a picture with wider aspect ratio should be displayed on a screen with narrower aspect ratio. User data allows a user to send private (not standardized by ISO/IEC and ITU-T committee) data. More detail is provided in Chapter 10.

5.2.15 Subjective Tests

In Figure 5.2, an example of a PSNR based comparison of AVC and MPEG-2 is provided. However, the human visual system perceives video quality differently than PSNR measurement. Therefore, the performance of a digital video processing system must also be evaluated based on visual/subjective tests. Several organizations, including the MPEG committee, have done subjective testing and comparison of AVC with MPEG-2 and other standards. Subjective tests done by the MPEG committee [15, 16] have been published. Those tests showed that the coding efficiency of AVC is about twice the coding efficiency of MPEG-2 also based on subjective tests, or in other words, AVC is able to provide a similar visual quality as MPEG-2 at about half the bit rate.

References

1. ISO/IEC JTC 1 and ITU-T, "Generic coding of moving pictures and associated audio information – Part 2: Video," ISO/IEC 13818-2 (MPEG-2) and ITU-T Rec. H.262, 1994 (and several subsequent amendments and corrigenda).
2. ISO/IEC JTC 1 and ITU-T, "Advanced video coding of generic audiovisual services," ISO/IEC 14496-10 (MPEG-4 Part 10) and ITU-T Rec. H.264, 2005 (and several subsequent amendments and corrigenda).
3. ISO/IEC JTC1, "Coding of audio-visual objects – Part 2: Visual," ISO/IEC 14496-2 (MPEG-4 visual version 1), 1999; Amendment 1 (version 2), 2000; Amendment 4, 2001 (and several subsequent amendments and corrigenda).
4. ITU-T, "Video coding for low bit rate communication," ITU-T Rec. H.263; v1: Nov. 1995, v2: Jan. 1998, v3: Nov. 2000.
5. P. Kuhn, "Algorithms, complexity analysis and VLSI architectures for MPEG-4 motion estimation," Kluwer Academic Publishers, 1999.
6. M. Flierl and B. Girod, "Video coding with superimposed motion-compensated signals – applications to H.264 and beyond," Kluwer Academic Publishers, 2004.
7. H. S. Malvar, A. Hallapuro, M. Karczewicz and L. Kerofsky, "Low-complexity transform and quantization in H.264/AVC," *IEEE Trans. Circuits and Systems for Video Technology*, Vol. 13, No. 7, pp. 598–603, July 2003.
8. A. Hallapuro, M. Karczewicz and H. Malvar, "Low complexity transform and quantization," JVT-B038, 2nd JVT meeting: Geneva, CH, Jan 29 – Feb 1, 2002.
9. I. Witten, R. Neal, and J. Cleary, "Arithmetic coding for data compression," Comm. of the ACM, Vol. 30, No. 6, pp. 520–540, 1987.
10. W. B. Pennebaker and J. L. Mitchell, "JPEG still image compression standard," Van Nelson Reinhold, 1993.
11. D. Marpe, H. Schwarz and T. Wiegand, "Context-based adaptive binary arithmetic coding in H.264/AVC video compression standard," *IEEE Trans. Circuits and Systems for Video Technology*, Vol. 13, No. 7, pp. 620–636, July 2003.
12. A. Puri, X. Chen, and A. Luthra, "Video coding using the H.264/MPEG-4 AVC compression standard", *Signal Processing: Image Communication*, Vol. 19, pp. 793–849, June 2004.
13. I. Moccagatta and K. Ratakonda "A performance comparison of CABAC and VCL-based entropy coders for SD and HD sequences," JVT-E079r2, 5th JVTMeeting: Geneva, CH, 9-17 October, 2002

14. M. Karczewicz and R. Kurceren, "The SP- and SI-Frames Design for H.264/AVC," *IEEE Trans. Circuits and Systems for Video Technology*, Vol. 13, No. 7, pp. 637–644, July 2003.
15. ISO/IEC JTC 1/SC 29/WG 11 (MPEG), "Report of the formal verification tests on AVC/H.264," MPEG document N6231, Dec., 2003 (publicly available at http://www.chiariglione.org/mpeg/quality_tests.htm).
16. C. Fenimore, V. Baroncini, T. Oelbaum, and T. K. Tan, "Subjective testing methodology in MPEG video verification," SPIE Conference on Applications of Digital Image Processing XXVII, July, 2004.

6

Content Analysis for Communications

6.1 Introduction

Content analysis refers to intelligent computational techniques which extract information automatically from a recorded video sequence/image in order to answer questions such as when, where, who and what. From such analysis, the machine is able to discover what is presented in the scene, who is there and where/when it occurs. There is no question about its importance as a general method, but why it is important here? Does it have something to do with the 4G wireless communications, our focus of the book?

The answer is absolutely! As multimedia communication applications such as mobile television and video streaming continue to be successful and generate tremendous social impact, and the communication networks continue to expand so as to include all kinds of channels with various throughputs, quality of services and protocols, as well as heterogeneous terminals with a wide range of capabilities, accessibilities and user preferences, the gap between the richness of multimedia content and the variation of techniques for content access and delivery are increasing dramatically. This trend demands the development of content analysis techniques in order to understand media data from all perspectives, and use the knowledge obtained in order to improve communication efficiency by considering jointly network conditions, terminal capabilities, coding and transmission efficiencies and user preferences.

Let's kick off this chapter with a few examples. As people have become more and more enthusiastic about watching multimedia content using mobile devices they personalize the content, for example, by summarizing the video for easy retrieval, or for easy transmission. Figure 6.1 provides an example of a still-image storyboard composed of a collection of salient images extracted automatically from the captured video sequence by a mobile phone camera, also referred to as *video summary*, to be delivered to remote friends so as to share the exciting experience of watching a tennis game. The summarizer identifies the exciting moments of the video based on certain analysis criteria so that the content coverage is guaranteed while a great deal of bandwidth and battery power for transmitting the entire video are saved.

4G Wireless Video Communications   Haohong Wang, Lisimachos P. Kondi, Ajay Luthra and Song Ci
© 2009 John Wiley & Sons, Ltd

Figure 6.1   An example of a 10-frame video summary of the Stefan sequence

(a) Original Image   (b) ROI of the image

Figure 6.2   An example of ROI

Another example is Region-of-interest (ROI) video coding for low bitrate applications, which can improve the subjective quality of an encoded video sequence by coding certain regions, such as facial image region, which would be of interest to a viewer at higher quality. As shown in Figure 6.2, the facial region in Figure 6.2(b) is normally considered as an ROI for the original image shown in Figure 6.2(a). Many analysis approaches, such as skin detection, face tracking and object segmentation, have been proposed so as to enable the accurate location and specification of the ROI.

The final example concerns privacy protection for video communications, especially for people who like to view head-and-shoulder videos in order to obtain more information from facial expressions and gestures, but are not interested in sharing other personal information, such as the surrounding environment and current activities. As shown in Figure 6.3, object segmentation can play an effective role in extracting the object from the current captured image and thus enabling the system to embed the object into a pre-specified background environment in order to protect the user's privacy. The aforementioned examples are part of the Universal Multimedia Access (UMA) family, an emerging next generation multimedia application which allows for universal access to the multimedia content with some or no user interaction.

This chapter follows the natural video content lifecycle from acquisition, analysis, compression and distribution to consumption. We start with video content analysis, which covers high-level video semantic structure analysis, low-level feature detection and object

Figure 6.3   An example of privacy protection

segmentation; after that we introduce the scalable content representation approach intro-duced by the MPEG-4 video standard, and then the coding and communication techniques used in the content-based video domain. Finally, we demonstrate the standardized con-tent description interface contained in MPEG-7 and MPEG-21 for multimedia content consumption and management purposes.

6.2   Content Analysis

It is interesting to observe that neither of the MPEG standards specifies the content anal-ysis methodologies other than specifying the syntax and semantics of the representation model. In other words, MPEG-4 codes the visual objects that compose a scene while not considering how the objects are obtained or specifying the criteria for composition. Similarly, MPEG-7 indexes visual data with feature descriptors whatever the methods that are employed to generate them. While content analysis is undoubtedly the critical part in the lifecycle of content from creation to consumption, and the foundation for support-ing both the MPEG-4 and MPEG-7 standards, the technique itself is still evolving. Its characteristics of being highly application-oriented make the current practice of leaving analysis methodologies out of the standards more beneficial.

As shown by the examples in section 6.1, visual content analysis covers some very wide topics and can be used for many different applications. The key is to extract the most important semantic characteristics for any given application. For example, the presence of a human-being and his/her movement is the most useful semantic information for the tennis game video clip; thus, the useful visual features are those which are capable of distinguishing these semantic meaningful objects/characteristics. Such an approach relies on both low-level features and high-level semantics, consequently the task of content analysis is to join the two techniques so as to understand the content within the scope of the specific application. Although it does not matter if we choose the top-down or bottom-up approach, in this chapter we start from the bottom with a few low-level features such as shape, color and texture, and then we demonstrate how these features can be used in image and video segmentation so as to separate different objects in the same frame, such as human being, sky, grass etc. Once the semantic meaningful objects are extracted, this information along with other cues can be used for the understanding of video structure. Clearly the most challenging part of the task is to diminish the gap between the high-level semantic concepts that are in a human being's mind and the low-level visual features that can be implemented in the algorithms executed by the machine.

6.2.1   Low-Level Feature Extraction

Low-level features such as edge, shape, color, texture and motion are the foundation of visual analysis. MPEG-7 (a.k.a. Multimedia Content Description Interface) provides descriptors in order to represent a few commonly used visual features. We will examine this in greater detail in section 6.5. In this section we focus on low-level features and their main usage in image and video analysis.

6.2.1.1   Edge

Edge detection [1] is one of the most commonly used operations performed in image analysis, in which the goal is to detect the boundary between overlapping objects or between a foreground object and the background, and locate the edge pixels on the boundary. The edges in an image reflect important structural properties. However in the practical world, the edges in an image are normally not an ideal step edge owing to many factors, such as the non-smooth object outline, focal blurriness and noises in the image-acquisition process.

The edges mostly appear as discontinuities in luminance (or chrominance) level (as shown in Figure 6.4). The derivative operator has been identified as the most sensitive operator for such change. Clearly, the 1st order derivative of the pixel intensity represents the intensity gradient of the current pixel, and the 2nd order derivative represents the change rate of the intensity gradient. To simplify the process, the difference between neighboring pixels is used as the approximation of the derivative operation, where the 1st order derivative is approximated by $I(x + 1, y) - I(x - 1, y)$ and $I(x, y + 1) - I(x, y - 1)$ for horizontal and vertical directions, respectively (where I is the pixel intensity function and (x, y) are coordinate of the current pixel), and the 2nd order derivative is approximated by $I(x + 1, y) - 2I(x, y) + I(x - 1, y)$ and $I(x, y + 1) - 2I(x, y) + I(x, y - 1)$. Consequently, the edge detector can be implemented by a small and discrete filter, and the filter has many variations according to the approximation. The famous *Sobel edge detector* is a 1st order dirivative operation. It uses the following filters:

$$S_x = \begin{bmatrix} -1 & 0 & 1 \\ -2 & 0 & 2 \\ -1 & 0 & 1 \end{bmatrix}, \ S_y = \begin{bmatrix} -1 & -2 & -1 \\ 0 & 0 & 0 \\ 1 & 2 & 1 \end{bmatrix}$$

where S_x is the horizontal component of the Sobel operator, and S_y is the vertical component. Clearly the filter is an approximation of the gradient at the central pixel. The gradient G_x and G_y (with G denotes gradients) for pixel (x, y) then becomes:

$$G_x = (I[x - 1, y + 1] + 2I[x, y + 1] + I[x + 1, y + 1]) - (I[x - 1, y - 1]$$
$$+ 2I[x, y - 1] + I[x + 1, y - 1])$$

and

$$G_y = (I[x + 1, y - 1] + 2I[x + 1, y] + I[x + 1, y + 1]) - (I[x - 1, y - 1]$$
$$+ 2I[x - 1, y] + I[x - 1, y + 1])$$

(a) Original image

(b) Edge map detected

Figure 6.4   An example of edge detection

After the operator components are calculated for each pixel in the image, the pixels with magnitude (or called edge response) of $\sqrt{G_x^2 + G_y^2}$ (or $|G_x| + |G_y|$) depending on the implementation) above a pre-assigned threshold are labeled as edge pixels and the edge direction can be calculated as the $\arctan(G_y/G_x)$. The determination of the threshold is not trivial; the lower the threshold, the more detected edges and the more susceptible are results to the noise; while, the higher threshold tends to miss details in the image and generate fragmented edges. A practical solution is to use two thresholds (also called hysteresis): first use the higher threshold to get the starting point of a new edge, and then keep track the edge using the lower threshold until the next pixel's magnitude is lower than the lower threshold which ends the current line. Such an algorithm assumes continuous edges.

The *Canny edge detector* (proposed by John Canny in 1986) is one of the most commonly used edge detectors. It assumes that the step edge is subject to white Gaussian noise and seeks for an optimized convolution filter in order to reduce the noise and the distance of the located edge from the true edge. It is found that an efficient approximation of the optimized filter would be the first derivative of a Gaussian function. Thus, the edge detection starts with the convolution of the raw image with a Gaussian filter in order to reduce noise. The result is then convolved with a first derivative operator (e.g. sobel) in both horizontal and vertical directions, and the edge response and direction are obtained. After that, a 'non-maximum suppression' process is carried out in order to remove those pixels whose gradient magnitudes are smaller than the gradient magnitude of their neighbors along the gradient direction. In other words, the pixels on edges are expected to be local maxima in gradient magnitude. Finally, the same hysteresis thresholding process as mentioned above is used to link the edge pixels.

6.2.1.2   Shape

Shape features [2] provide a powerful clue to pattern recognition, object tracking and object identification. The fact that human beings can recognize the characteristics of objects solely from their shape suggests that shape carries with it semantic information. MPEG-7 provides a set of shape descriptors for shape representation and matching for the content analysis tools to use as cues to in order to enhance performance. Clearly for applications like e-commerce, shape feature is superior as it is not so trivial to specify the shape of an object by text index or other means.

Many researchers characterize object shape with a set of function values which are independent of the geometric transformation such as translation and rotations. This practice helps to maintain small data size (i.e., a few function parameters) and makes feature description and usage convenient and intuitive. Listed below are the commonly used shape features:

- Geometric parameters, such as object bounding box, best-fit ellipse, the object eccentricity, area and shape roundness characteristics.
- Moment invariants, which are determined uniquely by contour function, and vice versa they can reconstruct the original function.
- Series expansions of the contour function, such as Fourier transform, discrete Sine (or Cosine) transform, discrete Walsh-Hadamard transform and discrete wavelet transform. In the transform coefficients, the mean value of the function reveals the center of gravity of the object shape.
- Shape curvature scale space image, which represents a multi-scale organization of the invariant geometrical features such as the curvature zero-crossing point of a planar curve. This method is very robust to noise.
- Mapped 1D function, such as a list of distances from the centroid to the object boundary, or the distances from object skeleton to the boundary, or a list of lengths and heights of the decomposed arc segments corresponding to the shape boundary.
- Bending energy, the energy derived from the shape curvature.
- Skeleton, which is derived from the medial axis transform or other means and reflects the local symmetry properties of the object shape.

As with any features, a similarity metric needs to be defined in order to match a shape with a shape template. Various similarity measurements have to be proposed. For example, the similarity of two vectors can be calculated via the Minkowski distance as:

$$L_p(x, y) = \left(\sum_{i=0}^{k} |x_i - y_i|^p \right)^{1/p},$$

where x, y are the two vectors in R^k space, and choosing different p value interprets different meanings for the distance, for example, p = 2 yields the Euclidean distance. When the total number of elements is different for the two sets in comparison, Hausdorff distance and bottleneck distance are widely used. Let us denote by A and B the feature vector sets used in shape matching, bottleneck distance refers to the minimum over all the correspondences f between A and B of the maximum distance d(a, f(a)), where a belongs to A and f(a) belongs to B, and d() can be Minkowski distance. Hausdorff distance H(A, B) is the maximum between h(A, B) and h(B, A), where h(C, D) is defined as the lowest upper bound over all points in C of the distance to D.

6.2.1.3 Color

Color has been used extensively as a low-level feature for image analysis. Despite the variety of color spaces available, RGB, HSV, CIE XYZ just to name a few, the color histogram remains the simplest and most frequently used means for representing color visual content. It represents the joint probability of the intensities of the three color channels and captures the global color distribution in an image. Since the human eye cannot perceive a large number of colors simultaneously, in other words, our cognitive space has the limitation of allowing around 30 colors to be represented internally and identified at one time; a compact color representation to characterize the domain color property has been widely used as a practical feature, which contains a number of 4-tuple elements with the 3D color and percentage of each color in the image.

Numerous similarity/dissimilarity measurement schemes have been proposed including the often used Euclidean distance, the one most commonly used for feature vectors is based on the Minkowski metric. In order to distinguish the impact of different color components, the quadratic distance metric measures the weighted similarity between histograms, and can provide more reasonable judgments.

6.2.1.4 Texture

Texture [1] normally reflects the visual property of an object's surface; therefore it can be a very useful cue for distinguishing objects with different texture patterns, for example, separating a tiger from the surrounding trees. Multi-scale frequency decomposition is the most popular technique for texture analysis, which includes wavelet transform, Gabor transform and steerable pyramid decomposition. The local energy of transformed frequency coefficients is the most commonly used texture feature. Some variations include using the mean, standard deviation, first and second order statistics functions of the frequency coefficients. It has been observed that the wavelet energy feature has an advantage

Figure 6.5   An example of object movement in various video frames

over other transforms such as DCT, DST, etc. On the other hand, some features for characterizing texture pattern and appearance such as repetition, directionality, regularity, and complexity, are also widely used. In section 6.2.2.4, we discuss the color and texture related features further when demonstrating a perceptually adaptive color-texture based segmentation approach.

6.2.1.5   Motion

Motion [3] reflects activities no matter whether the source of motion is from the camera or foreground objects, in other words, motion has very strong connection with high-level semantics, and thus needs special attention in content analysis (see an example of object movements in Fig. 6.5). Sometimes the video bitstream contains motion information that is obtained when the encoder conducts the motion estimation during encoding, however, the motion vectors may not be accurate owing to the fact that the basic operation unit, 8×8 (or other size depending on the codec and coding mode) block may contain components from more than one object, thus making the detected motion vector a compromised direction. The information is still very helpful when considering frame-level motion statistics, although it might be misleading for object segmentation tasks.

6.2.1.6   Mathematical Morphology

The dictionary definition of Morphology is 'the study of the form and structure of an object'. It is another effective way to describe or analyze the shape of an object. The operations normally involve two operands: an input image and a structuring element, which is a mask of predefined shape to form a pattern. There are four commonly used operations: dilation, erosion, opening and closing. Let us denote by A the input image, S the structure element matrix, then the results B after the mathematical morphology can be represented as follows:

- Dilation

$$B = A \oplus S = \bigcup_{s \in S} \{b \mid b = a + s, a \in A\}$$

- Erosion

$$B = A \circ S = \{b \mid any\, b + s \in A, s \in S\}$$

- Opening

$$B = (A \circ B) \oplus B$$

- Closing

$$B = (A \oplus B) \circ B$$

Clearly, the operations above can be combined so as to form higher-level operations, such as object boundary detection. By using a simple structuring element, for example a 2×2 matrix filled with 1, the boundary stripped can be described as:

$$Boundary\ B = A - (A \circ S).$$

6.2.2 Image Segmentation

The goal of image segmentation is to partition an image into visually distinct and uniform regions with respect to certain criteria such as color and texture. Still-image segmentation is a special case of video segmentation, which does not contain motion information. In other words, all the low-level features mentioned above other than motion feature can be used here for image segmentation purposes.

Before we move on to the details of image segmentation techniques, I would like to devote a few words to how to judge the quality of a segmentation result. Interestingly there is no single grand truth for natural scene segmentation, although the senses of human beings are quite consistent. Figure 6.6 [4] supports this argument by demonstrating six human manual segmentation results for the same picture. All of the segmentation results separate the tree from the stone, sky and ground. However, different users show their preference in the details, some users segment the tree and the ground into sub-regions, while others choose to keep them as one. Researchers in Berkeley proposed a segmentation evaluation scheme based on such an observation, which compares a segmentation result with the reference segmentation, which evaluates the consistency between them and tends to tolerate the refinement. The measurement works reasonably well, even if sometimes the better solution is only slightly better (see Figure 6.7(a)). However, there are also occasions when the evaluation result is not highly consistent with subjective evaluations or at least arguable (see Figure 6.7(b)).

In rest of this section, we will cover a few commonly used segmentation algorithms, including the threshold based approach, boundary-base approach, clustering based approach and region based approach, and highlight a new approach proposed recently by a group in the Northwestern University [5] which combines human perceptual models with principles of texture and color processing during the segmentation process. It demonstrates good performance even for low-resolution and degraded images.

Figure 6.6   Segmentation results from various people (source: the Berkeley Segmentation Dataset, http://www.eecs.berkeley.edu/Research/Projects/CS/vision/bsds/)

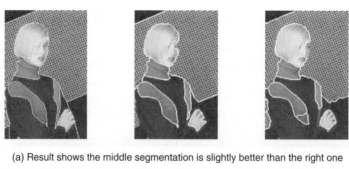

(a) Result shows the middle segmentation is slightly better than the right one

(b) Result shows the middle segmentation is much better than the right one

Figure 6.7   The evaluation based on Berkeley's measurement (Left: Human segmentation, Middle: Solution 1, Right Solution 2)

6.2.2.1   Threshold and Boundary Based Segmentation

Threshold-based segmentation uses a fixed or adaptive threshold to divide the image pixels into classes with the criterion that pixels in the same class must have features (color, grayscale or others) which lie within the same range. This approach is very straightforward and fast, and it can achieve good results if the image includes only two opposite components; however, it does not take advantage of spatial correlations in the image and is not robust when noise is present.

Boundary-based segmentation takes advantage of the edge features, and tends to form closed region boundaries from these discrete edges. However, owing to the fact that the edges detected by using Sobel or other operators are discontinuous and most of the time are over-detected, it takes a great deal of effort to merge them into closed contours, which make this approach very time-consuming.

6.2.2.2   Clustering Based Segmentation

K-means algorithm is one of the most popular clustering techniques which are widely used in image segmentation, although there are many variations using different feature vectors, intermediate processing operations and pixel classification criteria. Typically, clustering segmentation uses an iterative refinement heuristic known as Lloyd's algorithm, which starts by partitioning the image pixels into a number of clusters. The centroid of each cluster is calculated, then all the pixels are reassigned into clusters in order to ensure that all the pixels go to the cluster with the closest centroid point, and then the new cluster centroid is re-calculated. The algorithm iterates until it converges or a stop criteria is met. The similarity calculation is conducted in the selected feature spaces, for example, the color space or joint (weighted) color and texture space, thus the distance refers to the distance in the feature space. To speed up the performance, one way to obtain the initial clusters is by using the connected component labeling approach in order to obtain a number of connected regions (depending on the threshold pre-assigned for the current feature space). There are many studies on how to decide the optimal number of clusters, but different results may be obtained if the intial points are different. Therefore finding a good initialization point is very critical.

6.2.2.3   Region Based Approach

Split-and-merge and region growing are two representative methods for region based segmentation. In the split-and-merge method, the pixels are first split into a number of small regions (either use a quadtree approach or other sub-division method) with all the pixels in the same region homogeneous according to color, texture or other low level features; then all adjacent regions with similar attributes are merged until convergence. An example of a split-and-merge algorithm is shown in Figure 6.8, where the image is initially split into 1195 regions, then the regions are merged with each other, and finally 22 regions are generated.

The region growing algorithm starts by choosing a seed pixel as starting point and the growing is conducted by adding the neighboring pixels that meet the homogeneity criterion into the current region until the region cannot be further expanded (no homogenous neighboring pixels are available). This approach often takes advantage of the edge

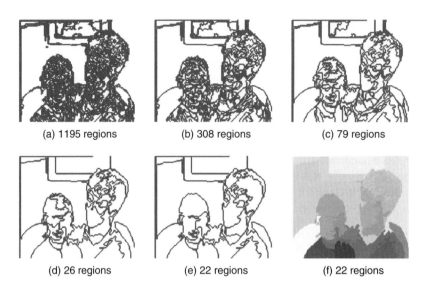

(a) 1195 regions   (b) 308 regions   (c) 79 regions

(d) 26 regions   (e) 22 regions   (f) 22 regions

Figure 6.8   Region growing (a-e: region boundary maps, f: filled region map)

information in order to stop the growing process by assuming that the intensity gap (or edge) is natural divider between separate regions.

6.2.2.4   Adaptive Perceptual Color-Texture Segmentation

So far none of the afore-mentioned methods has taken the human perceptual factor into account. For applications where human beings are the ultimate consumer of the segmentation results, the key to success is to take advantage of human perceptual related information and diminish the gap between low-level feature and high-level semantics. A natural desire is to bring perceptual models and principles into the visual feature and segmentation techniques. In this section, we demonstrate one of the representative works in this direction, the adaptive perceptual color-texture segmentation (APS) algorithm proposed by Chen *et al.* [5]. Although their efforts are targeted mainly at natural images, the concepts can be extended to other scenarios with extended perceptual models.

The APS algorithm proposes two spatial adaptive low-level features, one is a local color composition feature, and the other is a texture component feature. These features are first developed independently, and are then combined so as to obtain an overall segmentation.

A. Color Features

In APS, a new color feature called the spatially adaptive dominant color has been proposed in order to capture the adaptive nature of the human vision system and to reflect the fact that the region's dominant colors are spatially varying across the image. It is based on the observation that the human eye's perception of an image pixel's color is highly influenced by the colors of its surrounding pixels. This feature is represented by:

$$f_c(x, y, N_{x,y}) = \{(c_i, p_i), i = 1, 2, \ldots, M, p_i \in [0, 1]\}$$

where (x,y) is the pixel location in the image, $N_{x,y}$ is the neighborhood around the pixel, M is the total number of dominant colors in the neighborhood with a typical value of 4, c_i is the dominant color and p_i is the associated percentage. All p_i sums to 1. The proposed method uses the Adaptive Clustering Algorithm (ACA) to obtain this feature set. The ACA is an iterative algorithm that can be regarded as a generalization of the K-means clustering algorithm, as we mentioned in section 6.1.2.2. Initially it segments the image into a number of clusters using k-means and estimates the cluster centers by averaging the colors of the pixels in the same class over the whole image. Then the iteration starts and the ACA uses spatial constraints in the form of the Markov random field (MRF) to find the segmentation that maximizes the *a posteriori* probability density function for the distribution of regions given the observed image.

In order to compare the similarity between color feature vectors, a method called OCCD (optimal color composition distance) is used. The basic idea can be described as a ball matching process. Imagining the feature vectors in comparison as two bags and each bag has 100 balls with different colors inside, for example, bag 1 has M kinds of color balls including p_1*100 balls with color c_1, p_2*100 balls with color c_2, and so on, and similarly, bag 2 has M' kinds of color balls including p'_1*100 balls with color c'_1, p'_2*100 balls with color c'_2, and so on. In the matching process, we iteratively pick one kind of ball from bag 1 with color c_i, and find one color c'_j in bag 2 which is the closest color to c_i, then the $\min(p_i^*100, p_j'^*100)$ number of balls with color c_i for bag 1 and c'_j for bag 2 are taken out and the difference of $|c_i - c'_j|* \min(p_i^*100, p_j'*100)$ are accumulated as the matching difference. This action is repeated until all the balls are taken out from the bags and the bags are empty. The final accumulated matching difference is the difference in value that we are looking for.

So, mathematically the OCCD distance between the two feature vectors f_c and f'_c can be calculated as:

$$D_c(f_c, f'_c) = \sum_{i=0}^{m} dist(c_i, c'_i)^* p_i$$

where m is the total number of matching times, c_i, c'_i and p_i are the colors and percentage occurs in each matching, and dist(.) is the distance measurement for color space, and typically the Euclidean distance is used.

B. Texture Features

Among many texture analysis methods, APS demonstrates the methodology of steerable filter decomposition with four orientation sub-bands (horizontal, vertical, ±45 degree) with coefficients $s_0(x,y)$, $s_1(x,y)$, $s_2(x,y)$, and $s_3(x,y)$ according to pixel (x, y). The sub-band energy is first boosted with median operation, which preserves the edges while boosting low energy pixels in texture regions. Since a pixel in a smooth region would not contain substantial energy in any of the four orientation bands, so a threshold-based detection on the sub-band coefficients can determine whether a pixel is located in the smooth region.

For pixels in the non-smooth region, they are further classified into orientation categories (including horizontal, vertical, ±45 degree) or complex category (when no dominant orientation is found). The classification is based on the local histogram of the orientation in the neighborhood, and the orientation of a pixel is obtained by the maximum of the four sub-band coefficients.

The similarity measurement between texture feature vectors f_t and f'_t is defined as:

$$D_t(f_t, f_{t'}) = \begin{cases} 0 & if\ f_t = f_{t'} \\ t_{i,j} & otherwise \end{cases}$$

where $t_{i,j}$ is a pre-assigned threshold varying for different texture classes.

C. Segmentation Algorithm

The APS algorithm is based on both of the color and texture features mentioned above. In the process the smooth and non-smooth regions are considered using different approaches. For the smooth region, where the colors are varying slowly, it relies on ACA output with a simple region merging process where all the connected neighboring smooth regions are merged if the average color difference across the border is below a pre-assigned threshold. An example of the processing flow is demonstrated in Figure 6.9.

For a non-smooth region, it first obtains a crude segmentation, via a multi-grid region growing algorithm. In order to calculate the distance between the feature vectors of neighboring pixels so as to tell if they belong to the same region, APS uses the following formula which joins the color and texture feature distance together:

$$D(f, f') = D_c(f_c, f'_c) + D_t(f_t, f'_t)$$

In addition, the MRF spatial constraints are used here as well, in which the energy function to be minimized is defined as:

$$D(f, f^i) + S(N_d^i - N_s^i) \text{ for all } i$$

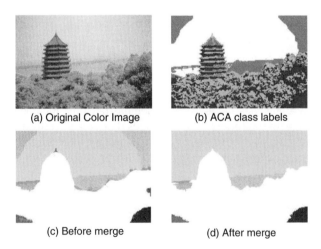

(a) Original Color Image   (b) ACA class labels

(c) Before merge   (d) After merge

Figure 6.9   Color segmentation for smooth regions (Images courtesy of Dr Junqing Chen, Aptina Imaging)

Where f is the feature vector for current pixel, f^i the feature vector for its ith neighboring pixel, S is the strength of the spatial constraint, N_s^i is the number of non-smooth neighbors that belongs to the same class as the ith neighbor, and N_d^i is the number of non-smooth neighbors that belongs to different class from the ith neighbor. S represents the strength of the MRF, Clearly this function punishes a scenario in which the neighboring pixels go to many different classes, while encouraging the current pixel to join the class that most of its neighbors belong to.

After the crude segmentation result is obtained, a refinement process follows in order to adjust adaptively the pixels on the non-smooth region boundary using the color composition features. For these boundary pixels, we determine which region it really belongs to by using the OCCD criterion in order to compare its color feature vector with an averaged feature vector within a larger window that reflects the localized estimation of the regions characteristics. In order to assure the region s smoothness, an MRF constraint similar to that above has been applied in this process. An example of the flow is demonstrated in Figure 6.10.

A few other segmentation examples are shown in Figure 6.11 by using the APS algorithm.

6.2.3 Video Object Segmentation

Video object segmentation differs from image segmentation owing to the additional information obtained in the temporal domain. Although all of the approaches mentioned in last section can be applied to video frames, video segmentation has much higher performance requirements especially for real-time applications. In [6], the video segmentation scenarios are classified into four scenarios: the first one is offline user interactive segmentation, which normally refers to media storage related applications, such as DVD or Blu-ray movies, where high segmentation quality is required but there is no limitation

Figure 6.10 APS segmentation flow (a) Original color image, (b) ACA color segmentation, (c) texture classes, (d) crude segmentation, (e) final segmentation, (f) final segmentation boundaries shown on top of the original image) (Images courtesy of Dr Junqing Chen, Aptina Imaging)

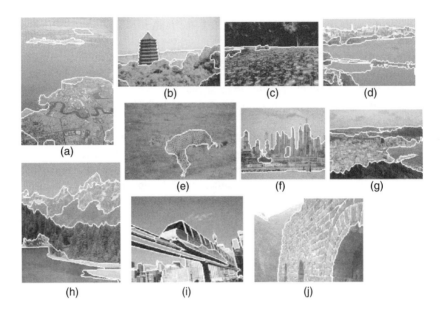

Figure 6.11 Adaptive perceptual color-texture segmentation results (Images courtesy of Dr Junqing Chen, Aptina Imaging)

for segmentation performance and content complexity; the second is the offline non-user interactive scenario, which allows for high content complexity but the segmentation quality requirement is a little lower than in the first category. The third category is real-time user interactive applications, such as video telephony or video conference, which allows for medium content complexity but with medium to high segmentation quality. The last category is real-time non-user interactive applications such as video surveillance or video telephony, which has low to medium content complexity and requires medium segmentation quality. Clearly the real-time non-user interactive scenario is the most challenging although certain constraints are used in order to achieve reasonable performance. As an example, indoor surveillance applications normally let the camera monitor a given area in order to detect intruders, so most of the time the camera captures static scene images with illumination changes. A real-time change detection algorithm can be deployed in order to segment out the interested objects (intruders) in such a scenario.

As usual we start from the back end: given that video segmentation results have been obtained, how can we evaluate segmentation quality? Is there a stand alone objective evaluation scheme for this purpose? We begin this section by introducing the efforts made in this area by the IST research team in Portugal [6], where an objective evaluation of segmentation quality is proposed, in particular when no ground truth segmentation is available to be used as a reference. In general, the proposed system is based on *a priori* information, such as intra-object homogeneity and inter-object disparity; that is, the interior pixels of an object are expected to have a reasonably homogeneous texture, and the objects are expected to show a certain disparity to their neighbors. The evaluator first checks each individual object. The intra-object features include shape regularity (compactness, circularity and elongation), spatial perceptual information, texture variance, and motion uniformity, etc. The inter-object features cover local contrast to neighbors, and the difference between neighboring objects. After all of the objects have been evaluated, the evaluator produces an overall quality measurement which combines individual object quality, their relevance and the similarity of object factors.

In this section, we describe the basic spatial-temporal approaches, the motion object tracking approaches and the head-and-shoulder object segmentation methods. We provide a detailed example in order to show the incorporation of face-detection and spatial segmentation in object segmentation for mobile video telephony applications.

6.2.3.1   COST211 Analysis Model

COST (Coopération Européenne dans le recherché scientifique et technique) is a project initiated by the European Community for R&D cooperation in Europe, and COST 211 is a collaborative research forum facilitating the creation and maintenance in Europe of a high level of expertise in the field of video compression and related activities. The current focus of the COST 211quat project is the investigation of tools and algorithms for image and video analysis in order to provide emerging multimedia applications with the means of availing themselves of the functionalities offered by MPEG-4 and MPEG-7. The objective is to define an Analysis Model (AM). The AM is a framework for image analysis and includes a complete specification of the analysis tools used within the framework. Figure 6.12 shows the current AM model. Clearly, other than the texture analysis module, change detection and motion vector segmentation are the two main techniques for dealing with the correlation between neighboring video frames. We will cover motion related segmentation approaches in the next few sections.

6.2.3.2   Spatial-Temporal Segmentation

Spatial-temporal segmentation is the most popular category in video segmentation, which incorporates the spatial segmentation methods demonstrated in section 6.2.2 with motion information, obtained crossing the video frames. The motion based segmentation approach relies on the observation that a natural object normally has a coherent motion pattern which is distinct from the background, thus motion can be used to group various regions into a semantic object. There has been a great deal of work in this area based on various motion models. One example is to consider video clips as a 3D data field and apply the optical flow equations and fit with affine motion models. The motion estimation in optical flow method is based on the invariance of luminance hypothesis, which assumes that a pixel conserves its intensity along its trajectory within the 3D data field if there is no noise and lighting changes. Clearly the assumption is never satisfied in a video application owing to noise and other factors during the image capturing process. area great deal of work has been presented on estimating or correcting the displacement field in the optical flow and simplifying the computation involved in motion estimation. On the other hand, direct or parametric motion estimation has become popular owing to the much lower computation burden. This kind of approach is not targeted at finding the true motions in the video data field, instead it uses a block-based matching method in order to find the best motion vectors that can match blocks between the current and next frame.

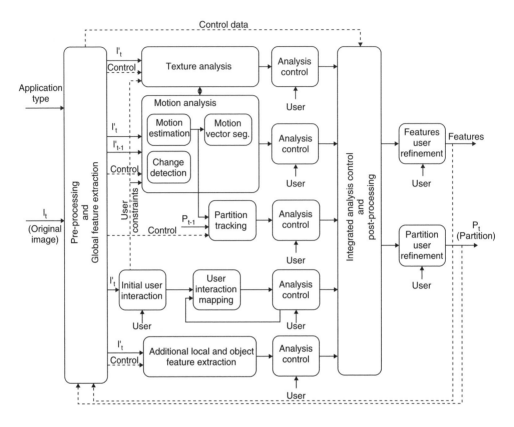

Figure 6.12   The European COST 211 quat AM model

Change detection is based on such a simplified motion model, which calculates the pixel intensity differences between neighboring frames so as to determine if a pixel is moving or not. In [7], a pixel is determined if it is within the change region by checking its fourth-order statistics moment:

$$sm(x, y) = \frac{1}{9} \sum_{(s,t) \in W(x,y)} (d(s, t) - m(x, y))^4$$

where d(.) is the inter-frame difference, W(x,y) is a small neighborhood window centered at (x,y) and m(x,y) is the sample mean of inter-frame difference within the window. Clearly this statistic would detect the boundary between moving objects. After that, the spatial segmentation approach, such as the color-based method, can merge the homogeneous regions separated by the boundary and complete the segmentation.

6.2.3.3   Moving Object Tracking

Moving object tracking as a specific category of video segmentation has gained more popularity in recent years after video surveillance became a hot topic. The key goal of such scenarios is to segment the moving object from a static or smoothly moving background. Considering the strong continuity of the background in neighboring video frames, the problem can be converted to the opposite side, that is to detect background instead of foreground moving objects, and subtracting the background from the original video frame will generate the objects of interest. This technique is called background subtraction, and the key technology is to set up and preserve an effective background model, which represents the background pixels and the associated statistics.

It is very natural to model background pixel intensity over time as a Gaussian distribution, however, this would only work for ideally static scenes, while not fitting for a real environment, considering uncertain factors such as sudden light change, confusion caused by a moving background (such as moving tree branches) and the shadows cast by foreground objects. In [3], a mixture Gaussian model is proposed for background modeling. In [8], non-parametric models are proposed.

In the following, we demonstrate a simplified version of the mixture Gaussian model proposed in [3] that was incorporated in a real-time moving object tracking system. The intensity of each pixel in the video frame is modeled and represented by a mixture of K (for example K = 5) Gaussian distributions, where each Gaussian is weighted according to the frequency with which it represents the background. So the probability that a certain pixel has intensity Xt at time t is estimated as:

$$P(X_t) = \sum_{i=1}^{K} w_{i,t} \frac{1}{\sqrt{2\pi}\sigma_i} e^{-\frac{1}{2}(X_t-\mu_i,t)^T \Sigma^{-1}(X_t-\mu_i,t)},$$

where $w_{i,t}$ is the normalized weight, μ_i and σ_i are the mean and the standard deviation of the ith distribution. As the parameters of the mixture model of each pixel vary across the frame, our task is to determine the most influential distributions in the mixture form for each pixel and use them to determine if the current pixel belongs to background or not. Heuristically, we are mostly interested in the Gaussian distributions with the most supporting evidence and the least variance, thus we sort the K distributions based on the value of w/σ and maintain an ordered list, to keep the most likely distributions on top and the leave less probable transient background distributions at the bottom.

In [3], the most likely distribution models for a pixel are obtained by:

$$B = \arg\min_b \left(\sum_{j=1}^{b} w_j > T \right)$$

where the threshold T is the fraction of the total weight given to the background. The current pixel in the evaluation is checked against the existing K Gaussian distributions in order to detect if the distance between the mean of a distribution and the current pixel intensity is within 2.5 times the standard deviation of this distribution. If none of the K distributions succeeds in the evaluation, the least probable distribution which has the smallest value of w/σ is replaced by a new Gaussian distribution with the current pixel value as its mean, and a pre-assigned high variance and low prior weight. Otherwise if the matched distribution is one of the B background distributions, the new pixel is marked as background, otherwise foreground.

To keep the model adaptive, the model's parameters are updated continuously using the pixel intensity values from the next frames. For the matched Gaussian distribution, all the parameters at time t are updated with this new pixel value Xt. In addition, the prior weight is updated by:

$$w_t = (1 - \alpha)w_{t-1} + \alpha$$

and the mean and variance are updated by:

$$\mu_t = (1 - \rho)\mu_{t-1} + \rho X_t$$

and

$$\sigma_t^2 = (1 - \rho)\sigma_{t-1}^2 + \rho(X_t - \mu_t)^2$$

where α is the learning rate controlling adaptation speed, $1/\alpha$ defines the time constant which determines change, and ρ is the probability associated with the current pixel, scaled by the learning rate α. So ρ can be represented by:

$$\rho = \alpha \cdot \frac{1}{\sqrt{2\pi}\sigma_t} e^{-\frac{(X_t - \mu_t)^2}{\sigma_t^2}}.$$

For unmatched distributions, the mean μ_t and variance σ_t remain unchanged, while the prior weight is updated by:

$$w_t = (1 - \alpha)w_{t-1}$$

One advantage of this updating method is that, when it allows an object to become part of the background, it doesn't destroy the original background model. In other words, the original background distribution remains in the mixture until it becomes the least probable distribution and a new color is observed. So if this static object happens to move again, the previous background distribution will be reincorporated rapidly into the model. In Figure 6.13, an example is given in order to demonstrate this approach. Although the movement of the foreground objects from frame 8 to 10 (in the video clip) is rather small, the portion of the movement (human heads) in the frames are extracted successfully.

6.2.3.4 Head-and-Shoulder Object Segmentation

Automatic head-and-shoulder video object segmentation is very attractive for applications such as video telephony, where the video encoder can allocate more resources to the human object regions, which might be of interest to a viewer, in order to code them at higher quality so as to achieve better perceptual feelings. As another example, for video surveillance applications, the segmented object/face can be inputted into a (face) database system in order to match with target objects. The traditional automatic object segmentation research activities were focused on motion analysis, motion segmentation and region segmentation [10, 11]. In [12], a statistical model-based video segmentation algorithm is presented for head-and-shoulder type video, which abstracts the human object into a blob-based statistical region model and a shape model, thus the object segmentation problem is converted into a model detection and tracking problem. Challapali [13] proposes

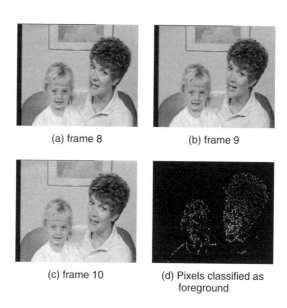

(a) frame 8   (b) frame 9

(c) frame 10   (d) Pixels classified as
                                                    foreground

Figure 6.13   Mixture of Gaussian model for background subtraction

to extract the foreground object based on a disparity estimation between two views from a stereo camera setup, and also proposes a human face segmentation algorithm by using a model-based color and face eigenmask matching approach. Cavallaro [14] proposes a hybrid object segmentation algorithm between region-based and feature-based approaches. It uses region descriptors to represent the object regions which are homogeneous with respect to the motion, color and texture features, and track them across the video sequence.

Clearly, face detection can provide tremendous help in head-and-shoulder object segmentation. There is a great deal of research which has been published in the literature on this area (see [15, 16] for reviews). Wang [17] and Chai [18] propose a simple skin-tone based approach for face detection, which detects pixels with a skin-color appearance based on skin-tone maps derived from the chrominance component of the input image. Hsu [19] proposes a lighting compensation model in order to correct the color bias for face detection. In addition, it constructs eye, mouth and boundary maps in order to verify the face candidates. Wong [20] adopts a similar approach but uses an eigenmask, which has large magnitude at the important facial features of a human face, in order to improve the accuracy of detection.

In the following, we provide a detailed example of head-and shoulder object segmentation deployed in mobile video telephony applications [21], where a hybrid framework which combines feature-based and model-based detection with a region segmentation method is presented. The algorithm flowchart is shown in Figure 6.14, where segmentation is carried out in four major steps and eight sub-tasks. First, color based quick skin/face detection is applied in order to remove those pixels which cannot be facial pixels, and facial features, such as eye and mouth, are localized and verified; then, the detected features are classified into various groups (according to faces) and a head-and-shoulder geometric model is used to find the approximate shape of the object for each face. After

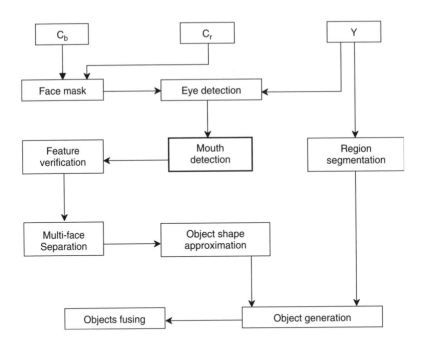

Figure 6.14 Flowchart of the header-and-shoulder object segmentation algorithm

that, a split-and-merge region growing approach is used to divide the original image into homogeneous regions. Finally, an automatic region selection algorithm is called upon in order to form these regions into a segmented object based on the approximated shape of the head-and-shoulder object, and the objects are fused into a final output image.

A. Skin-Color Based Quick Face Detection

It has been found in [17] that a skin-color region can be identified by the presence of a certain set of chrominance values distributed narrowly and consistently in the YCbCr color space. In addition, it has been proved in [18] that the skin-color map is robust as against different types of skin color. Clearly, the skin color of all races which are perceived as apparently different is due mainly to the darkness or fairness of the skin, in other words, the skin color is characterized by the difference in the brightness of the color and is governed by Y but not Cr or Cb. Therefore, an effective skin-color reference map can be achieved based on the Cr and Cb components of the input image. In this work, we use the CbCr map recommended in [18] which has the range of $Cr \in [133, 173]$ and $Cb \in [77, 127]$ for skin detection. After the skin mask is obtained, we use the mathematical morphological operations [22] of dilation and erosion to remove the noise and holes caused by facial features such as the eye and mouth regions in the mask. An example of quick face detection in the 'Foreman' sequence is shown in Figure 6.15.

As may be expected for some clips, skin-color based quick detection is unable to obtain the human face exclusively. As shown in Figures 6.16 and 6.17, according to the 'Mother and Daughter' and 'Akiyo' sequences, the clothes regions appear to have a similar tone to

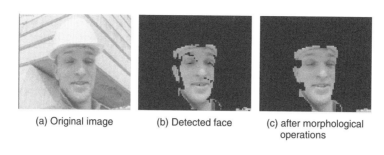

(a) Original image   (b) Detected face   (c) after morphological
                                                            operations

Figure 6.15   An example of quick face detection

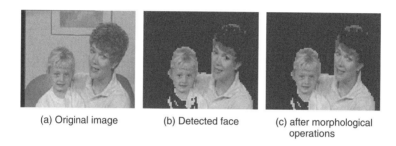

(a) Original image   (b) Detected face   (c) after morphological
                                                            operations

Figure 6.16   Quick face detection results for the 'Mother and Daughter' sequence

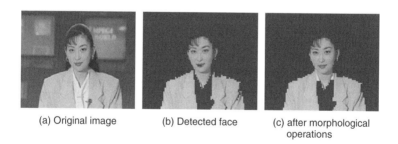

(a) Original image   (b) Detected face   (c) after morphological
                                                            operations

Figure 6.17   Quick face detection results for the 'Akiyo' sequence

the skin tone, and are therefore selected falsely as being inside the skin region. Therefore, it is important to realize that this quick face detection step does help to remove some non-face regions, but further processes are needed in order to find the exact face region and verify its correctness.

B. Facial Feature Localization

We build a facial filter based on the common knowledge of human faces and their features, such as the elliptical shape of the facial region and the overall spatial relationship constraints among these facial features. Thus, locating these facial features is helpful in deriving the approximate face location. In the case where there is more than one semantic object involved in the frame, the problem becomes more complicated.

Eye Detection

In this book, eye detection is based on two observations in [19] with a little simplification. The steps are:

a. The chroma components around the eyes normally contain high Cb and low Cr values. Therefore, an eye map can be constructed by:

$$C = \frac{Cb^2 + (255 - Cr)^2 + (Cb/Cr)}{3}.$$

When the eye map is obtained, a threshold is used to locate the brightest regions for eye candidates and morphological operations are applied in order to merge close regions into a connected one (as shown in Figure 6.18).

b. The eyes usually contain both dark and bright pixels in the luma component. Therefore, grayscale morphological operators (as shown in Figure 6.19(a) and (b)) can be used to emphasize brighter and darker pixels in the luma component around eye regions. An eye map can be constructed by:

$$L = \frac{Dilation(Y)}{Erosion(Y) + 1}$$

When the eye map is obtained, a threshold is used to locate the brightest regions for eye candidates and morphological operations are applied to merge close regions into a connected one (as shown in Figure 6.19(d)).

Finally, these two eye maps are joined in order to find the final eye candidates, as illustrated in Figure 6.19(e). In Figure 6.19(e), there is an eye candidate (the rightmost region) which is not correct. It will be removed later at the feature verification stage.

Mouth Detection

Normally, the color of the mouth region contains a stronger red component and a weaker blue component than other facial regions [19]. Hence, the chrominance component Cr is great than Cb in the mouth region. On the other hand, the mouth has a relative low

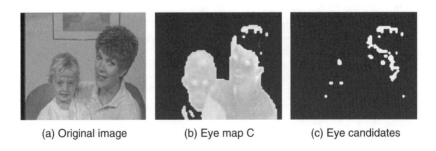

(a) Original image   (b) Eye map C   (c) Eye candidates

Figure 6.18   Eye map C generation

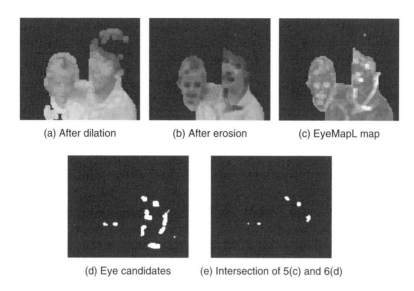

(a) After dilation	(b) After erosion	(c) EyeMapL map
(d) Eye candidates	(e) Intersection of 5(c) and 6(d)	

Figure 6.19   Eye map L generation

response in the Cr/Cb feature, but it has a high response in Cr2. So a mouth map can be constructed as:

$$M = Cr^2 \left(Cr^2 - \lambda \frac{Cr}{Cb} \right)^2$$

where

$$\lambda = 0.95 \frac{\displaystyle\sum_{(x,y) \in SkinMask} C_r(x, y)^2}{\displaystyle\sum_{(x,y) \in SkinMask} \frac{C_r(x, y)}{C_b(x, y)}}$$

As processed in the eye detection and shown in Figure 6.20, thresholding and morphological processes are applied on the obtained the M map in order to generate mouth candidates.

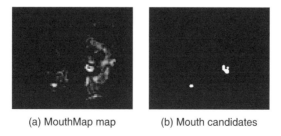

(a) MouthMap map	(b) Mouth candidates

Figure 6.20   Mouth map M generation

C. Feature Verification

Feature verification is very critical in face detection in order to assure its robustness. The detected facial features go through three verification steps so as to remove false detections.

Valley Detection

Normally facial features are located in valley regions (characterized by high intensity contrast inside the region). These regions can be detected by grayscale-close and dilation morphological operations. As shown in Figure 6.21, the valley regions of 'Mother and Daughter' and 'Akiyo' sequences are detected. If a feature candidate has no overlapping areas with the detected valley regions, it will be removed from the candidate list.

Eye Pair Verification

The eyes are verified based on a set of common knowledge [23] listed as follows:

a. Two eyes are symmetric with respect to the major axis of the face (as shown in Figure 6.22), which means $|AO1| = |AO2|$, both eyes have similar area, and the shape similarity can be compared by projecting to the axis OA.
b. Eye is symmetric with respect to the PCA (Principle Component Analysis) axis.
c. For most cases, an eyebrow can be detected above the eye.

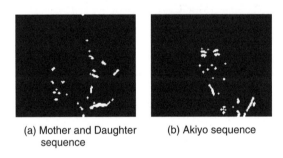

(a) Mother and Daughter   (b) Akiyo sequence
sequence

Figure 6.21   Valley regions for various video clips

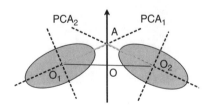

Figure 6.22   The symmetric property of eye pairs

A weighted score-system is used for verification. In other words, we check each criterion listed below and give a score for the result. Later on, all the scores obtained are accumulated in order to make a final decision.

a. The eye centroid location is inside a valley region.
b. The locations of the eye centroid and detected iris are close enough; (the iris location is found by projecting the intensity value in an eye to horizontal and vertical axis and finding the point corresponding to the minimum accumulated total intensity value).
c. Eyebrow is found above the eye.
d. The PCA axis of the eye is within a range of reasonable directions.
e. The eye candidate is able to find its second-half within a reasonable distance.
f. The pair of eyes has a symmetric PCA axis according to the axis OA.
g. The pair of eyes has a symmetric shape according to the axis OA.

Clearly, based on the score system, those false detections obtaining scores below a preset threshold are removed from the candidate list.

Eye-Mouth Triangle Verification
Every eye-mouth triangle of all possible combinations of two eye candidates and one mouth candidates has to be verified. The geometry and orientation of the triangle is first reviewed and unreasonable triangle pairs are removed from further consideration. As shown in Figure 6.23, only two possible eye-mouth triangles (in dash-lines) are kept for further verification.

A template is used to verify the triangle area gradient characteristics. The idea is that the human face is a 3D object, thus the luminance throughout the facial region is non-uniform, and the triangle area contains the nose, which should make the gradient information more complicated than for other facial areas like chins.

D. Multi-Face Separation
In video clips with more than one object in a video frame, it is critical to separate the set of detected eyes and mouths into groups corresponding to different faces. Clearly, the difficulties of this task are threefold: first of all, the total number of faces existing in the frame is not available; second, some features might be missing during the detection

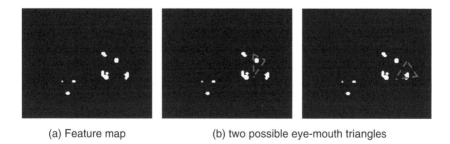

      (a) Feature map   (b) two possible eye-mouth triangles

Figure 6.23   An example of eye-mouth triangle verification

process; third, an exhaustive check of all the potential combinations has exponential computational complexity.

By treating the problem as pairing eyes with mouth, the original problem is mapped into a graph theory problem, that is, considering a bipartite graph G = (V, E) with vertices set V = {mouth} + {eye pairs} and edge set E = {(vi, vj) vi and vj belong to different sets, and the distance between the node vi and vj is within a reasonable range}. If we define a matching S as a subset of E such that no two edges in S are incidental to the same vertex or directly connected vertices, then the problem becomes one of finding the maximum matching. It is a variant of the famous maximum matching problem, because in the original maximum matching problem definition, the constraint on the matching only requires that 'no two edges in S are incidental to the same vertex'. For the original maximum matching problem, the solution can be found either in [24] or [25], and both of them provide a solution with a polynomial complexity.

It is important to note the possibility of converting our problem into the original maximum matching problem. We define an edge set E' = {(vi, vj)| there exits vk such that (vi, vk) ∈ E, (vj, vk) ∈ E but (vi, vj) ∉ E}, so after we expand the edge set from E to $E \cup E'$, our problem becomes the original maximum matching problem except that an additional constraint must be included that 'the results matches must be a subset of E instead of $E \cup E'$'. Clearly, the constraint would not affect the usage of the solution in [24, 25]. Therefore, the multi-face separation problem can be solved in polynomial time complexity.

E. Head-and-Shoulder Object Approximation

After the eye-mouth triangle is found, it is not hard to build a head-and-shoulder model based on the geometry relationship between the nodes of the triangle. Clearly, a better shape model results in more accurate approximation. In this work, to speed up the performance, we use a simple rectangular model to approximate the head-and-shoulder shape. As an example, the head-and-should area for the 'Akiyo' sequence is shown in Figure 6.24.

For video sequences containing more than one object, after separating the eyes and mouths into different groups, the head-shoulder area for each object is generated. As shown in Figure 6.25, the child and the mother have different head-shoulder areas for further processing.

(a) Original image   (b) eye-mouth triangle   (c) head-shoulder area

Figure 6.24   An example of head-and-shoulder model

(a) Original image   (b) child head-shoulder   (c) mother head-shoulder

Figure 6.25   Head-shoulder model for sequence containing more than one object

F. Split-and Merge Region Growing

Split-and-merge region growing approach is an important step in this process. The basic idea is to separate the relationship between neighboring pixels inside an image into two classes: similar and dissimilar, then cluster the connected similar pixels into small regions, and then form the meaning image components by merging these regions. Please revisit section 6.2.2.3 for more detail on this approach and see Figure 6.8 for sample results.

G. Single Object Segmentation by Automatic Region Selection

In this book, we propose an automatic region selection algorithm for object segmentation. The algorithm considers only regions (after split-and-merge region growing) inside the object bounding box, which is a rectangular area that contains the object, and the size of the box can be estimated from the eye-mouth triangle. Thus, further processing can be conducted within the bounding box, instead of the original frame. In the region selection procedure, we assume that all the regions that are inside the approximated head-and-shoulder shape belong to the foreground object, and those regions which have more than a certain percentage (for example, 60%) of their total pixels inside the approximated shape also belong to the foreground object. In this way, we can extract the foreground object from the original image. As shown in Figure 6.26, the object is extracted from the 'Akiyo' sequence.

(a) Object bounding box   (b) segmented regions   (c) extracted foreground object

Figure 6.26   Object segmentation of the 'Akiyo' sequence

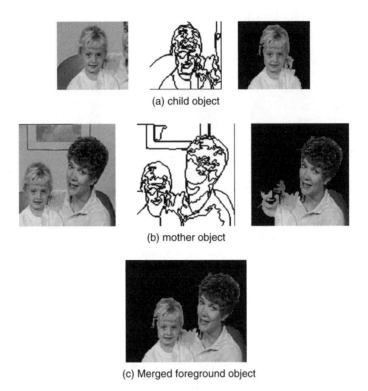

(a) child object

(b) mother object

(c) Merged foreground object

Figure 6.27   Object segmentation of 'Mother and daughter' sequence

H. Object Fusing

For video sequences containing more than one object, similar procedures are applied, and the extracted objects are merged to form the final foreground (as shown in Figure 6.27).

6.2.4 Video Structure Understanding

In the real world, people generally characterize video content by high-level concepts such as the action, comedy, tragedy or romance in a movie, which does not link directly to the pixel level attributes in the video. In this section, we study the video structure, which plays a significant role in characterizing a video and helps video understanding. It can be observed that certain video clips exhibit a great deal of content structure information, such as newscast and sports video, while others do not. There are many studies on newscast video understanding which take advantage of the spatial and temporal structure embedded in the clip. Spatial structure information includes the standardized scene layout of the anchor person, the separators between various sections, etc. Temporal structure information refers to the periodic appearances of the anchor-person in the scene, which normally indicates the starting point of a piece of news. In [26], the stylistic elements, such as montage and mise-en-scene, of a movie are considered as messengers for video structure understanding. Typically, montage refers to the special effect in video editing that composes different shots (shot is the basic element in videos that represents the continuous frames recorded from the moment the camera is on to the moment it is off) to form the movie scenes; and it conveys temporal structure information, while the mise-en-scene relates to the spatial structure. Statistical models are built for shot duration and activity, and a movie feature space consisting of these two factors is formed which is capable of classifying the movies in different categories.

Compared to other videos, sports video has well-defined temporal content structure, in which a long game is divided into smaller pieces, such as games, sets and quarters, and there are fixed hierarchical structures for such sports. On the other hand, there are certain conventions for the camera in each sports, for example, when a service is made in a tennis game, the scene is usually presented by an overview of the field. In [27], tennis videos are analyzed with a hidden Markov model (HMM). The process first segments the player using dominant color and shape description features and then uses the HMMs to identify the strokes in the video. In [28, 29], the football video structure is analyzed by using HMMs to model the two basic states of the game, 'play' and 'break', thus the whole game is treated as switching between these two states. The color, motion and camera view features are used in these processes, for example, the close-ups normally refer to break state while global views are classified as play.

6.2.4.1 Video Abstraction

Video abstraction is a compact representation of a video clip and is currently widely used by many communication applications, as we mentioned at the beginning of this chapter. In general, there are three types of video abstraction formats:

- *Video highlights (or video skimming):* a concatenation of a number of video segmentations representing the most attractive parts of the original video sequence. This is often seen in a newscast when a preview is shown.
- *Video summary:* a collection of salient still images extracted from the underlying video sequence in order to represent the entire story. This can be used in multimedia messaging services for people to communicate by sharing key frames in a video story.
- *Video structure map (or data distribution map):* a hyperlinked structural format that organizes the video into a hierarchical structure. This way the video is like a book with chapters and sections listed in the content index, so users can locate a certain video frame by going quickly through this structure.

Zhu [30] presents an effective scheme that obtains the video content structure along with the hierarchical video summary in the same framework. As shown in Figure 6.28, this detects the video content structure by taking the following steps:

1. *Video shot detection and key frame extraction:* initially, the video shots are separated according to the distribution of activities within the video frames; then within each shot the key frame is selected according to the visual similarity between frames measured by color histogram, coarseness and directionality texture, and the one selected generally exhibits sufficient dissimilarity and captures the most visual variances of the shot.

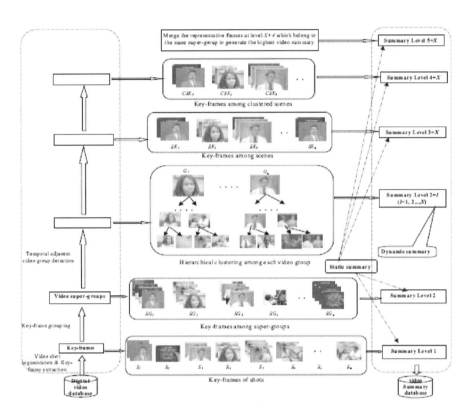

Figure 6.28   Video content structure (Courtesy of Dr Xingquan Zhu, FAU) (Copyright © 2004, Springer Berlin/Heidelberg)

2. *Video super group detection:* by checking the similarity between the extracted key frames using a self-correlation matrix mechanism, the shots with visual similarity are clustered into super groups.

3. *Video group detection:* the task is to divide the shots within a super group into a number of groups according to the temporal distance between neighboring shots. The underlying assumption is that the shots with a large temporal distance are more likely to belong to different scenarios. Therefore, a video group contains visually similar and temporally close shots, and acts as the intermediate bridge between shots and semantic scenes.

4. *Video scene detection:* the video groups are merged into scenes according to the characteristics of different types of scenes. In one kind, the scene consists of frames with consistency of chromatic composition and lighting conditions; in another, called the montage scene, a large visual variance is found among the frames although the scene consists of temporally interlaced frames from different groups in order to convey the same scenario; There are also hybrid scenes that are a combination of these.

5. *Video scene clustering:* a clustered scene represents a more compact level in the structure, in which visually similar scenes are grouped together according to the super group information. Specifically, for each scene, the union of the super groups that contains the shots in the current scene is obtained so as to become a potential cluster candidate.

The final clusters are obtained by removing the redundancy among these potential candidates.

Clearly the cluster->scene->video group->shot->key frame structure can serve as a video structure map, although Zhu [30] uses it for a further process designed to seek hierarchical video summarization. In the next section, we discuss video summary extraction further and provide an example of real-time video summarization for video communications applications.

6.2.4.2   Video Summary Extraction

There are many published works on video summarization. In [31], a video sequence is viewed as a curve in a high dimensional space and a video summary is represented by the set of control points on that curve which meet certain constraints and best represent the curve. Clustering techniques [32–37] are widely used in video summarization. In [37] an un-supervised clustering is proposed in order to group the frames into clusters based on the color histogram features in the HSV color space. The frames closest to the cluster centroids are chosen as the key frames. In [33] cluster-validity analysis is applied in order to select the optimal number of clusters. Li [38] and Lu [39] use graph modeling and optimization approaches to map the original problem into a graph theory problem. In [38], video summarization is modeled into a temporal rate-distortion optimization problem by introducing a frame distortion metric between different frames which is determined by the number of missing frames and their location in the original sequence. Although color and motion were used to represent distortion in [40] and a weighted Euclidean distance between two frames in the Principal Component Analysis (PCA) subspace was used in [38], the proposed approaches only consider the content coverage aspect of the selected frames, while failing to take into account their representation and the visual quality. Lu [39] uses a similar technique which defines a spatial-temporal dissimilarity function between key frames and thus by maximizing the total dissimilarity of the key frames ensures good content coverage of the video summary. However, Lu [39] uses the temporal distance between key frames to represent the temporal dissimilarity, which might undermine the content representation of the generated summary.

In this section, we introduce a practical and low-complexity video summary extraction algorithm for real-time wireless communications application that is proposed in [41]. This considers jointly summary representation, content variation coverage and key frame visual quality in a general framework and constructs a new cost function for optimizing the summary indices. In addition, it takes advantage of the hinting information stored during the video encoding process, which saves a large volume of memory access in order to analyze the original video frames.

Figure 6.29 shows the overall system architecture of a video processor in a wireless multimedia chip. In the system, when a video sequence (raw image data) enters the front end module, the incoming image is first pre-processed to auto focus, noise removal and so on, and then the data enters the video encoder for compression. Some visual statistics obtained during this process are stored for future use. If the video summarization function is turned on, the stored information along with the user inputs and interactions are used as side-information in order to enhance summarization performance. Once the summary

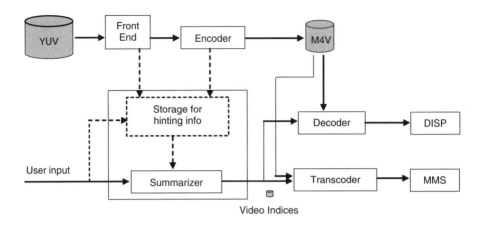

Figure 6.29   System architecture of the video summarization system and potential applications

indices are extracted, they are either sent to the decoder and the multimedia display processor for displaying and video editing purposes, or are sent to a transcoder and then to the multimedia messaging service (MMS) for transmitting to remote mobile devices. These are two of the many potential applications that may make use of the results of video summarization. The system allows the summarization to be programmed into either off-line or real-time mode. The difference between them is that in off-line processing the summarizer has a global view of the sequence. Therefore it will help to generate a summary with better global representation. However, the cost is the large storage request for storing the hinting information for each frame and generating the summary, and it is sometimes unaffordable for very long video clips. For real-time processing, a window-based processing strategy is applied, that is, once the hinting information for a window of frames, for example 1000 frames, is ready, the window of frames will be summarized, and in the meantime, the new incoming frames are processed and encoded and the statistics are recorded in the memory for future summarization. The advantage of real-time processing is that it has less extra storage request, but clearly the summary generated might lose global representation, which is a natural trade off that can be determined by users.

A. Hinting Information
In front end module, the hinting information is generated during the processing of automatic white balancing, automatic exposure control and automatic focus control. An auto focus process consists of two sub-processes: 1) determining the focus value for a given focus lens position; and 2) the algorithm for determining efficiently the optimal focusing position based on a series of focus values. The focus value is computed from luminance signal Y by:

$$F = \sum_i MAX_j \{[Y(i, j) - Y(i, j + 2)]^2 + [Y(i, j) - Y(i + 2, j)]^2$$

$$+ [(Y(i, j) - Y(i + 2, j + 2)]^2\}$$

where j = I*2, J*2+2, J*2+4, ..., 2*M-2; I = I*2, I*2+2, I*2+4, ..., 2*L-2; I = starting row of focus window in sub-sampled – by-2 domain, J = starting column of focus window in sub-sampled-by-2 domain, L = ending row of focus window in sub-sampled-by-2 domain (L-I< = 508), M = ending column of focus window in sub-sampled-by-2 domain (M-J< = 508), and (M-J) shall be even. Clearly, a higher value of F might correspond to a lower chance of blurriness for the image.

An auto exposure process involves three steps: light metering, scene analysis and exposure compensation. In this process, the input image is divided into 256 regions, and each of these regions is further subdivided into four sub-regions. The sum of luminance value of the pixels in the region, the minimum local sum luminance value in the region, the maximum local sum luminance value in the region and the maximum absolute delta local sum luminance value in the region are generated. From these data, we can estimate approximately the sum of luminance value of the pixels in each sub-region, and thus generate a 64-bin luminance histogram of the frame. In the meantime, we obtain a down sampled 8 × 8 luminance image of the frame.

An auto white balance system computes the gains on the red, green and blue channels that are necessary in order to compensate for the color shift in the white color due to scene illumination. This process involves three steps: pixel color metering, illumination estimation and white balancing. A 128-point histogram for the chrominance components for the frame can be obtained.

During the encoding process, statistics are obtained such as motion vectors, SAD (sum of absolute difference) and prediction mode of macroblocks. From this information, we are able to obtain a value to represent the degree of motion activity of the frame. On the other hand, the user's interaction brings useful side-information as well, for example when the user zooms in the camera and stays for a short period, which means that something interesting might occur. This will bring us certain semantic information for summarization.

B. Summary Extraction

In this section, we demonstrate how to convert the video summary extraction problem into an optimization problem and solve it with the shortest path algorithm. Let us denote by N the number of total frames in the sequence, and by M the length of the expected video summary, then the problem is to find the indices of selected M frames $\{a_i\}$ (i = 1,..., M, and $a_0 = 0$) which can best summarize the sequence. Here the 'best' means that the summary frames would have good local representation, covering content variation, and have good visual quality.

Good local representation means that the selected frame would have good local similarity among its neighbor frames, in other words, the key frames would be similar enough to its neighbor frames to represent them in the final summary clips. We use the color similarity here to evaluate the similarity of neighbor frames. Let us denote by $\{H_i\}$ the YCbCr color histogram obtained from the front end auto exposure and auto white balance processes, then we can define the frame local representation of the ith frame by:

$$A(i) = \begin{cases} Sim(H_{i-1}, H_i) & if\ i = N \\ \dfrac{Sim(H_{i-1}, H_i) + Sim(H_i, H_{i+1})}{2} & otherwise \end{cases}$$

where Sim(.) is the similar function in comparing two 1-D vectors, and it can be defined by:

$$Sim(\vec{x}, \vec{y}) = \frac{\vec{x} \cdot \vec{y}}{||\vec{x}|| \cdot ||\vec{y}||}$$

Covering content variation can be interpreted by the fact that that consecutive frames in the selected summary frames have large dissimilarity. Let us denote by $\{L_i\}$ the down sampled 8×8 luminance image obtained from the front end auto exposure process, then we define the similarity of summary frames by:

$$B(i, j) = \begin{cases} 0 & \text{if } i = 0 \\ \gamma Sim(H_i, H_j) + (1 - \gamma)Sim(L_i, L_j) & \text{otherwise} \end{cases}$$

where γ is a weighting factor with its value between $[0, 1]$. Here the luminance similarity is also considered in order to detect the cases where object movements occur against a still or stable background.

Good visual quality can be interpreted in that the selected frame will have less blurriness (caused by the shifting of the camera) and the object/background in the selected frame have relatively low movement compared to its neighbor frames. It is important to notice that PSNR (peak signal-to-noise ratio) has not been used here to evaluate the visual quality of a frame, because it might mislead the key frame selection. Let us denote by $\{||MV_i||\}$ the total length of the macroblock motion vectors, $\{S_i\}$ the total macroblock SAD in the frame, and $\{F_i\}$ the focus of the image obtained from front end auto focus processing, then we define the visual quality of the image by:

$$C(i) = \begin{cases} \eta||MV_i||S_i^2 + (1 - \eta)(F_{MAX} - F_i) & \text{if } i = N \\ \eta \dfrac{||MV_i||S_i^2 + ||MV_{i+1}||S_{i+1}^2}{2} + (1 - \eta)(F_{MAX} - F_i) & \text{otherwise} \end{cases}$$

where η is a weighting factor with its value between $[0, 1]$, and F_{MAX} is a pre-assigned upper bound of the focus value.

Clearly, a good summary would require having larger $\sum_{i=1}^{M} A(a_i)$, smaller $\sum_{i=1}^{M} B(a_{i-1}, a_i)$ and smaller $\sum_{i=1}^{M} C(a_i)$. It would be natural to convert the summary selection into an optimization problem that:

$$\text{Minimize } T(a_1, a_2, \ldots a_M) = \sum_{i=1}^{M} \{\alpha[1 - A(a_i)] + \beta B(a_{i-1}, a_i) + (1 - \alpha - \beta)C(a_i)\}$$

where α and β are weighting parameters between $[0, 1]$.

In the following, we will provide an optimal solution for the problem above and also provide a sub-optimal solution which takes into account constraints from storage, delay and computation complexity.

C. Optimal Solution

We first create a cost function:

$$G_k(a_k) = \underset{a_1,a_2,\ldots,a_{k-1}}{Minimize} \ T(a_1, a_2, \ldots, a_k)$$

which represents the minimum sum up to and including frame a_k. Clearly

$$G_M(a_M) = \underset{a_1,a_2,\ldots,a_{M-1}}{Minimize} \ T(a_1, a_2, \ldots, a_M)$$

and

$$\underset{a_M}{Minimize} \ G_M(a_M) = \underset{a_1,a_2,\ldots,a_M}{Minimize} \ T(a_1, a_2, \ldots, a_M)$$

The key observation for deriving an efficient algorithm is the fact that given cost function $G_{k-1}(a_{k-1})$, the selection of the next frame index a_k is independent of the selection of the previous decision vectors $a_1, a_2, \ldots, a_{k-2}$. This is true since the cost function can be expressed recursively as:

$$G_{k+1}(a_{k+1}) = \underset{a_k}{Minimize} \ \{G_k(a_k) + \alpha[1 - A(a_{k+1})] + \beta B(a_k, a_{k+1})$$

$$+ (1 - \alpha - \beta)C(a_{k+1})\}$$

The recursive representation of the cost function above makes the future step of the optimization process independent from its past step, which is the foundation of dynamic programming. The problem can be converted into a graph theory problem of finding the shortest path in a directed acyclic graph (DAG). The computational complexity of the algorithm is O(NM^2).

D. Sub-Optimal Solution

In some cases, the optimal solution mentioned above may not be feasible, for example when N is too large for the memory storage or the computational complexity is higher than the allocated power and CPU time. Therefore, a sub-optimal solution is required in order to deal with such circumstances. In real-time processing, the video sequence will be forced into dividing into groups with a fixed window size, which keeps N at a relatively acceptable magnitude. However, the sub-optimal solution is still useful for speeding up performance.

The solution consists of three major steps: shot boundary detection, shot compression ratio calculation and optimized shot key frame selection. The basic idea is to divide the clip into a number of shots and then find the optimal key frame location inside each shot. Since the algorithm of shot key frame selection has been demonstrated above, we focus on the first two steps in the following.

Our shot boundary detection algorithm is a color-histogram based solution in the YCbCr color space. The basic idea is to check the similarity of the color-histogram of consecutive frames, once the similarity is below a pre-set threshold, which means a scene change might happen, the current location will be recorded as a shot boundary. If the number of shot boundaries obtained is larger than the summary length, than the boundary with the minimum location similarity will the selected as the summary frame indices. Otherwise, we need to calculate the summary length for each shot.

In our system, a motion activity based shot compression ratio calculation method is proposed. Let us denote by P the total number of divided shots, by $\{N_i\}$ the length of each shot, and by $\{M_i\}$ the summary length for each shot to be calculated, then:

$$M_i = 1 + \frac{\displaystyle\sum_{j=1+\sum_{k=1}^{i-1} N_k}^{\sum_{k=1}^{i} N_k} (\|MV_j\|S_j^2)}{\displaystyle\sum_{j=1}^{N} (\|MV_j\|S_j^2)} (M - P).$$

Clearly, the algorithm assigns a longer summary to shots with high-volume activities, and assigns fewer summary frames to shots with lower motion and activities. This content-based strategy is closer to a human being's natural logic, and will generate totally different video summary frames compared to the approach of adopting uniform sampling key frame selection especially for clips with high contrast in content activity.

6.2.5 Analysis Methods in Compressed Domain

In real-time wireless applications, computation, power and storage are major constraints for video communications chips, which motivate the study of video analysis techniques in compressed video. In this way, the analysis is conducted on the parsed syntax extracted from the bitstream, such as motion vectors, DCT coefficient, etc, while not waiting until the video pixel intensity is fully decoded. In this scenario, the full decoding may not even be happening, for example, if the analysis tools are linked with a video transcoder that enables compressed domain transcoding, a video frame after analysis is either dropped or transcoded using the parsing syntax data. In this section, we introduce a few useful analysis methods in the DCT domain.

In compressed video, the following features are often used for analysis purposes:

- DC value of a macroblock, which reflects the energy of the coded macroblock, however, it is mostly useful for I-frame, but quite difficult for P- and B-frames, where the coded data are residual data instead of image pixels. In [42], an approximation approach is proposed in order to obtain a DC image corresponding to the average pixel intensity value of P-frames from the previous I-frame.
- Prediction mode of a macroblock, which reflects whether there is a reference in the previous frame with similar pixel intensities compared to the current macroblock that was detected in the block-based motion estimation process.
- Skip mode of a macroblock, which means the there is a reference in the previous frame which matches perfectly the current macroblock so the encoder chooses to not code this macroblock at all.
- Motion vectors, which indicate the movement direction and distance of the macroblock from its reference. The magnitude, homogeneity and variance of the motion vectors are often used as motion intensity features.

- Reference frames of B-frame, which consist of four types: intro-coded, forward prediction, backward prediction and bidirectional prediction. Clearly the reference direction can be used to predict if a scene change occurs in the video, and the number of bi-directionally predicted macroblocks in a B frame can be used as a measurement of the correlation between the past and future I and P-frames.
- Edge detected in the compressed domain: In [43] and [44], edge detection algorithms were proposed by using DCT coefficients with a certain amount of calculations.

In the literature all the features above have been used in scene change detection. The percentage of macroblocks in a frame coded by skip mode was often used as a measure for scene motion intensity change. The higher the percentage, the lower the scene motion intensity change. On the other hand, the DC-image was often derived for comparing the neighboring picture intensity histogram. Furthermore, the reference relationships between adjacent frames are often analyzed based on the assumption that if two frames are strongly referenced, then most of the macroblocks in the frame will not be coded with intra-coded type. Therefore a measure can be constructed to detect the percentage of macroblocks in a frame that are referenced to a certain type in order to determine whether a scene change occurs.

6.3 Content-Based Video Representation

So far, we have demonstrated many key technologies in the content analysis area, thus it is reasonable to imagine that if the system provides certain computations ahead of time either online or offline, the video should be represented after segmentation in a way that supports convenient content access, interactivity and scalability. Obviously the impact of such technology on our everyday life is amazing, for example this content interactivity would change watching TV from a passive activity to a brand new user experience called 'Interactive TV'. With the widespread use of the Internet, the TV user can customize content by manipulating the text, image, audio, video and graphics. The user can even adjust the multi-window screen formats so as to allow for the display of sports statistics or stock quotes. For example, when motorcycle racing is shown on interactive TV, besides watching the scene, the viewer can access information on each rider or team through the Internet, and can locate the current position of the rider on the track, and even obtain the telemetry data sent from the motorcycles such as how fast the motorcycle is and how many revolutions per minute the engine is achieving.

It is common sense that the conventional video representation by a sequence of rectangular frames or fields, also referred to as 'frame-based video', lacks content representation capability. Within such a demanding background of content-based interactivity functionality, the concept of object-based video representation is proposed, in which the video information is represented in terms of its content. In particular, object-based representation divides the video content into a collection of meaningful objects. This approach offers a broad range of capabilities in terms of access, manipulation and interaction with the visual content. MPEG-4 is the first International video standard that supports object-based video.

In object-based video, the video object is represented by the object's shape and texture (as shown in Figure 6.30). Shape information is either provided by the studio or extracted

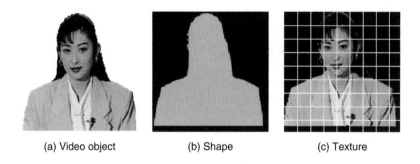

(a) Video object   (b) Shape   (c) Texture

Figure 6.30   Example of video object composed of shape and texture

from a frame-based video by using image/video segmentation. In the architecture of a typical object-based video coding and communication system (as shown in Figure 6.31), at the transmitter, the objects in the scene is extracted (if not provided) by video segmentation and are encoded separately by the source encoder. The compressed elementary streams are then multiplexed so as to form a single bitstream. The channel encoder helps to add redundant information in order to protect the bitstream before the packet is transmitted. At the receiver side, channel coded data are decoded and the received bitstream is then demultiplexed in order to obtain the elementary streams for source decoding. The decoded video objects are then composed to form the displayed scene based on a composition scenario that is either transmitted from the encoder or determined at the decoder. The decoder has the flexibility to determine which objects are used to compose the scene and in what order.

A virtue of the object-based video model is that it has all the benefits of frame-based video, if you can imagine that a frame-based video is only a subset of object-based

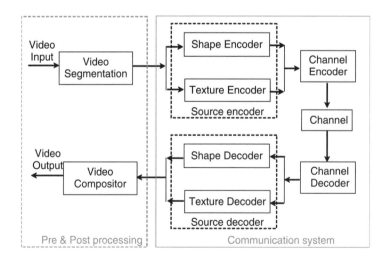

Figure 6.31   Architecture of object-based video communications system

video when we treat the rectangular frame as an object. However, object-based video representation provides a set of new functionalities, especially for the increasing demands of interactive multimedia applications. The advantages of this model can be classified in four categories:

1. *Content-based interactivity:* with object-based video representation, the powerful support for object manipulation, bitstream editing and object-based scalability allows for new levels of content interactivity. Owing to the fact that the objects are encoded independently, the user can interact with the scene in many ways. For example, the user can move, flip, rotate the object, modify the attribution of the object and even let objects interact with each other. The users can modify scene composition flexibly, for example, they can add new objects to the scene and remove existing objects, and they can decide the display order of the objects (for example, flip the display order of background and foreground).

2. *Extended content creation ability:* with object-based video representation, it is possible and easier to create content by reusing existing objects stored in a large database, thus everyone can become a content creator. This saves a lot of time and energy in repeatedly creating new and similar films in traditional studio work. Furthermore, the object-based model supports the integration of all types of objects, such as the integration of synthetic and natural content, which provides powerful content creation ability. Thus, it is possible to have scenes where cartoon characters coexist with live actors, and the synthetic character can even be coded as a 3D wire frame model, which supports more flexible interactions such as 3D rotation and animations.

3. *Object-based selective processing ability:* with object-based video representation, it is possible to encode and process different objects with selective tools that are adaptive to their different natures. This ability will bring significant coding efficiency gains and bit rate reduction, because traditional image/video coding approaches are not optimal for non-image type data, such as text and graphical elements. With this new representation, instead of treating text characters as image data, we could use Unicode to code them in a very efficient fashion and obtain even better visual quality.

4. *Universal access:* with object-based video representation, the content is accessible over a wide range of media, such as mobile network and Internet connections. Content scalability enables heavier protection of transmission of more important objects to the receiver in order to guarantee that the received content has an acceptable quality. For end users with lower computational ability and sparse resources, this content scalability effectively enhances the existing spatial, temporal and SNR scalability in its dynamic resource allocation.

The generality of the object-based video representation model enables it to support a wide range of applications; it goes from low bit rate mobile video telephony to high bit rate consumer electronic applications such as DVD and HDTV. Recently, the enormous popularity of cellular phones and palms indicates the interest in mobile communication. With the emergence of 4G mobile communication systems, it is expected that mobile video communications such as mobile video telephony and video conferencing will become even more popular. However, in wireless networks the widespread communication of multimedia information is always hampered by limitations such as bandwidth, computational capability and reliability of the transmission media. Object-based video representation is a very good fit for such cases. In typical mobile video communications, it is expected that the background will change much faster than the person actually speaking, thus the encoder has to expend a lot of bits on the background, which may not be the focus of user's interest. However, by coding the person and the background as separate objects, the user can decide upon the most interesting object (for example, the person) and let the encoder expend most of the bits on this object. When the bit rate is rather low, the object which is not interesting (for example, the background) can even be simply discarded or not transmitted at all.

6.4   Content-Based Video Coding and Communications

6.4.1   Object-Based Video Coding

Compared to conventional frame-based video coding, which is represented by encoding a sequence of rectangular frames, object-based video coding is based on the concept of encoding arbitrarily shaped video objects. The history of object-based video coding can be traced back to 1985, when a region-based image coding technique was first published [45]. In it, the segmentation of the image is transmitted explicitly and each segmented region is encoded by transmitting its contour as well as the value for defining the luminance of the region. The underlying assumption is that the contours of regions are more important for subjective image quality than the texture of the regions. The concept was also extended to video encoding [46]. The coder is very well suited for the coding of objects with flat texture; however, the texture details within a region may not be preserved. In 1989, an object-based analysis-synthesis coder (OBASC) was developed [47]. The image sequence is divided into arbitrarily shaped moving objects, which are encoded independently. The motivation behind this is that the shape of moving objects is more important than their texture, and that human observers are less sensitive to geometric distortions than to coding artifacts of block-based coders. OBASC was mainly successful for simple video sequences. As mentioned in the previous section, object-based video coding was first standardized by MPEG-4, where the shape and texture information of video objects is encoded by separate schemes [48, 49]. In this section, we will demonstrate in detail the shape and texture coding approaches adopted by the MPEG-4 standard, because most of the work in this thesis is implemented on the MPEG-4 standard.

As expected, video object (VO) is the central concept in MPEG-4, and it is characterized by intrinsic properties such as shape, texture and motion. For each arbitrarily shaped video object, each frame of a VO is called a video object plane (VOP). The VOP encoder consists essentially of separate encoding schemes for shape and texture. The purpose of the use of shape is to achieve better subjective picture quality, increased coding efficiency, as well as object-based video representation and interactivity. The shape information for a VOP, also referred to as an alpha-plane, is specified by a binary array corresponding to the rectangular bounding box of the VOP specifying whether an input pixel belongs to the VOP or not, or a set of transparency values ranging from 0 (completely transparent) to 255 (opaque). In this thesis, only a binary alpha plane is considered, although the proposed algorithm can be extended to the grayscale alpha plane cases. The texture information for a VOP is available in the form of a luminance (Y) and two chrominance

(U, V) components. It is important to point out that the standard does not describe the method for creating VOs, that is, they may be created in various ways depending on the application.

The Context-based Arithmetic Encoding (CAE) method [50] is the shape coding approach adopted by the MPEG-4 standard. CAE is a bitmap-based method, which encodes for each pixel whether it belongs to the object or not; it is also a block-based method, where the binary shape information is coded utilizing the macroblock structure, by which binary alpha data are grouped within 16×16 binary alpha blocks (BAB) (see Figure 6.32). Each BAB (if neither transparent nor opaque) is coded using CAE. A template of 10 pixels for intra mode (9 pixels for inter mode) is used to define the context for predicting the alpha value of the current pixel. The templates for Intra and Inter BAB encoding are shown in Figure 6.33. A probability table is predefined for the context of each pixel. After that, the sequence of pixels within the BAB drives an arithmetic encoder with a pair of alpha value and its associated probability. Owing to the support of the models in Figure 6.32, the encoding of a BAB depends on its neighbors to the left, above, above-left and above right.

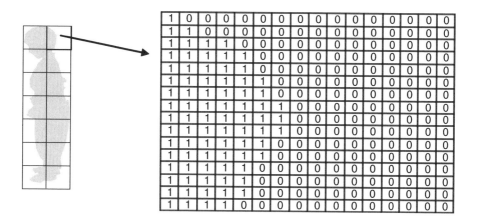

Figure 6.32 Block-based shape representations (the pixels marked 0 are transparent pixel and those marked 1 are opaque pixels)

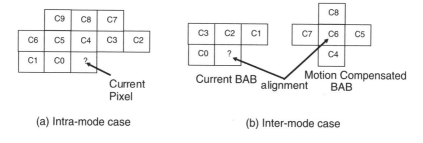

Figure 6.33 Context for CAE encoding

BABs are classified into transparent, opaque and border macroblocks using a test imposed on the target BAB containing only zero-valued pixels and BAB containing only 255-valued pixels. In the classification, each BAB is subdivided into 16 elementary sub-blocks, each of size 4×4, and the BAB is classified as transparent or opaque only if all of these sub-blocks are of the same classification. The sum of absolute differences between the sub-block under test and the target sub-block is compared to 16*Alpha_TH, where the Alpha_TH can take values from the set $\{0, 16, 32, 64, \ldots, 256\}$.To reduce the bit-rate, a lossy representation of a border BAB might be adopted. According to that, the original BAB is successively down-sampled by a conversion ratio factor (CR) of 2 or 4, and then up-sampled back to the full-resolution. In Inter-mode, there are seven BAB coding modes. The motion vector (MV) of a BAB is reconstructed by decoding it predictively as MV = MVP + MVD, where MVP is the motion vector predictor (see Figure 6.34) and MVD (motion vector differences) lies in the range of $[-16, +16]$. The MVP is chosen from a list of candidate motion vectors, MVs1, MVs2, MVs3, which represent shape motion vectors for 16×16 macroblocks, and mv1, mv2, mv3, which represent texture motion vectors for 8×8 blocks. They are considered in the specific order MVs1, MVs2, MVs3, mv1, mv2 and mv3. If no candidate MV is valid among them, MVP is set to $(0, 0)$. Upon the selection of MVP, the motion compensated (MC) error is computed by comparing the BAB predicted by MVP and the current BAB. If the computed MC error is smaller or equal to 16*Alpha_TH for all 4×4 sub-block, the MVP is directly utilized as MV and the procedure is terminated. Otherwise, MV is determined by searching around the MVP while computing the 16×16 MC error using the sum of absolute differences (SAD). The search range is $[-16, +16]$ pixels around the MVP along both the horizontal and vertical directions. The motion vector that minimizes the SAD is taken as MV, and MVD is coded as MVD = MV-MVP.

The texture coding approaches in MPEG-4 are similar to most of the existing video coding standards. The VOPs are divided into 8×8 blocks followed by 2D 8×8 discrete cosine transforms (DCT). The resulting DCT coefficients are quantized and the DC coefficients are also quantized (QDC) using a given step size. After quantization, the DCT coefficients in a block are zigzag scanned and a 1D string of coefficients is formed for entropy coding. In MPEG-4, the texture content of the macroblock's bitstream depends to a great extent on the reconstructed shape information. Before encoding,

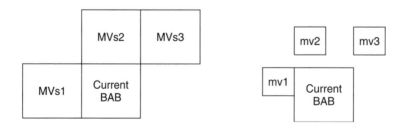

Figure 6.34 Candidates for motion vector predictors (MVPs): MVs1, MVs2, MVs3 represent shape motion vectors for 16×16 macroblocks, while mv1, mv2, and mv3, are texture motion vectors for 8×8 blocks. The order for selecting MVPs is MVs1, MVs2, MVs3, mv1, mv2, and mv3

low-pass extrapolation (LPE) padding [51] (non-normative tool) is applied to each 8×8 boundary (non-transparent and non-opaque) blocks. This involves taking the average of all luminance/chrominance values over all opaque pixels of the block, and all transparent pixels are given this average value. If the adaptive DC prediction is applied, the predicted DC selects either the QDC value of the immediately previous block or that of the block immediately above it. With the same prediction direction, either the coefficients from the first row or the first column of a previously coded block are used to predict the coefficients of the current blocks. For predictive VOPs (P-VOPs), the texture data can be coded predictively, and a special padding technique called macroblock-based repetitive padding is applied to the reference VOP before motion estimation. After motion estimation, a motion vector and the corresponding motion-compensated residual are generated for each block. The motion vectors are coded differentially (see Figure 6.35), while the residual data are coded as intra-coded texture data, as mentioned above.

On the decoder side, the shape and texture data are composed to reconstruct the original video data. Since in this thesis we assume that the shape information is represented by a binary alpha plane, after composition, only the corresponding texture within the shape boundary is kept (an example is shown in Figure 6.36).

6.4.2   Error Resilience for Object-Based Video

Since object-based video has a different video representation model from frame-based video, new approaches for error resilience in such video have to be pursued. So far, this

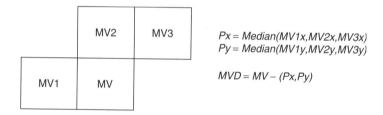

$$Px = Median(MV1x, MV2x, MV3x)$$
$$Py = Median(MV1y, MV2y, MV3y)$$

$$MVD = MV - (Px, Py)$$

Figure 6.35   Motion vector prediction

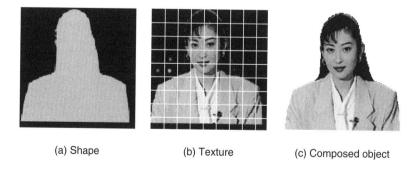

(a) Shape   (b) Texture   (c) Composed object

Figure 6.36   Example of composition of shape and texture

is a relatively new area in which much work still remains to be done. The existing work can be classified into three directions:

- *Evaluation metrics for object-based video:* there are two major works on this direction. In [9, 52, 53], a set of metrics was defined in order to evaluate the error resilient need of each object in the form of intra coding refreshment. The shape (or texture) refreshment need metric is based on the probability of losing part of the shape (or texture) data, and the difficulty in concealing the lost shape (or texture). In this thesis, an estimation of expected distortion at the decoder is derived based on the assumption that the encoder has the knowledge of transmission channel characteristics and the decoder error concealment strategies.
- *Error resilient encoding:* this refers to the choice of optimal coding parameters for shape and texture [54], or optimal intra refreshment time instants [55], based on the evaluation metrics mentioned above. In this thesis, we also propose an error resilient encoding method by using data hiding. The metric mentioned above is used to find the optimal source coding parameters and embedding level for hiding shape and motion data in texture.
- *Shape error concealment:* although some of the texture error concealment approaches mentioned in section 2.3 can be used in shape concealment, a great deal of work published recently was based on the characteristics of shape itself. The work can be classified simply as motion compensated concealment [56–59] and spatial concealment [60, 61]. The former method estimates the motion vector of missing BAB and uses the estimated vectors to replace the missing shape data from previous frame. The latter method uses the local neighbor information to conceal the missing shape. In [61], a maximum *a posteriori* (MAP) estimator is designed by modeling the binary shape as a Markov random filed (MRF). Each missing pixel is estimated as a weighted median of the pixels in its clique, where the weight corresponding to the difference between the pixels is estimated and another pixel in its clique is selected based on the likelihood of an edge passing through in the direction of the subject pair of pixels. The motivation behind this is to weigh the difference between the current pixel and a pixel in its clique in a direction along which pixels have a tendency to be the same. In [60, 62], the geometric boundary information of the received shape data is used to recover the missing BAB, and high order curves such as Bezier curves and Hermite splines are used to model the missing boundary. Clearly, motion compensated concealment performs poorly when objects appear/disappear or rotate, and the spatial concealment performs poorly when packet loss is high. To take advantage of both and avoid the problems above, a spatial-temporal technique was proposed recently [53].

In MPEG-4, a data partition scheme is adopted for protecting shape data, where a motion mark is used to separate the macroblock header, shape information and motion vectors from the texture information. The advantages are twofold. First, an error within the texture data does not affect the decoding of shape. Second, data partition facilitates unequal error protection, which can increase protection of shape and motion vectors.

There are many recent works on content-based video communications, including network-adaptive video coding, joint source-channel coding and joint coding and data hiding, but owing to limitations of space, we will move on to these topics in Chapter 9.

6.5   Content Description and Management

So far we have demonstrated content analysis and representation techniques. However, from the professional user's or consumer's point of view, the fundamental problem of how to describe the content for consumption has not been addressed. Furthermore, the content exchange between databases across various systems raises another issue of interoperability. The MPEG-7 and MPEG-21 standards have recently come into being in order to address these tasks.

6.5.1   MPEG-7

MPEG-7, formally named the 'Multimedia Content Description Interface', is a standard for describing the multimedia content data that supports some degree of interpretation of the meaning of the information, which can be passed onto, or accessed by, a device or a computer code [64]. It provides a comprehensive set of multimedia description tools in order to generate descriptors which are to be used for content access, and supports content exchange and reuse across various application domains. In this section, we focus on the visual descriptors covered in MPEG-7, while not covering the overall system of MPEG-7. Interested readers should refer to [63, 64] and [65] for more detail.

In MPEG-7, there are seven color descriptors, that is:

- *Color space:* supports six spaces including RGB, YCbCr, HSV, HMMD, linear transformation matrix with reference to RGB and monochrome.
- *Color quantization:* defines a uniform quantization of a color space, and can be combined with dominant color descriptor for usage.
- *Dominant color:* specifies a set of colors that is sufficient to characterize the color information of a region of interest.
- *Scalable color:* is an HSV color histogram encoded by a Harr transform, which is scalable in terms of accuracy of representation.
- *Color layout:* specifies the spatial distribution of color in a compact form; It is defined in terms of frequency domain and supports scalable representation.
- *Color structure:* captures both the color content similar to the color histogram and the structure of the content. It is different from the color histogram in that this descriptor can distinguish between images with an identical amount of color components as long as the structure of the groups of pixels having these colors is different.
- *Group of Frames/Group of Picture color:* it is a extension of scalable color in that it represents the color histogram of a collection of still images.

There are three texture descriptors, that is:

- *Homogeneous texture:* represents the first and the second moments of the energy in the frequency domain.
- *Texture browsing:* uses 12-bit (maximum) to characterize the texture's regularity, directionality and coarseness.
- *Edge histogram:* represents the spatial distribution of five edges in different orientations.

There are three shape descriptors, that is:

- *Region shape:* makes use of all pixels constituting the shape within a frame, and uses a set of Angular Radial Transform coefficients to represent the region.
- *Contour shape:* captures characteristic shape feature of an object, and uses a curvature scale-space (CSS) representation to capture the perceptually meaningful features of a shape. CSS is compact and robust to non-rigid motion, partial occlusion of the shape and perspective transforms (see Figure 6.37 for an example).
- *Shape 3D:* it provides an intrinsic shape description of 3D mesh models, based on the histogram of 3D shape indices representing the local attributes of the 3D surface.

In MPEG-7, there are 4 motion descriptors:

- *Camera motion:* it represents the 3D camera motion parameters information that is extracted automatically by capture devices, such as horizontal rotation (fixed, panning), horizontal transverse movement (tracking), vertical rotation (tilting), vertical transverse movement (booming), translation along the optical axis (dollying), and rotation around the optical axis (rolling). An example of camera motion is shown in Figure 6.38.
- *Motion trajectory:* defines the spatial and temporal location of a representative point of the object.
- *Parametric motion:* describes the object motion in video sequences as a 2D parametric model, such as translation, rotation, scaling, perspective projection, etc.
- *Motion activity:* captures the intuitive notion of 'intensity of action' or 'pace of action' in a video segment. In general, motion activity includes attributes of intensity, direction, spatial and temporal distribution of activity.

Figure 6.37   Example of shapes where a contour shape descriptor is applicable [64]

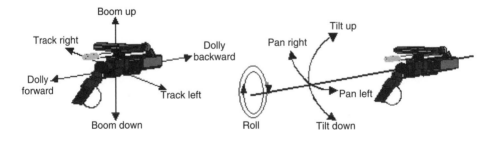

Figure 6.38   Types of MPEG-7 camera motion [64]

In addition, there are two location descriptors, that is:

- *Region locator:* this enables one to locate a region within a video frame by a box or polygon.
- *Spatio-temporal locator:* describes a 3D region in the video frames.

Finally, MPEG-7 provides a face recognition descriptor based on the principal components analysis (PCA) technique. The feature set is extracted from a normalized face image, and represents the projection of a face vector onto 49 basis vectors which span the space of all possible face vectors.

6.5.2 MPEG-21

MPEG-21 is initialized so as to enable the transparent and augmented use of multimedia data across the heterogeneous network facilities. It provides a general framework for all players in the content creation, delivery and consumption chain and ensures that they have equal opportunities.

There are two basic concepts in MPEG-21, that is:

- *Digital Items:* the fundamental unit of distribution and transaction within the framework, representing the content to be searched and consumed; the digital item could be created with a very complex structure, such as a hyperlinked object composed of many sub-items in various media formats. MPEG-21 has a digital item declaration specification in order to provide flexibility in digital item representations.
- *User:* refers to any party that interacts with the digital items, including individuals, organizations, communities, corporations, government, etc. In addition, MPEG-21 treats all users equally; no matter who they are, in other words, it does not distinguish content provider from consumer. MPEG-21 defines a framework for one user to interact with another regarding a digital item. The interaction covers a very wide range including creating content, delivering content, rating content, selling content, consuming content, subscribing content, etc.

MPEG-21 specifies the model, representation and schema of digital item declaration, and also provides tools for digital item identification and adaptation. The standard covers very wide areas which even include intellectual property management and scalable coding. However, most of the materials are out of the scope of this chapter. Interested readers should refer to the MPEG-21 specs for more detail.

References

1. J. R. Parker, *"Algorithms for image processing and computer vision"*, Wiley Computer Publishing, 1997.
2. G. Stamous and S. Kollias, *"Multimedia Content and the Semantic Web"*, John Wiley & Sons, Inc, 2005.
3. C. Stauffer and W. E. L. Grimson, "Learning Patterns of Activity Using Real-Time Tracking", *IEEE Transactions on Pattern Analysis and Machine Intelligence*, August 2000. 47.
4. D. Martin, C. Fowlkes, D. Tal, and J. Malik, "A database of human segmented natural images and its application to evaluating segmentation algorithms and measuring ecological statistics," in *ICCV*, (Vancouver, Canada), July 2001.

5. J. Chen, T. N. Pappas, A. Mojsilovic, B. E. Rogowitz, "Adaptive perceptual color-texture image segmentation", IEEE Trans. Image Processing, Vol. 14, No. 10, Oct. 2005.

6. P. L. Correia and F. B. Pereira, "Object evaluation of video segmentation quality", IEEE Trans. Image Processing, Vol. 12, No. 2, pp. 182–200, Feb. 2003.

7. MPEG4 video group, "Core experiments on multifunctional and advanced layered coding aspects of MPEG-4 video", Doc. ISO/IEC JTC1/SC29/WG11N2176, May 1998.

8. A. Elgammal, R. Duraiswami, D. Harwood and L. S. Davis, "Background and Foreground Modeling Using Nonparametric Kernel Density Estimation for Visual Surveillance", *Proceedings of IEEE*, July 2002.

9. L. D. Soares and F. Pereira, "Texture refreshment need metric for resilient object-based video", in *Proc. IEEE International Conference on Multimedia and Expo*, Lausanne, Switzerland, Aug. 2002.

10. M. M. Chang, A. M. Tekalp and M. I. Sezan, "Simultaneous motion estimation and segmentation", *IEEE Trans. Image Processing*, Vol. 6, pp. 1326–1333, Sept. 1997.

11. J. G. Choi, S. W. Lee, and S. D. Kim, "Spatio-temporal video segmentation using a joint similarity measure", *IEEE Trans. Circuits and System for Video Technology*, Vol. 7, pp. 279–285, 1997.

12. H. Luo and A. Eleftheriadis, "Model-based segmentation and tracking of head-and-shoulder video objects for real time multimedia services", IEEE Trans. Multimedia, Vol. 5, No. 3, p. 379–389, Sept. 2003.

13. K. Challapali, T. Brodsky, Y. Lin, Y. Yan, and R. Y. Chen, "Real-time object segmentation and coding for selective-quality video communications", *IEEE Trans. Circuits and Systems for Video Technology*, Vol. 14, No. 6, pp. 813–824, June 2004.

14. A. Cavallaro, O. Steiger, and T. Ebrahimi, "Tracking video objects in cluttered background", *IEEE Trans. Circuits and Systems for Video Technology*, Vol. 15, No. 4, pp. 575–584, April 2005.

15. E. Hjelmas, "Face detection: a survey", *Computer Vision and Image Understanding*, Vol. 83, pp. 236–274, 2001.

16. M. Yang, D. Kriegman, and N. Ahuja, "Detecting faces in images: a survey", *IEEE Trans. Pattern Analysis and Machine Intelligence*, Vol. 24, No. 1, pp. 34–58, Jan. 2002.

17. H. Wang, and S. Chang, "A highly efficient system for automatic face region detection in MPEG video", *IEEE Trans. Circuits and Systems for Video Technology*, Vol. 7, No. 4, pp. 615–628, August 1997.

18. D. Chai, and K. N. Ngan, "Face segmentation using skin-color map in videophone applications", *IEEE Trans. Circuits and Systems for Video Technology*, Vol. 9, No. 4, pp. 551–564, August 1999.

19. R. Hsu, M. Abdel-Mottaleb, and A. K. Jain, "Face detection in color image", *IEEE Trans. Pattern Analysis and Machine Intelligence*, Vol. 24, No. 5, pp. 696–706, Jan. 2002.

20. K. Wong, K. Lam, and W. Siu, "A robust scheme for live detection of human faces in color images", *Signal Processing: Image Communication*, Vol. 18, pp. 103–114, 2003.

21. H. Wang, G. Dane, K. El-Maleh, "A multi-mode video object segmentation scheme for wireless video applications", in *Proc. 1*st *International Workshop on Multimedia Analysis and Process*, Honolulu, USA, August 2007.

22. J. Serra, *Image Analysis and Mathematical Morphology*, New York: Academic, 1982.

23. Y. Tian, T. Kanade, and J. F. Cohen, "Multi-state based facial feature tracking an detection", *Carnegie Mello University Technical Report CMU-RI-TR-99-18*, August, 1999.

24. J. Edmonds, "Maximum matching and polyhedron with 0, 1-vertices", *J. Res. Nat. Bur. Standards 69B*, pp. 125–130, 1965.

25. H. N. Gabow, "An efficient implementation of Edmonds' algorithm for maximum matching on graphs", J. ACM, Vol. 23, No. 2, pp. 221–234, April 1976.

26. N. Vasconcelos and A. Lippman, "Statistical models of video structure for content analysis and characterization", *IEEE Trans. Image Processing*, Vol. 9, No. 1, Jan. 2000.

27. M. Petkovic, W. Jonker, Z. Zivkovic, "Recognizing strokes in tennis videos using hidden markov models", in *Proc. IASTED International Conference on Visualization, Imaging and Image Processing*, Marbella, Spain, 2001.

28. L. Xie, S-F. Chang, A. Divakaran, H. Sun, "Structure analysis of soccer video with hidden Markov Models", In Proc. International Conference on Acoustic, Speech, and Signal Processing (ICASSP), 2002.

29. P. Xu, L. Xie, S-F. Chang, A. Divakaran, A. Vetro, H. Sun, "Algorithms and system for segmentation and structure analysis in soccer video", In Proc. International Conference on Multimedia and Exposition (ICME), Tokyo, August 2001.

30. X. Zhu, X. Wu, J. Fan, A. K. Elmagarmid, W. G. Aref, "Exploring video content structure for hierarchical summarization", *Multimedia Systems*, 10: 98–115, 2004.

31. D. DeMenthon, V. Kobla and D. Doermann, "Video summarization by curve simplification", in Proc. ACM Multimedia, Bristol, U. K., 1998.

32. Y. Gong and X. Liu, "Video summarization using singular value decomposition", Computer Vision and Pattern Recognition, Vol. 2, pp. 13–15, June 2000.

33. A. Hanjalic and H. Zhang, Än integrated scheme for automatic video abstraction based on unsupervised cluster-validity analysis", IEEE Trans. Circuits and Systems for Video Technology, Vol. 9, Dec. 1999.

34. S. Lee and M. H. Hayes, "A fast clustering algorithm for video abstraction", in Proc. International Conference on Image Processing, Vol. II, pp. 563–566, Sept. 2003.

35. Y. Rao, P. Mundur, and Y. Yesha, "Automatic video summarization for wireless and mobile network", in Proc. International Conference on Communications, Paris, June 2004.

36. B. L. Tseng and J. R. Smith, "Hierachical video summarization based on context clustering", in Proc. SPIE IT 2003 – Internet Multimedia Management Systems, SPIE, 2003.

37. Y. Zhuang, Y. Rui, T. S. Huang, and S. Mehrotra, "Adaptive key frame extraction using unsupervised clustering", in Proc. International Conference on Image Processing, Chicago, pp. 866–870, Oct. 1998.

38. Z. Li, G. M. Schuster, and A. K. Katsaggelos, "Rate-distortion optimal video summary generation", IEEE Trans. Image Processing, to appear, 2005.

39. S. Lu, I. King, and M. R. Lyu, "Video summarization by video structure analysis and graph optimization", in Proc. IEEE International Conference on Multimedia and Expo, Taipei, Taiwan, June 2004.

40. Z. Li, A. K. Katsaggelos, and B. Gandhi, "Temporal rate-distortion optimal video summary generation", in Proc. International Conference on Multimedia and Expo, Baltimore, USA, July 2003.

41. H. Wang and N. Malayath, "Selecting Key Frames From Video Frames", US patent pending, WO/2007/120337, to appear.

42. B. Yeo and B. Liu, "Rapid scene analysis on compressed video", *IEEE Trans. Circuits and systems for Video Technology*, Vol. 5, No. 6, pp. 533–544, Dec. 1995.

43. H. S. Chang and K. Kang, "A compressed domain scheme for classifying block edge patterns", *IEEE Trans. Image Processing*, Vol. 14, No. 2, pp. 145–151, Feb. 2005.

44. B. Shen and I. K. Sethi, "Direct feature extraction from compressed images", in Proc. SPIE: Storage and Retrieval for Still Image and Video Databases IV, vol. 2670, Mar. 1996, pp. 404–414.

45. M. Kunt, "Second-generation image coding techniques", *Proceedings of the IEEE*, Vol. 73, pp. 549–574, Apr. 1985.

46. W. Li and M. Kunt, "Morphological segmentation applied to displaced difference coding", *Signal Processing*, Vol. 38, pp. 45–56, July 1994.

47. H. G. Musmann, P. Pirsch, and H. Grallert, "Advances in picture coding", *Proceedings of the IEEE*, Vol. 73, pp. 523–548, April 1985.

48. A. K. Katsaggelos, L. P. Kondi, F. W. Meier, J. Ostermann, and G. M. Schuster, "MPEG-4 and Rate-Distortion-Based Shape-Coding Techniques", *IEEE Proc., special issue on Multimedia Signal Processing*, Vol. 86, No. 6, pp. 1126–1154, June 1998.

49. G. Melnikov, G. M. Schuster, and A. K. Katsaggelos, "Shape Coding using Temporal Correlation and Joint VLC Optimization", *IEEE Trans. Circuits and Systems for Video Technology*, Vol. 10, No. 5, pp. 744–754, August 2000.

50. N. Brady, F. Bossen, and N. Murphy, "Context-based arithmetic encoding of 2D shape sequences", in *Special session on shape coding, ICIP'97*, Santa Barbara, USA. pp. 29–32, 1997.

51. Andre Kaup, "Object-based texture coding of moving video in MPEG-4", *IEEE Trans. Circuits and System for Video Technology*, Vol. 9, No. 1, pp. 5–15, Feb. 1999.

52. L. D. Soares and F. Pereira, "Shape refreshment need metric for object-based resilient video coding", in Proc. *IEEE International Conference on Image Processing*, Vol. 1, pp. 173–176, Rochester, Sept. 2002.

53. L. D. Soares and F. Pereira, "Refreshment need metrics for improved shape and texture object-based resilient video coding", *IEEE Trans. Image Processing*, Vol. 12, No. 3, pp. 328–340, March 2003.

54. H. Wang, A. K. Katsaggelos, "Robust network-adaptive object-based video encoding", in *Proc. IEEE Int. Conf. Acoustics, Speech, and Signal Processing*, Montreal, Canada, May 2004.

55. L. D. Soares and F. Pereira, "Adaptive shape-texture intra coding refreshment for error resilient object-based video", in *Proc. IEEE Workshop on Multimedia Signal Processing*, St. Thomas, USVI, Dec. 2002.

56. M-J. Chen, C-C Cho, and M-C. Chi, "Spatial and temporal error concealment algorithms of shape information for MPEG-4 video", in *Proc. IEEE International Conference on Consume Electronics*, June 2002.

57. M. R. Frater, W. S. Lee, and J. F. Arnold, "Error concealment for arbitrary shaped video objects", in *Proc. Int. Conf. Image Processing*, Vol. 3, Chicago, Oct. 1998. pp. 507–511.

58. P. Salama and C. Huang, "Error concealment for shape coding", in *Proc. IEEE Conference on Image Processing*, pp. 701–704, Rochester, Sept. 2002.

59. G. M. Schuster and A. K. Katsaggelos, "Motion Compensated Shape Error Concealment", *IEEE Trans. Image Processing*, Vol. 15, No. 2, pp. 501–510, February 2006.

60. G. M. Schuster, X. Li, and A. K. Katsaggelos, "Shape error concealment using Hermite splines", *IEEE Trans. Image Processing*, 2005.

61. S. Shirani, B. Erol, and F. Kossentini, "Concealment method for shape information in MPEG-4 coded video sequences", *IEEE Trans. Multimedia*, Vol. 2, No. 3, Sept. 2000, pp. 185–190.

62. X. Li, A. K. Katsagglos, and G. M. Schuster, "A recursive line concealment algorithm for shape information in MPEG-4 video", in *Proc. International Conference on Image Processing*, Rochester, Sept. 2002

63. ISO02 ISO/IEC 15938-3: 2020, *Information Technology-Multimedia Content Description Interface – part3: Visual*, Version 1, ISO, Geneva, 2002.

64. MPEG Requirements Group, "MPEG-7 Overview", *ISO/IEC JTC1/SC29/WG11N6828 Document*, Palma de Mallorca, October 2004.

65. B. S. Manjunath, P. Salembier, T. Sikora (eds), "*Introduction to MPEG-7 Multimedia Content Description Interface*", John Wiley & Sons Ltd, Chichester, 2002.

7

Video Error Resilience and Error Concealment

7.1 Introduction

Traditionally, the goal of video compression is to represent the video data with as few bits as possible for a given video quality. If the video is to be stored or transmitted via an error-free channel, this is fine. However, if the only goal is compression efficiency, the resulting bitstream is very sensitive to bit errors. A reason for this is the extensive use of Variable Length Codes (VLC) in video compression. Thus, even a single bit error can cause a loss of synchronization between encoder and decoder, which may even cause the rest of the bitstream to be lost. Therefore, if the video is to be transmitted over a lossy channel, further considerations are necessary beyond pure compression efficiency.

There are three main categories of techniques that help to provide reliable video communications: *error resilience, channel coding* and *error concealment*. These techniques are complementary to each other. In order to establish reliable video communications, it is necessary to trade off compression efficiency with resistance to channel errors. *Error resilience* deliberately keeps some redundant information in the video bitstream in order to make it more resilient to channel losses. Thus, with the use of error resilience, losses in the channel will not have a catastrophic effect and video communication will not be interrupted, although video quality may drop. Some error resilience techniques that will be discussed later in detail are: resynchronization markers, Reversible Variable Length Coding (RVLC), independent segment decoding and the insertion of intra blocks or frames. Scalable and multiple description coding can also be seen as error resilience techniques.

Error resilience alone is usually not enough to cope with the high error rates associated with wireless channels. In this case, *channel coding*, also known as *Forward Error Correction* (FEC) may be used to reduce the error rate. Channel coding adds additional bits to the video bitstream so as to enable error detection and/or correction. Thus, while source coding aims at reducing redundancy in the video signal, channel coding adds some redundancy. As will be discussed in detail later, there are two main types of channel codes: *block codes* and *convolutional codes*. Furthermore, there are channel codes which can

4G Wireless Video Communications   Haohong Wang, Lisimachos P. Kondi, Ajay Luthra and Song Ci
© 2009 John Wiley & Sons, Ltd

cope with bit errors (for example, BCH codes, Rate Compatible Punctured Convolutional (RCPC) code and Turbo codes), as well as codes that can cope with packet losses (for example, Reed-Solomon (RS) codes).

For most cases of wireless video transmission, error resilience and channel coding are not enough to provide loss-free communications. Most likely, some data will be lost and will not be available to the receiver. In that case, *error concealment* is used at the decoder in order to estimate the missing information and attempt to conceal the losses from the viewer. Some error concealment techniques that will be discussed in this chapter are: motion-compensated temporal interpolation, spatial interpolation and recovery of coding modes and motion vectors.

7.2 Error Resilience

When designing codecs for video communications over lossy channels, compression efficiency cannot be the only consideration. The resulting video bitstream must also be resilient to channel errors. Although error resilience techniques cannot guarantee error-free video transmission, they prevent catastrophic failures that could completely disrupt video communications. If the channel bit error rate or packet loss rate is low enough, error resilience itself can provide communications of acceptable quality. Otherwise, error resilience is used in conjunction with channel coding and/or error concealment. Error resilience makes the video bitstream robust to channel errors at the expense of compression efficiency. Depending on the error resilience technique used, this can happen directly, by inserting extra bits in the bitstream, or indirectly, by imposing limitations in the predictors that can be used by the encoder. We will next review some common error resilience techniques [1–3].

7.2.1 Resynchronization Markers

In general, it is possible to encode data using fixed length coding or variable length coding. If fixed length coding is used, a bit error will affect only a single code word. The decoder is always able to know where each code word starts, so it will be able to decode the next code word. If variable length coding is used, however, the beginning of each code word is not known *a priori*. Thus, a single bit error can cause incorrect decoding of the rest of the bitstream. If the bit error causes an invalid code word to form, then the error will be detected immediately. However, if the bit error causes the emulation of a valid code word, the error will not be detected until a subsequent code word is deemed to be invalid. In either case, once an error is detected, the rest of the bitstream cannot be decoded reliably. It becomes clear that in video compression, where variable length codes are heavily used, a single bit error can completely disrupt decoding if additional measures are not taken.

Resynchronization markers provide a means of preventing bit errors from causing catastrophic effects. Resynchronization markers are inserted periodically into the bitstream and they are relatively long sets of bits that cannot be emulated by the code words used by the codec or by small perturbations of those code words. If an error is detected in VLC decoding, then the decoder will search for the next resynchronization marker, which is designed so that it can be detected with high probability, even in the presence of bit errors. Once the next resynchronization marker is located, variable length decoding can start again. Of course, this 'skipping' of data between the point where the error is detected and the next resynchronization marker causes information to be lost. Furthermore, the decoder needs to know information about the data that are encoded after the next resynchronization marker (for example, which frame and macroblocks they correspond to, which quantizer is used, etc.) Thus, the resynchronization marker must be followed with a header that provides this information. Clearly, the more closely spaced the resynchronization markers are, the better the error resilience, since the distance between a bit error and the next resynchronization marker will be smaller and less information will be lost. Using more resynchronization markers in a bitstream reduces compression efficiency since resynchronization markers are extra bits in the bitstream that do not provide any extra information. Compression efficiency is further reduced because data before the resynchronization marker cannot be used for the prediction of data after the resynchronization marker. Thus, the encoding of motion vectors and dc coefficients may be less efficient due to the use of resynchronization markers. The determination of the interval between the resynchronization markers requires the exploration of a tradeoff between error resilience and compression efficiency.

Figure 7.1 illustrates the use of the resynchronization markers. Once a bit error is detected, the decoder searches for the next resynchronization marker and resumes decoding from there. All bitstream data between the bit error and the next resynchronization marker are discarded.

7.2.2 Reversible Variable Length Coding (RVLC)

As mentioned previously, the use of variable length coding can lead to complete disruption of video communications if bit errors occur and the insertion of resynchronization markers is a means of combating that. Even with the use of resynchronization markers, the number

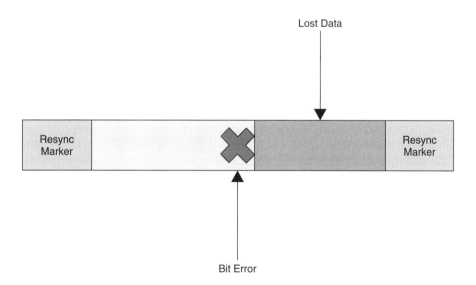

Figure 7.1 Illustration of the use of resynchronization markers

of bits that can be lost due to a single bit error can be almost as high as the interval between two consecutive markers, depending on the location of the bit error. Reversible Variable Length Coding (RVLC) is a way of reducing the amount of lost information. RVLC creates code words that can be decoded in both the forward and backward direction.

Figure 7.2 illustrates the use of RVLC in conjunction with resynchronization markers. The decoder starts the decoding in the forward direction, as usual. If an error is detected, the decoder searches for the next resynchronization marker. Then, it begins decoding starting at the next resynchronization marker and in the backward direction. Since there is at least one bit error in the interval between the two resynchronization markers, an error (an invalid code word) will be detected and decoding will stop again. Thus, there will still be some lost information (shaded area in Figure 7.2). However, the lost information will be less than in the case where standard non-reversible VLC codes are used.

RVLC has been adopted by both MPEG-4 and H.263. It can be shown that the use of RVLC only causes a small reduction in compression efficiency [3].

7.2.3   Error-Resilient Entropy Coding (EREC)

Video coders encode the video data using Variable Length Codes (VLC). Thus, in an error prone environment, any error would propagate throughout the bit stream unless we provide a means of resynchronization. The traditional way of providing resynchronization is to insert special synchronization code words into the bit stream. These code words should have a length that exceeds the maximum VLC code length and also be robust to errors. Thus, a synchronization code should be able to be recognized even in the presence of errors. The Error-Resilient Entropy Coding (EREC) [4] is an alternative way of providing synchronization. It works by rearranging variable length blocks into fixed length slots of data prior to transmission.

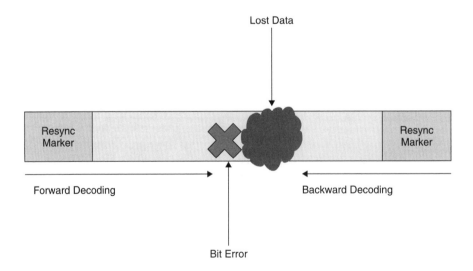

Figure 7.2   Illustration of the use of Reversible Variable Length Coding in conjunction with resynchronization markers

EREC is applicable to coding schemes where the input signal is split into blocks and these blocks are coded using variable-length codes. For example, these blocks can be the macroblocks in H.263. Thus, the output of the coding scheme is variable-length blocks of data. Each variable-length block must be a prefix code. This means that in the absence of errors the block can be decoded without reference to previous or future blocks. The decoder should also be able to know when it has finished decoding a block. The EREC frame structure consists of N slots of length s_i bits. Thus, the total length of the frame is $T = \sum_{i=1}^{N} s_i$ bits. It is assumed that the values of T, N and s_i are known to both the encoder and the decoder. Thus, the N slots of data can be transmitted sequentially without risk of loss of synchronization. Clearly, each EREC frame can accommodate up to N variable-length blocks of data, each of length b_i assuming that the total data to be coded $\left(\sum_{i=1}^{N} b_i \right)$ is not greater than the total bits available (T). The quantity $R = T - \sum_{i=1}^{N} b_i \geq 0$ represents redundant information. In order to minimize R, a suitable value of T needs to be chosen. It can be agreed upon in advance or transmitted. The slot lengths s_j can be predefined as a function of N and T and do not have to be transmitted. Usually, N is also agreed upon (for example, in an H.263 based codec, it can be 99, the number of macroblocks in a frame).

EREC reorganizes the bits of each block into the EREC slots. The decoding can be performed by relying on the ability to determine the end of each variable-length block. Figure 7.3 shows an example of the operation of the EREC algorithm. There are six blocks of lengths 11, 9, 4, 3, 9, 6 and six equal length slots with $s_i = 7$ bits.

In the first stage of the algorithm, each block of data is allocated to a corresponding EREC slot. Starting from the beginning of each variable-length block, as many bits as possible are placed into the corresponding slot. If $b_i = s_i$, the block will be coded completely leaving the slot full. If $b_i < s_i$, the block will be coded completely, leaving $s_i - b_i$ unused bits in the slot. Finally, if $b_i > s_i$, the slot is filled and $b_i - s_i$ bits remain to be coded. In the example in Figure 7.3, blocks 1, 2 and 5 have $b_i > s_i$ whereas blocks 3, 4 and 6 have $b_i < s_i$. Therefore, at the end of the first stage, slots 1, 2 and 5 are full, whereas slots 3, 4 and 6 have free space which will be used to code data from blocks 1, 2 and 5.

In the following stages of the algorithm, each block with data yet to be coded searches for slots with space remaining. At stage n, block i searches slot $i + \phi_n$ (modulo N), where ϕ_n is a predefined offset sequence. If there is space available in the slot searched, all or

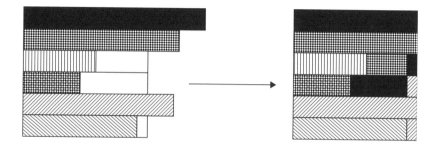

Figure 7.3 Illustration of the EREC algorithm

as many bits as possible are placed into that slot. In Figure 7.3, the offset sequence is 0, 1, 2, 3, 4, 5. Therefore, in stage 2, the offset is 1 and each block searches the following slot. Thus, in stage 2, one bit from block 5 is placed in slot 6 and two bits from block 2 are placed in slot 3.

Clearly, if there is enough space in the slots (i.e., if $R \geq 0$), the reallocation of the bits will be completed within N stages of the algorithm. Figure 7.3 shows the final result of the EREC algorithm.

In the absence of errors, the decoder starts decoding each slot. If it finds the block end before the slot end, it knows that the rest of the bits in that slot belong to other blocks. If the slot ends before the end of the block is found, the decoder has to look for the rest of the bits in another slot. Where to look for is clear since the offset sequence ϕ_n is known to the decoder. Since s_i is known to the decoder, it knows the location of the beginning of each slot. Thus, in case one slot is corrupted, the location of the beginning of the rest of the slots is still known and the decoding of them can be attempted. It has been shown that the error propagation is quite low when using the EREC algorithm.

As mentioned earlier, EREC eliminates the need for transmission of synchronization code words. However, the value of T should be transmitted. Clearly, EREC has the potential of saving bits over the traditional synchronization methods using synchronization code words. However, it also increases computational complexity and is not used currently in any video compression.

7.2.4 Independent Segment Decoding

It is clear that, owing to the motion compensation used in video coding, an error (for example, a corrupted block) in a frame can affect a much larger region in subsequent frames. *Independent segment decoding* is a means of isolating error propagation due to channel errors within specific regions of the video frames. A frame is partitioned into regions (segments) and only information from the corresponding spatial location of the previous frame can be used to encode a segment in the current frame. Thus, only pixels that belong to the corresponding segment in the previous frame can be used for predicting a segment. This leads to a reduction in compression efficiency, but clearly limits the effects of error propagation to within the segment. Independent segment decoding is supported by standards such as H.263.

7.2.5 Insertion of Intra Blocks or Frames

Video compression standards support the encoding of specific video frames or blocks within frames in intra mode, i.e., without using interframe prediction. Usually, intra coding requires significantly more bits than inter coding. Thus, if compression efficiency is our only goal, intra blocks or frames should only be used when prediction fails and encoding in inter mode would require more bits than encoding in intra mode. This would be the case in the event of a scene change, too much motion or new objects appearing or disappearing from the scene.

Intra blocks or frames can also be used for error resilience purposes. As mentioned previously, due to the motion compensation used in inter coding, an error in a frame will propagate to subsequent frames. If independent segment decoding is used, the effects of error propagation will be isolated within a specific region of the video frame. However, all subsequent frames will continue to be corrupted. Error propagation will stop entirely when a subsequent frame is encoded in intra mode. If a macroblock is encoded in intra mode, error propagation will stop in the corresponding spatial location. Thus, it makes sense to insert intra blocks or frames periodically in an encoded video sequence in order to improve error resilience.

The insertion of intra blocks or frames in a video sequence involves a trade off between compression efficiency and error resilience. Periodic encoding of video frames in intra mode will significantly limit error propagation if the period is sufficiently small. However, this will result in a significant decrease in compression efficiency, since intra frames typically require many more bits than inter frames. For relatively low bit rate communications, it is better to insert intra blocks for error resilience instead of whole frames. As mentioned previously, intra macroblocks will stop error propagation in their location. The number of intra macroblocks as well as their location needs to be determined. The percentage of intra macroblocks in a video sequence should depend on the channel quality. Clearly, a bad channel will require a higher intra macroblock rate. Regarding the placement of intra macroblocks, it is possible to use heuristic methods. For example, the time of the last intra update of a given macroblock can be considered. The placement of the intra macroblocks may also be determined by solving a rate-distortion optimization problem, which takes into account the channel conditions and the content of the video sequence.

7.2.6   Scalable Coding

As mentioned in Chapter 3, a *scalable* encoder produces a bitstream that can be partitioned into layers. One layer is the *base* layer and can be decoded by itself and provide a basic signal quality. One or more *enhancement* layers can be decoded along with the base layer to provide an improvement in signal quality. There are three main types of scalability in video coding. In *SNR scalability*, inclusion of enhancement layers in decoding increases the PSNR of the video frames (reduces the quantization noise). In *spatial scalability*, the enhancement refers to an increase in the spatial resolution of the video frames, whereas in *temporal scalability*, it refers to an increase in the temporal resolution of the video sequence (frame rate).

A traditional use of scalability is in video transmission over heterogeneous networks. Let us assume that a video server wishes to transmit the same video to several classes of video users, which are connected to the network at different bit rates. Using scalable coding, the base layer may be transmitted to the low bit rate users, while the base plus one or more enhancement layers may be transmitted to the higher bit rate users. Thus, each user will receive video with a quality that corresponds to the user's connection speed, and the video server will only need to compress the video data once. Had scalable video coding not been used, the server would have to compress the video several times, once for each target bit rate.

Scalable layers form a hierarchy. In order for an enhancement layer to be useful in the decoding, the base layer and all previous enhancement layers need to be available to the decoder. Thus, the earlier layers in the hierarchy are more important than the later ones. The most important layer is the base layer, since it can be decoded by itself and produces a video sequence of a certain quality. Also, if the base layer is not received, the video sequence cannot be decoded and any received enhancement layers are useless.

Scalability can be considered an error concealment method when used in conjunction with *Unequal Error Protection* (UEP). Since the base layer is more important than the enhancement layers, it should receive stronger protection from channel errors. Unequal error protection may be accomplished using channel coding with different channel coding rates for each scalable layer. Next generation networks will also support channels that offer different qualities of service.

Scalability is supported by several video compression standards, such as MPEG-2, MPEG-4 and H.263, as well as the scalable extension of H.264 [5]. Also, the current state-of-the-art wavelet based video codecs offer scalability. Classic SNR scalability, as supported in the earlier standards, is implemented by encoding the base layer as in non-scalable encoders and then re-encoding the quantization error to produce enhancement layers. This will result in a decrease in compression efficiency for two main reasons: since the enhancement layers are not always available at the receiver, in order to avoid drift problems, as discussed in Chapter 3, motion compensation should be based on base layer information only. Prediction will therefore be based on worse quality frames and the prediction error will contain more information, thus requiring more bits for its encoding, compared to non-scalable coding. The second main reason why classic SNR scalability leads to reduced compression efficiency is the fact that the encoding of the enhancement layers requires the transmission of more headers, resynchronization markers, etc. More recent scalable codecs, such as the scalable extension of H.264 and wavelet-based codecs, encode a group of frames at a time instead of a frame at a time. As discussed in Chapter 3, this avoids drift problems and leads to scalability without the penalty of reduced compression efficiency. The trade off is increased computational complexity and delay, since a number of frames need to be available at the encoder before the group of frames can be encoded. Also, the decoder needs to receive the information bits for the whole group of frames before any frames can be decoded.

7.2.7 Multiple Description Coding

Multiple description coding (MDC) [6–12] is another coding paradigm which, like scalable coding, produces a bitstream which can be partitioned into layers. However, unlike scalable coding, the layers in multiple description coding do not form a hierarchy. Any one of the layers can be decoded by itself. The more layers (descriptions) are available to the receiver, the better the decoded video quality. In order for every description to be decodable by itself, descriptions should share some important information about the video. Thus, descriptions need to be correlated.

Figure 7.4 shows a generic multiple description coding system. The encoder in the figure produces two descriptions, although encoders with any number of descriptions may be designed. The two descriptions have bit rates R_1 and R_2, respectively and they are assumed to be transmitted over two separate communication channels. In this example, 'on/off' channels are assumed, i.e. each channel may be either 'on' and provide lossless transmission of a description, or 'off' (completely broken). At the decoder, if only description 1 is received, Decoder-1 is used and produces video with distortion D_1. If only description 2 is received, Decoder-2 is used and produces video with distortion D_2. If both descriptions are received, Decoder-0 is used and produces video with distortion D_0, where $D_0 < \min\{D_1, D_2\}$.

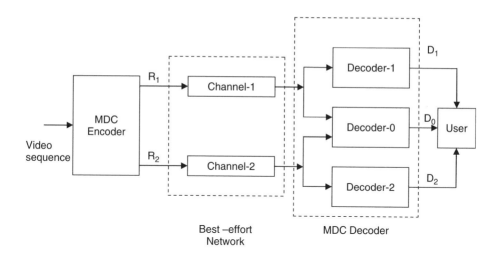

Figure 7.4 Generic multiple description coding system

A key concept in multiple description coding is the concept of *redundancy*. For a two description MDC codec, the redundancy ρ is defined as:

$$\rho = \frac{R_1 + R_2 - R^*}{R^*}$$

where R^* is the bit rate that would be required to encode the video sequence using a regular non-layered codec. Since the two descriptions are correlated, the numerator of the above equation will always be positive. ρ expresses the percentage of redundant bits that are transmitted using an MDC codec compared to a regular non-layered codec.

Although MDC is considered frequently in the context where two or more separate communication channels are available, descriptions may also be transmitted over the same channel. The more descriptions are correctly received by the decoder, the better the video quality. In contrast with scalable coding, there is no need for unequal error protection of the multiple descriptions, since they are equally important (assuming balanced multiple description coding, although unbalanced multiple description coding, where the descriptions are not equally important, is also possible).

Comparing multiple description coding with scalable coding, the former typically requires a larger amount of redundancy, since descriptions need to be correlated. The redundancy is deliberately added in multiple description coding, whereas, in scalable coding, the redundancy (increase in the number of bits required to represent the video compared to non-scalable coding) is due to a reduction in compression efficiency, as explained earlier in this chapter. In scalable coding, the layers form a hierarchy, so the level of error protection for each layer should be commensurate with its position in the hierarchy, with the base layer receiving the most protection. If the base layer is lost, no video can be produced at the receiver. MDC descriptions do not form a hierarchy, so any received layer will produce video. Thus, MDC descriptions may be transmitted over a 'best effort' channel one after another without any UEP or other kind of prioritization.

This is precisely the main advantage of multiple description coding over scalable coding, which comes at the expense of increased redundancy (reduced compression efficiency).

So far, we have used the term 'multiple description coding' to refer to 'SNR multiple description coding'. Thus, inclusion of additional MDC layers in the decoding results in an increase of the PSNR of the decoded video sequence. This type of MDC may be accomplished using several approaches including overlapping quantization, correlated predictors, correlating linear transforms, block wise correlating lapped orthogonal transforms, correlating filter banks and DCT coefficient splitting.

It is also possible to have 'temporal' MDC, where additional received descriptions improve the frame rate of the video sequence. This may be accomplished, for example, by partitioning the video sequence into subsets (threads) and encoding each one of them independently. For example, assuming two descriptions, frame 0 is a 'sync frame', which is encoded in intra mode and belongs to both threads. Then, frames $1, 3, 5, 7 \ldots$ form the first description, where frame 1 is predicted from frame 0, frame 3 from frame 1, frame 5 from frame 3, etc. Similarly, frames $2, 4, 6, 8 \ldots$ form the second description. Then, if a frame in one description is corrupted due to channel errors, there will be no error propagation to frames belonging to the other description. Thus, the frames in the other description may continue to be displayed at the receiver. This will reduce the frame rate by half; however, there will be no corruption. The reduced frame rate will continue until another sync frame (a frame that belongs to both descriptions) is encountered. This type of multiple description coding is known as *video redundancy coding* in H.263.

7.3   Channel Coding

Error resilience techniques enable video communications using compressed data, even in the presence of channel errors. However, the error rates introduced by wireless channels are usually too high for error resilient codecs to cope with. *Channel coding* or *Forward Error Correction (FEC)* is used to reduce the error rates seen by the video codec. Channel coding adds redundancy (extra bits) in the video bitstream in order to help detect and/or correct errors. Channel coding is not specific to video communications and may be used to protect any kind of bitstream. A comprehensive review of channel coding is beyond the scope of this book. More information can be found in [13] and [14].

There are two classes of channel codes, *block codes* and *convolutional* codes. In block coding, therefore, the rate of the resulting code (ratio of number of source output bits over the number of channel input bits) is k/n.

A block code is systematic if the k source output bits (information bits) are concatenated with $n-k$ *parity* bits to form an n-bit code word. The parity bits are related algebraically to the information bits. Thus, in a systematic block code of length n, the first k bits are the information bits and the remaining $n-k$ bits are the parity bits, whereas in a non-systematic block code, information bits and parity bits may be interlaced. Figure 7.5 shows an example of a systematic block code.

In convolutional coding, source output bits of length k_0 are mapped into n_0 channel input bits, but the channel input bits depend not only on the most recent k_0 source bits but also on the last $(L-1)k_0$ inputs of the encoder. Therefore, the convolutional encoder has the structure of a finite-state machine where at each time instance, the output sequence depends not only on the input sequence, but also on the state of the encoder, which is

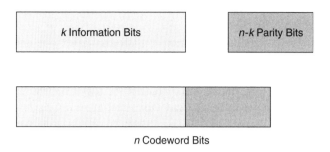

Figure 7.5 A example of a systematic block code

determined by the last $(L-1)k_0$ input bits of the encoder. The parameter L is called the *constraint length* of the convolutional code.

In Figure 7.6 we can see a convolutional encoder with $k_0 = 1, n_0 = 2$ and $L = 3$. We can see that the structure of the convolutional encoder is very similar to that of a Finite Impulse Response (FIR) digital filter. Each box in Figure 7.6 corresponds to a flip-flop (one-bit memory). Thus, the set of the three boxes corresponds to a shift register. Convolutional codes got their name from the fact that they perform a convolution of the input sequence with a sequence (impulse response), which depends on the connections made between the shift register and the output.

The state of the encoder is determined by the contents of the first two flip-flops, since the content of the third flip-flop will be pushed out of the shift register when the next input bit arrives and the next output will not depend on it.

We can see that in the convolutional encoder of Figure 7.6 we get two output bits for each input bit. There are two outputs which are multiplexed into a single output bit stream. Therefore, $k_0 = 1$ and $n_0 = 2$. Also, the output of the encoder also depends on the previous two input bits. Thus, $(L-1)k_0 = 2$ and since $k_0 = 1$, we conclude that the constraint length of the encoder is $L = 3$, which in this case also coincides with the number of flip-flops in the shift register.

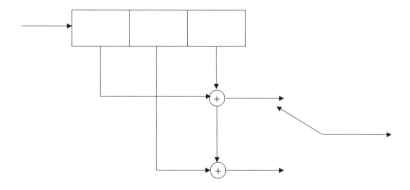

Figure 7.6 An example of a convolutional code

7.4   Error Concealment

Channel coding is very effective in reducing the channel error rate and error resilience can enable video communications in the presence of errors. However, it is impossible to guarantee that no information will be lost. The goal of *error concealment* is to estimate the missing information at the decoder and conceal the loss from the viewer. Error concealment is a similar concept to image and video restoration. Classic image and video restoration attempts to reduce the effects of additive noise and/or blurring due to motion and the optics. Error concealment, on the other hand, aims to restore a specific type of artifact, which occurs when parts of a compressed video bitstream are corrupted. If the video bitstream consisted of completely uncorrelated data, it would be impossible to estimate the missing data from the received information. However, all video bitstreams contain some redundant information, which the encoder was unable to completely remove, or was deliberately placed by error resilience techniques. Still, the redundancy in the video bitstream is low, so error concealment cannot completely recover the lost information.

Error concealment techniques can be classified into *intra* techniques and *inter* techniques [1]. Intra techniques conceal a missing part of a video frame using successfully received information in adjacent parts of the frame. Intra techniques are basically spatial interpolation techniques. Inter techniques utilize information from the previous frame in order to conceal a missing part of the current video frame. Intra error concealment techniques are usually used for intra frames/blocks and inter techniques are usually used for inter frames/blocks, but this is not necessary. Inter techniques are usually more effective, provided that the previous frame has enough correlation with the current frame. For this reason, MPEG-2 allows for the transmission of motion vectors for intra frames in order to aid in the application of inter error concealment techniques. However, if inter error concealment is applied in cases of low correlation between two successive video frames, the results may be catastrophic. Thus, if the mode information the receiver has is not reliable, it is a more conservative choice to use intra error concealment.

7.4.1   Intra Error Concealment Techniques

Intra error concealment techniques include spatial interpolation, Maximally Smooth Recovery (MSR) and Projection onto Convex Sets (POCS). The problem of intra error concealment is a hard one, since it is very difficult to guess the contents of a region of an image, given the contents of surrounding pixels. The results of intra techniques are reasonably good if the missing area is relatively small. For larger areas, our expectations should not be very high.

7.4.2   Inter Error Concealment Techniques

The simplest inter error concealment technique is to replace the missing area in the current video frame with the corresponding area in the previous frame. This corresponds to assuming that both the motion vector and prediction error are zero and may work satisfactorily if there is very little motion in the scene. However, if there is significant motion, the macroblocks that were copied from the previous frame will look 'out of place' and will be annoying to the viewer. A more effective technique is to replace a missing macroblock in the current frame with the corresponding motion-compensated macroblock in the previous frame, i.e. the macroblock the corresponding motion vector points to. This technique is known as *motion-compensated temporal interpolation* and is widely used in practice. If the motion vector has not been received successfully, it must be estimated. The quality of the error concealment greatly depends on the estimation of the motion vectors.

Several techniques have been proposed for the estimation of motion vectors. These include:

- assuming that the lost motion vectors are zero;
- using the motion vectors of the corresponding macroblock in the previous frame;
- using the average of the motion vectors from spatially adjacent macroblocks;
- using the median of the motion vectors from spatially adjacent macroblocks;
- re-estimating the motion vectors [15].

Taking the average or the median of the motion vectors from adjacent macroblocks works satisfactorily in the presence of low to moderate motion. It should be emphasized that, since motion vectors are typically encoded differentially in the standards, if a motion vector is lost, we cannot assume that the motion vectors of all adjacent macroblocks will be available, since the encoding of some of them will depend on the lost motion vector. Thus, the mean or the median should include only the motion vectors that can actually be decoded using the received data. Re-estimating the motion vectors may work better than the average and median techniques for higher motion, but this comes at the expense of increased computational complexity.

In [15], an overlapped region approach for the recovery of lost motion vectors is proposed. This approach has the advantage that edge or signal continuity between the missing block and its neighbors can be preserved. The motion vector is re-estimated without requiring any differential information from the neighboring blocks. Figure 7.7 depicts the idea behind the algorithm. Two frames are shown in it, the current *l*th frame on the right and the previous (*l*-1)th frame on the left. The gray region represents a missing block in the frame, and the band above and to the left of it the neighboring pixels to be used for the

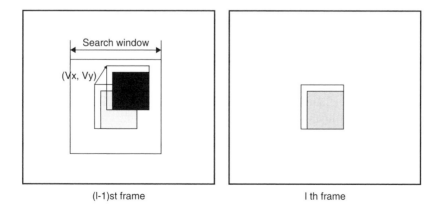

Figure 7.7   Motion vector re-estimation

estimation of the lost motion vectors. Such a band is used since only blocks to the left and above the current block have been decoded. The $(l$-1)st decoded frame is utilized as a reference frame, and the motion vector for the lost macroblock is determined as:

$$(Vx, Vy) = \arg \min_{(m,n)\in S_{mv}} \sum_i \sum_j |\hat{x}(i, j; l) - \hat{x}(i - m, j - n; l - 1)|,$$

where (i,j) represents the pixels inside the band to the left and above of the missing block, S_{mv} denotes the search region in the previous frame, and $|.|$ denotes the absolute value.

Since region matching assumes that the displacement within the region is homogeneous, support of this region is critical. A large support decreases the reliability of the estimated motion vectors, because the correlation between the missing block and the region around it used for matching is reduced. On the other hand, a small support region might be matched to various locations in the previous frame. In [15] it is reported that $4-8$ rows and columns of neighboring pixels result in good matching results.

7.5   Error Resilience Features of H.264/AVC

We next discuss the error resilience features of H.264/AVC in detail. Some of them are similar to features supported by earlier standards and some are completely new. More information can be found in [16]. It should be pointed out that H.264 is aimed for packet-based networks and thus copes mainly with packet losses instead of bit errors. Thus, it is assumed that packets that contain bit errors are discarded and not fed into the decoder.

7.5.1   Picture Segmentation

Picture segmentation is supported in H.264 in the form of slices. Each slice is formed by an integer number of macroblocks of one picture. Macroblocks are assigned into slices in raster scan order, unless FMO mode is employed, as will be discussed later.

Slices interrupt the in-picture prediction mechanisms, thus, the use of small slices reduces compression efficiency. Slices are self-contained, thus, if a coded slice is available to the decoder, all the macroblocks in that slice can be decoded.

7.5.2   Intra Placement

Intra placement (the insertion of intra blocks or frames) is also supported by earlier standards and was discussed in section 7.2.5. The following aspects of its application to H.264/AVC should be pointed out.

In order to improve coding efficiency, H.264 allows for intra macroblock prediction even from inter macroblocks. However, this means that intra macroblocks predicted in this fashion will not remove error propagation. Thus, H.264 also allows for a mode that prevents this form of prediction, so that intra macroblocks can offer true elimination of error propagation.

H.264 supports multiple reference picture motion compensation. A video frame is therefore not constrained to use the previous frame for motion compensation, but may also use earlier frames. Thus, encoding a slice or frame in intra mode will erase error propagation for the current frame but subsequent frames may again be corrupted because they would be predicted based on earlier frames that could be corrupted. Intra frames or slices may therefore not erase error propagation completely in H.264. Thus, H.264 also supports IDR slices in addition to intra slices. All slices in an IDR picture must be IDR slices. An IDR picture is also encoded in intra mode, but subsequent pictures are not allowed to use pictures before the IDR picture for motion compensation. The insertion of an IDR picture therefore removes error propagation completely in H.264.

7.5.3 Reference Picture Selection

H.264 allows for reference picture selection, which was also supported in H.263. If feedback is available, the decoder may signal to the encoder that a particular frame was received damaged. Then, the encoder will avoid using the damaged frame for the prediction of the next encoded frame. Instead, it will use a previous frame that is known to have been received correctly by the decoder. If feedback is not available, the reference picture selection capability of H.264 may be used to implement video redundancy coding, which was described in section 7.2.7.

7.5.4 Data Partitioning

Data partitioning allows for the grouping of the symbols within a slice into partitions, so that the symbols in each partition have a close semantic relationship with each other. Three different partition types are used in H.264:

- *Type A Partition.* This partition contains the most important data in the video bitstream, such as macroblock types, quantization parameters, motion vectors, and other header information. If this partition is lost, the data in the other partitions are useless.
- *Type B Partition.* This is the intra partition and contains intra Coded Block Patterns (CBP) and intra coefficients. This partition requires the Type A partition to be available at the receiver in order to be useful. This partition is more important than the inter partition, discussed below, because it can stop further drift.
- *Type C Partition.* This is the inter partition and contains only inter CBP and inter coefficients. In order for this partition to be useful in the decoding, The Type A partition should be available to the decoder. Since most data in a bitstream are usually coded in inter mode, the Type C partition is, in many cases, the largest partition in a coded slice. This partition is the least important partition because its information does not re-synchronize the encoder and decoder and does not remove drift.

At the decoder, if the Type B or Type C partition is missing, the header information in the Type A partition may be used in order to aid in error concealment. For example, motion vectors may be used to perform motion-compensated temporal interpolation (section 7.4.2).

7.5.5 Parameter Sets

The parameter sets contain header information which is necessary for the decoding of the video sequence. Multiple parameter sets may be stored at the decoder. The encoder just signals which parameters set is to be used instead of transmitting it every time. There are two kinds of parameter sets: The *sequence parameter set*, which contains header information related to a sequence of pictures (all pictures between two IDR frames), and the *picture parameter set*, which contains the header information for a single picture.

The parameter sets may be transmitted out-of-band using a reliable control protocol. Alternatively, they may be transmitted in-band, using strong error protection. It is also possible for the encoder and decoder to agree on a set of parameter sets to be used, so that these parameter sets are always available to the decoder and do not have to be transmitted. Thus, the use of parameter sets improves error resilience because it greatly increases the probability that important header information will be available to the decoder.

7.5.6 Flexible Macroblock Ordering

Normally, macroblocks are assigned to slices using the scan order. As the name implies, Flexible Macroblock Ordering (FMO) allows for a flexible grouping of macroblocks to form slices. In FMO, each macroblock is assigned to a slice group using a macroblock allocation map. Figure 7.8 shows an example of FMO. The picture in this example contains two slice groups. The macroblocks of the first slice group are shown in gray, whereas the macroblocks of the other slice group are shown in white. It is clear that, if one slice group is lost and the other one is correctly received, the macroblocks in the lost slice groups will have several neighboring blocks that have been received correctly. Thus, error concealment will be easier.

It has been shown [16] that in video conferencing applications with CIF-sized pictures using FMO, error concealment is so successful that it is very hard even for the trained eye to realize that information has been lost, even at packet loss rates of 10 %. The trade off is reduced compression efficiency because of the broken in-picture prediction mechanisms between non-neighboring macroblocks. Also, In highly optimized environments, use of FMO results in a somewhat higher delay.

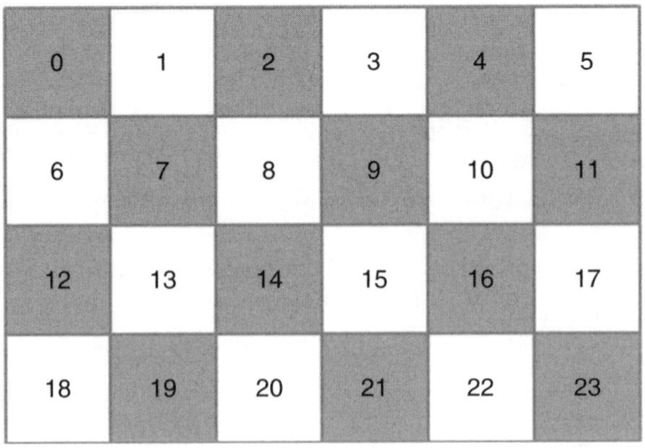

Figure 7.8 An illustration of Flexible Macroblock Ordering (FMO) for a picture with two slice groups and 6 × 4 macroblocks

7.5.7 Redundant Slices (RSs)

Redundant Slices (RSs) allow for the encoding of one or more redundant representations of a slice, in addition to the original representation (primary slice). The redundant representations may be encoded using a different (usually coarser) quantization parameter than the primary slice. Thus, the redundant slices will usually utilize fewer bits than the original representation. If the primary slice is available to the decoder, it will be used to reconstruct the macroblocks and the RSs will be discarded.

References

1. Y. Wang and Q. Zhu, "Error Control and Concealment for Video Communication: A Review", *Proceedings of the IEEE*, Vol. 86, pp. 974–997, May 1998.
2. Y. Wang, S. Wenger, J. Wen and A. K. Katsaggelos, "Error Resilient Video Coding Techniques," *IEEE Signal Processing Magazine*, pp. 61–82, July 2000.
3. Y. Wang, J. Ostermann and Y.-Q Zhang, *Video Processing and Communications*, Prentice-Hall, 2002.
4. D. W. Redmill and N. G. Kingsbury, "The EREC: An Error-Resilient Technique for Coding Variable-Length Blocks of Data," *IEEE Transactions on Image Processing*, Vol. 5, No. 4, pp. 565–574, April 1996.
5. H. Schwarz, D. Marpe and T. Wiegand, "Overview of the Scalable Video Coding Extension of the H.264/AVC Standard," *IEEE Transactions on Circuits and Systems for Video Technology*, Vol. 17, No. 9, pp. 1103–1120, September 2007.
6. A. E. Gamal and T. M. Cover, "Achievable Rates for Multiple Descriptions," *IEEE Transactions on Information Theory*, Vol. IT-28, No. 6, pp. 851–857, November 1982.
7. V. K. Goyal, "Multiple Description Coding: Compression Meets the Network," *IEEE Signal Processing Magazine*, Vol. 18, pp. 74–93, September 2001.
8. V. K. Goyal and J. Kovacevic, "Generalized Multiple Description Coding with Correlating Transforms", *IEEE Transactions on Information Theory*, Vol. 47, pp. 2199–2224, September 2001.
9. K. R. Matty and L. P. Kondi, "Balanced Multiple Description Video Coding Using Optimal Partitioning of the DCT Coefficients," *IEEE Transactions on Circuits and Systems for Video Technology*, Vol. 15. No. 7, pp. 928–934, July 2005.
10. L. Ozarow, "On a Source Coding Problem with Two Channels and Three Receivers," *Bell Systems Technical Journal*, Vol. 59, pp. 1901–1921, December 1980.
11. A. R. Reibman, H. Jafarkhani, Y. Wang and M. T. Orchard, "Multiple Description Video Using Rate-Distortion Splitting," in Proc. *IEEE International on Image Processing*, Thessaloniki, Greece, pp. 978–981, October 2001.
12. Y. Wang, M. T. Orchard, V. A. Vishampayan and A. R. Reibman, "Multiple Description Coding using Pairwise Correlating Transforms," *IEEE Transactions on Image Processing*, Vol. 10, ppp. 351–366, March 2001.
13. S. Haykin, *Digital Communications*, John Wiley & Sons, Inc, 1988.
14. J. G. Proakis, *Digital Communications*, 4th Edition, McGraw-Hill, 2001.
15. M. C. Hong, H. Schwab, L. P. Kondi and A. K. Katsaggelos, "Error Concealment Algorithms for Compressed Video", *Signal Processing: Image Communication*, No. 14, pp. 473–492, May 1999.
16. S. Wenger, "H.264/AVC Over IP," *IEEE Transactions on Circuits and Systems for Video Technology*, Vol. 13, No. 7, July 2003.

8

Cross-Layer Optimized Video Delivery over 4G Wireless Networks

8.1 Why Cross-Layer Design?

As we discussed in Chapter 2, a networking system can be designed by using the methodology of layering architecture such as Open Systems Interconnection (OSI) or TCP/IP, where each layer focuses on solving different design issues. For example, physical layer (PHY) minimizes bit errors during transmissions, medium access layer (MAC) deals with channel access, network layer handles routing issues, and transport layer manages issues of congestion control.

However, the original design goal of current layering network architecture is to support simple delay-insensitive and loss-intolerant data services with little QoS consideration, which cannot support delay-sensitive, bandwidth-intense and loss-tolerant multimedia services. Therefore, how to accommodate multimedia services into future broadband wireless network design becomes a very challenging problem.

Cross-layer design methodology provides a new concept, in which the interactions among different protocol layers are utilized to achieve the best end-to-end service performance. For example, the end-to-end delay in a multi-hop multimedia network can be represented by three components: 1) transmission delay, influenced by channel conditions, modulation and channel coding scheme, number of allowed packet retransmission, and source coding rate; 2) queuing delay, determined by the source rate, transmission rate, and the selected routing path; and 3) propagation delay, impacted by the number of hops of the selected path. As we analyzed above, these three components are impacted by the system parameters residing in various network layers, which imply that all related system parameters need to be considered to achieve the desired end-to-end delay performance.

So far, most of research efforts in cross-layer design are mainly focused on jointly considering a subset of network layers in the optimization of parameter selection and adaptation. This divide-and-conquer strategy is useful when a large number of system parameters are considered. However, by doing so, a large-scale optimization problem is divided into a number of smaller-scale optimization problems and is solved individually. An alternative is to shrink the state space of each layer thus to reduce the entire state space of the whole problem. For example, as mentioned in [1], using learning and classification techniques, a number of features of system parameters can be identified and then be used for the global optimization. However, these problem simplification strategies may lead to sub-optimal solutions. As pointed out in [2], how to do the problem simplification is very important design issue. We need to decide whether we should make the problem simplification in the problem modeling or in the problem solving phase.

Clearly, it is always beneficial to do the simplification to a later stage as long as the computational complexity is manageable. In cross-layer optimized multimedia communications, problem simplification in the modeling phase may either underestimate or overestimate the importance of certain system parameters, leading to sub-optimality. Furthermore, divide-and-conquer method may cause "Ellsberg Paradox", meaning that even each network layer is optimized for maximizing the global objective function, the overall system performance may not be able to achieve the global optimality due to the possible strong interdependency existing among different network layers. Therefore, it is necessary to build a theoretical framework for us to understand cross-layer behaviors and to formulate the problem systematically, and then we can find practical methods to reduce the complexity in the problem solving stage as well as guide the design to achieve the best end-to-end service quality.

8.2 Quality-Driven Cross-Layer Framework

In multimedia communications applications, quality refers to the user experience of network service, which is the main goal of system design. The quality degradation is general caused by many factors such as limited bandwidth, excessive delay, power constraints, computational complexity limitation. Thus, quality is the key performance metric that connects all related system parameters together. Hence, it is reasonable to conduct the cross-layer design in a quality-driven fashion. On the other hand, the goal of cross-layer system design is to find the optimal tradeoff within a N-dimensional space with given constraints, in which the dimensions include distortion, delay, power, complexity, etc. An example is shown in Figure 8.1, where the optimal rate-distortion curve of source coding is demonstrated to represent the optimal tradeoff between rate and distortion, the coarse image corresponds to a lower bit rate and a higher distortion (in the mean squared error sense), while the finer the image quality, the higher the bit rate. The 2-D case can be easily extended to N-D case, and thus the curve will become a convex hull. It is also important to realize that the curve in Figure 8.1 has a property that the rate is a non-increasing function of the distortion. The reason is that when more bits are used, the optimality of a given problem ensures that better or at least equal video quality will be achieved. This property can also be extended to the N-Dimensional case.

Lagrangian relaxation is a popular approach to for solve such problems. By using a Lagrange multiplier λ, the original constrained problem is converted into an unconstrained one. An intuitive geometric interpretation [3] for an unconstrained optimization problem

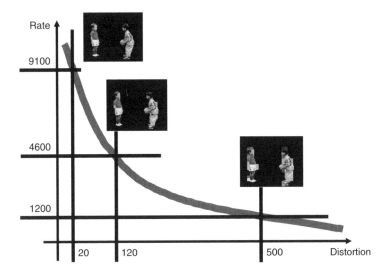

Figure 8.1 Optimal rate-distortion curve

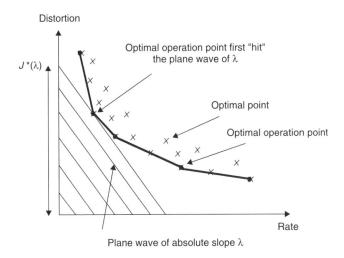

Figure 8.2 Geometric interpretation of the unconstrained optimization problem with one Lagrangian multiplier ($J^*(\lambda) = \min_x(D + \lambda R)$, where x denotes an operating point) [3]

with one Lagrangian multiplier is shown in Figure 8.2, where the Lagrangian cost $J(\lambda)$ associated with any admissible operating point is interpreted as the y-intercept of the straight line of slope $-\lambda$ passing through that point on the operational rate-distortion plane. Therefore, the minimum Lagrangian function is achieved for that point which is 'hit' first by the 'plane-wave' of slope $-\lambda$ emanating from the fourth quadrant of the R-D

plane towards the R-D curve in the first quadrant. For N-Dimensional cases, multiple λs are used and the optimal curve would become a manifold.

In the following, we will introduce quality-driven video delivery over 4G networks in a cross-layer design fashion.

8.3 Application Layer

To enhance the flexibility of video encoding, many coding parameters such as macro block modes, motion vectors and transform coefficient levels are adopted by most video standards [4–7]. To achieve a high rate-distortion efficiency, optimal real-time video encoding at the application layer has received significant research attentions [8–12]. To perform optimal video coding control, one important issue is the choice of the objective function for optimization. From a user's point of view, end-to-end video quality is the most straightforward and reasonable utility function in the optimization framework for wireless video communications. This is the concept of quality-driven video encoding which has also been widely accepted and implemented in practice.

One critical question is how to accurately estimate the end-to-end delivered video distortion on the decoder side? A great deal of research has taken the pre-calculated or some rate-distortion-based distortion as the optimization objective function [12–18]. However, the pre-calculated or rate-distortion-based distortion calculation fails to give an accurate estimation of the total distortion caused by different coding modes, quantization noise, error propagation, transmission error and error concealment in the process of video coding. As a result, this makes it difficult to adapt quality-driven video coding to the dynamic network conditions in wireless video communications.

Calculating the expected decoder distortion at the encoder has been addressed in literature [10–12, 19–22]. The most recognized method is called the recursive optimal per-pixel estimate (ROPE) proposed in [10]. ROPE can efficiently evaluate the overall video distortion caused by the quantization noise, video packetization, transmission packet loss and error concealment schemes in video coding. Considering the complicated characteristics of emerging video standards such as H.264/AVC [7] including the in-loop deblocking filter, fractional sample motion accuracy, complex intra prediction and advanced error concealment, A "K-decoder" algorithm was proposed in [19] which simulated K copies of random channel behavior and decoder operation. By the strong law of large numbers, the expected distortion at the decoder can be estimated accurately in the encoder if K is chosen large enough. The method has been introduced into the H.264/AVC reference software so as to achieve Lagrangian video coder control. In the rest of this chapter, the ROPE method will be used unless otherwise stated.

8.4 Rate Control at the Transport Layer

8.4.1 Background

The transmission rate of the dominant transport protocol TCP is controlled by a congestion window which is halved for every window of data containing a packet drop and increased by roughly one packet per window of data. This rate variability of TCP, as well as the additional delay introduced by reliability mechanisms, makes it not well-suited for real-time streaming applications, although it is efficient for bulk data transfer. Therefore, it is generally believed that the transport protocol of choice for video streaming should be UDP, on top of which several application-specific mechanisms can be built (error control, rate control, etc.) [23–27]. However, the absence of congestion control in UDP can cause performance deterioration for TCP-based applications if large-scale deployment takes place in wired/wireless networks.

Therefore, congestion control, typically in the form of rate-based congestion control for streaming applications, is still of primary importance for multimedia applications to deal with the diverse and constantly-changing conditions of the Internet [28]. In essence, the goal of congestion control algorithms is to prevent applications from either overloading or underutilizing the available network resources. Therefore, the Internet community is seeking for a good solution in order to mitigate the impact imposed by UDP-based multimedia applications.

The aforementioned reasons lead to the emergence of TCP-Friendly protocols, which are not only slow responsiveness in order to smooth data throughput, but are also TCP-friendly. The definition of "TCP-friendliness" is that a traffic flow, in steady state, uses no more bandwidth in a long term than what a conforming TCP flow would use under comparable conditions [29], which avoids the starvation or even congestion collapse of TCP traffic when the two types of traffic coexist. Being one of the most popular TCP-Friendly protocols, TFRC is an equation-based congestion control mechanism that uses the TCP throughput function to calculate the actual available rate for a media stream [28].

Owing to its well-known advantages and bright prospects, there has been a great deal of research [29–34] done on TFRC, since it was formally introduced in [30] and standardized in [35]. However, most of these research activities only focus on the performance of TFRC in different networking systems or the comparison with TCP. Some work on video streaming using TFRC focuses only on the TFRC performance with pre-encoded video streams as stated in [36, 37]. This section presents the first work that integrates TFRC with real-time video encoding in a cross-layer optimization approach. This scheme aims at setting up a bidirectional interaction between source online encoding and network TFRC. For example, when the network congestion imposed by heavy traffic takes place and deteriorates to a point where the rate control mechanism alone cannot make any improvement, the video encoder could accordingly adapt its coding parameters and decrease the coding rate output into the network so as to alleviate network congestion, while maintaining a best possible delivered video quality on the receiver side. On the other hand, when a large amount of data are generated in compression due to video clips with fast scence changes, the rate control at the transport layer could increase its sending rate so as to avoid the possible large transmission delay.

The framework is formulated to select the optimal video coding parameters and the transport layer sending rate in order to minimize the average video distortion with a predetermined delay bound. To provide a smooth video playback experience, the video delay bound refers to the video frame delay deadline. Different video frames are associated with different delay deadlines, which are determined by both the video decoding and display settings at the application layer. To solve the problem formulated above and implement the proposed framework, a controller is designed at the source node as shown in [17]. The responsibility of the controller is to: 1) communicate with each layer and obtain the corresponding information. For example, it retrieves the expected video distortion from the encoder and the network conditions from lower layers; 2) perform optimization and determine dynamically the optimal values of the corresponding parameters of each layer; and 3) pass the determined values back to each layer.

8.4.2 System Model

The system model of the proposed framework is shown in Figure 8.3, which consists of a three-layer structure and a controller. The controller is the core of the proposed framework, which is equipped with all possible values of the key parameters of each layer. With the feedback information from the network such as RTT, queue length, and packet loss rate, the controller performs optimization and chooses the optimal set of parameter values in a cross-layer approach, in order to achieve the best video distortion/delay performance.

8.4.3 Network Setting

The Application Layer: In video coding, the rate-distortion rule indicates that setting a finer coding parameter will directly improve the coded video quality. However, the resulting large data rate might increase transmission delay and increase the chance of transmission error and data loss due to network congestion. In the framework, let S be the video coding parameters of each video unit. A video unit can be a video frame, a video slice or a macro block (MB), which depends on specific complexity requirement. S can be the quantization step size (QP), the intra/inter prediction mode, or their combinations. Such coding parameters have significant impacts on the rate-distortion performance of video coding systems.

Assume a N-frame video sequence $\{f_1, f_2, \ldots, f_N\}$ to be encoded and transmitted from the source to the destination. The sequence is packetized into multiple data units. Each

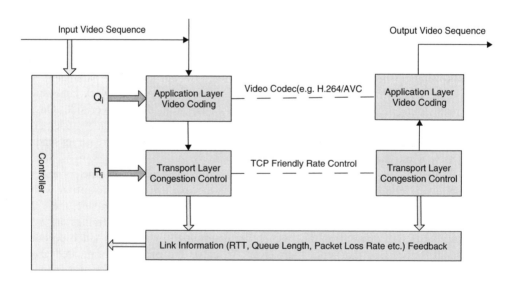

Figure 8.3 The system model for optimized real-time video communications based on TFRC

data unit/packet is independently decodable and represents a slice of the video. In recent video coding standards, such as H.264 [38], a slice could either be as small as a group of macro blocks (MBs), or as large as an entire video frame. Each slice header acts as a resynchronization marker, which allows the slices to be independently decodable and to be transported out of order, but still being decoded correctly at the decoder. Let I_n be the total number of slices in video frame f_n, $\xi_{n,i}$ the ith slice of frame f_n, and $J_{n,i}$ the total number of pixels in $\xi_{n,i}$. Let $f_{n,i}^j$ denote the original value of pixel j in the slice $\xi_{n,i}$, $\hat{f}_{n,i}^j$ the corresponding encoder reconstructed pixel value, $\tilde{f}_{n,i}^j$ the corresponding reconstructed pixel value at the decoder, and $E[d_{n,i}^j]$ the expected distortion at the receiver of pixel j in slice $\xi_{n,i}$. As in [10], the expected mean-squared error (MSE) is used as the distortion metric. Then the total expected distortion $E[D]$ for the entire video sequence can be calculated by summing up the expected distortion over all the pixels

$$E[D] = \sum_{n=1}^{N} \sum_{i=1}^{I_n} \sum_{j=1}^{J_{n,i}} E[d_{n,i}^j] \tag{8.1}$$

where

$$E[d_{n,i}^j] = E[(f_{n,i}^j - \tilde{f}_{n,i}^j)^2]$$
$$= (f_{n,i}^j)^2 - 2 f_{n,i}^j E[\tilde{f}_{n,i}^j] + E[(\tilde{f}_{n,i}^j)^2] \tag{8.2}$$

Since $\tilde{f}_{n,i}^j$ is unknown to the encoder, it can be thought as a random variable. To compute $E[d_{n,i}^j]$, the first and second moments of $\tilde{f}_{n,i}^j$ need to be calculated based on the used packetization scheme and error concealment scheme, and the feedback information on the end-to-end packet loss rate. Readers interested in the detailed calculation of the first and second moments are referred to [10].

The Transport Layer: At the transport layer, the TFRC mechanism is used to avoid the drawbacks of using TCP and to provide smooth throughput for real-time video applications. TFRC is designed for applications that use a fixed packet size and vary their sending rate in packets per second in response to congestion [35]. In TFRC, the average sending rate is given by the following equation:

$$R = \frac{l}{rtt\sqrt{\frac{2p}{3}} + t_{RTO}\left(3\sqrt{\frac{3p}{8}}\right) p(1 + 32p^2)} \tag{8.3}$$

where l is the packet size in bytes, rtt is the round-trip time estimate, t_{RTO} is the value of the retransmission timer, p is the packet loss rate and R is the transmit rate in bytes/second.

In this framework, we propose to optimize the sending rate from a set of optional rates. For each given sending rate, we can derive an inverse function of the equation (8.3) for p:

$$p = \overline{f}(RTT, R) \tag{8.4}$$

That is to say, the equivalent TFRC packet loss rate p can be estimated using the above inverse function (8.4) with the measured rtt. With a given p, the expected distortion can be estimated by the ROPE algorithm. The transmission delay of slice $\xi_{n,i}$ can be expressed as:

$$T_{n,i} = \left\lceil \frac{L_{n,i}}{l} \right\rceil \frac{l}{R} \qquad (8.5)$$

where $\left\lceil \frac{L_{n,i}}{l} \right\rceil$ is the number of the l-byte packets into which the total $L_{n,i}$ bytes of the slice $\xi_{n,i}$ are fragmented.

8.4.4 Problem Formulation

As discussed above, the source coding parameter and the sending rate can be jointly optimized within a quality-driven framework. The goal is to minimize the perceived video distortion within a given slice delay bound. To provide smooth video display experience to users, each frame f_n is associated with a frame delay deadline T_n^{max}. Thus, all the slices of video frame f_n have the same delay constraint T_n^{max}. Let $\vec{S}_{n,i}$ and $\vec{R}_{n,i}$ be the source coding parameter and the TFRC sending rate for slice $\xi_{n,i}$. Therefore, the problem can be formulated as:

$$\min_{\{\vec{S}_{n,i}, \vec{R}_{n,i}\}} \sum_{n=1}^{N} \sum_{i=1}^{I_n} E[D_{n,i}]$$

$$s.t. : Max\{T_{n,1}, T_{n,2}, \ldots T_{n,I_n}\} \leq T_n^{max}, \quad n = 1, 2, \ldots, N \qquad (8.6)$$

8.4.5 Problem Solution

For simplicity, let $v_w = \{S_{n,i}, R_{n,i}\}$ be the parameter vector to be optimized for slice $\xi_{n,i}$, where $1 \leq w \leq N \times I$ is the index of the slice $\xi_{n,i}$ over the whole video clip. According to (8.6), any selected parameter vector v_w which results in a single-slice delay larger than the constraint T_n^{max} cannot belong to the optimal parameter vector $v_w^* = \{S_{n,i}^*, R_{n,i}^*\}$. Therefore, we can make use of this fact by redefining the distortion as follows:

$$E[D_{n,i}'] = \begin{cases} \infty : & T_{n,i} > T_n^{max} \\ E[D_{n,i}] : & T_{n,i} \leq T_n^{max} \end{cases} \qquad (8.7)$$

In other words, the average distortion of a slice with a delay larger than the delay bound is set to infinity. This means that, if a feasible solution exists, the parameter vector which minimizes the average total distortion, as defined in (8.6), will not result in any slice delay larger than T_n^{max}. Therefore, the minimum distortion problem can be transformed into an unconstrained optimization problem using the above redefinition of the single-slice distortion.

Most decoder concealment strategies introduce dependencies between slices. For example, if the concealment algorithm uses the motion vector of the previous MB to conceal the lost MB, then it would cause the calculation of the expected distortion of the current slice to depend on its previous slices. Without losing the generality, we assume that due to the concealment strategy, the current slice will depend on its previous a slices ($a \geq 0$). To solve the optimization problem, we define a cost function $G_k(v_{k-a}, \ldots, v_k)$, which represents the minimum average distortion up to and including the kth slice, given that $v_{k-a}, \ldots, v_k$ are decision vectors for the $(k-a)$th to kth slices. Let O be the total slice number of the video sequence, and we have $O = N \times I$. Therefore, $G_O(v_{O-a}, \ldots, v_O)$ represents the minimum total distortion for all the slices of the whole video sequence. Thus, solving (8.6) is equivalent to solve:

$$\min_{v_{O-a}, \ldots, v_O} G_O(v_{O-a}, \ldots, v_O)) \tag{8.8}$$

The key observation for deriving an efficient algorithm is the fact that given $a+1$ decision vectors $v_{k-a-1}, \ldots, v_{k-1}$ for the $(k-a-1)$th to $(k-1)$th slices, and the cost function $G_{k-1}(v_{k-a-1}, \ldots, v_{k-1})$, the selection of the next decision vector v_k is independent of the selection of the previous decision vectors $v_1, v_2, \ldots, v_{k-a-2}$. This means that the cost function can be expressed recursively as:

$$G_k(v_{k-a}, \ldots, v_k) = \min_{v_{k-a-1}, \ldots, v_{k-1}} \{G_{k-1}(v_{k-a-1}, \ldots, v_{k-1}) + E[D_k]\} \tag{8.9}$$

The recursive representation of the cost function above makes the future step of the optimization process independent from its past step, which is the foundation of dynamic programming.

The problem can be converted into a graph theory problem of finding the shortest path in a directed acyclic graph (DAG) [39]. The computational complexity of the algorithm is $O(O \times |v|^{a+1})$ (where $|v|$ the cardinality of v), which depends directly on the value of a. For most cases, a is a small number, so the algorithm is much more efficient than an exhaustive search algorithm with exponential computational complexity.

8.4.6 Performance Evaluation

In this section, we present some simulation results to verify the efficient performance of the cross-layer framework. Video coding is performed using the H.264/AVC JM 12.2 codec. The first 100 frames of the QCIF sequence 'Foreman' are coded at frame rate 30 frames/second. All frames except the first are encoded as P frames. We assume that a video frame consists of only one slice. To avoid prediction error propagation, a 10% macro block level intra-refreshment is used. When a packet is lost during transmission, the decoder applies a simple error concealment scheme by copying the corresponding slice from the most recent, correctly received frame. In our experiments, we consider the QP of each slice as the video coding parameter to be optimized. The admissible set of QP values are 4, 6, 8, 10, 14, 18, 24, 30, 38, and 46. The expected video quality at the receiver is measured by the peak signal-to-noise ratio (PSNR).

For the network simulation, the used multihop topology is shown in the Figure 8.4. Each node has a queue with a queue length of 100 packets for packet buffering. As in [29], we set the default value of t_{RTO} as $t_{RTO} = 4 \times rtt$ in our NS-2 simulator. We

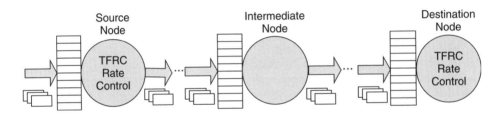

Figure 8.4 The network model used in the experiment

also set the link capacity as 2 Mbps and the optional TFRC sending rates are 1.0 Mbps, 1.2 Mbps, 1.4 Mbps, 1.6 Mbps, 1.8 Mbps, and 1.9 Mbps. The RTT at the transport layer is estimated by measuring the time elapsed between sending a data packet and receiving the acknowledgement. Figure 8.5 depicts the RTT estimate of the network with different number of hops when the packet size is 4000 bits.

Figure 8.6 shows the corresponding PSNR-delay performance comparisons with a different number of hops between the joint optimization scheme and the existing scheme with video coding and sending rate being optimized separately. As shown in the figure, the number of hops has a significant impact on the PSNR-delay performance of the proposed scheme as well as that of the existing scheme. This is because a smaller number of hops leads to smaller RTT's. Within a fixed video frame delay deadline, the smaller

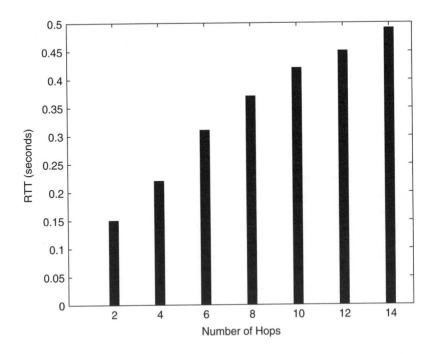

Figure 8.5 Measured RTT's vs. the number of hops

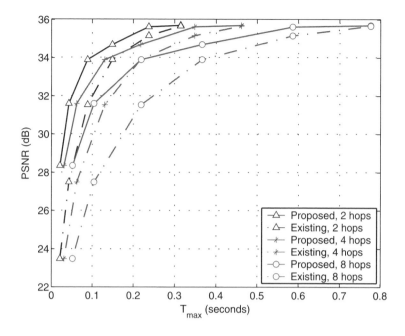

Figure 8.6   PSNR-delay performance comparisons with different number of hops

RTT allows finer video coding, which brings a better received video quality. As shown in the figure, the cross-layer scheme has a significant video quality improvement over the existing scheme, especially under tighter video frame delay deadlines.

Table 8.1 shows the statistics of the selections of QP and TFRC sending rate in the joint optimization scheme over the first 100 frames with the corresponding $T_n^{\max}$. The hop number is 4 and the packet size is 4000 bits. The values in brackets denote the the frequencies of the used QPs and TFRC sending rates in coding the 100 video frames, respectively. This table indicates the variations of QP and TFRC sending rate required to adapt the network traffic conditions. To illustrate the quality of video frames, we plot a sample video frame from the original video in Figure 8.7(a) and compare it to the

Table 8.1   Statistics of QP and TFRC sending rate

Scheme	$T_n^{\max} = 0.031$s PSNR = 28.3 (dB)		$T_n^{\max} = 0.031$s PSNR = 31.6 (dB)	
	QP	R_{TFRC} (Mbps)	QP	R_{TFRC} (Mbps)
Existing	46	1.4	38	1.4
Proposed	46 [21]	1.2 [5], 1.4 [65]	38[28]	1.2 [3], 1.4 [43]
	38 [60]	1.6 [26]	30 [56]	1.6 [32],1.8 [15]
	30 [19]	1.8 [4]	24 [16]	1.9 [7]

(a) Original   (b) Optimized   (c) Existing

Figure 8.7   Frame 82 from the reconstructed video

reconstructed video frames under the cross-layer scheme and the existing scheme for $T_n^{\max} = 0.062s$, respectively. The frame delivered by using the cross-layer scheme has a visual quality very close to the original frame, while the frame under the existing scheme is considerably blurry.

8.5   Routing at the Network Layer

8.5.1   Background

The increasing demand for video communication services is promoted by two facts: one is the pervasive use of various computing devices, and the other is the potential deployment of multi-hop wireless networks in order to connect these computing devices. However, transmitting video over multi-hop wireless networks encounters many challenges, such as unreliable link quality owing to multi-path fading and shadowing, signal interferences among nodes and dynamic connectivity outages. Therefore, routing performance affects the end-to-end quality-of-service (QoS) of video applications significantly. We need to answer a series of questions: How to find paths adaptively which can maximize received video quality under stringent delay constraint? How to determine paths adaptively so that network resource is utilized fully in good network conditions, How is a minimal QoS requirement guaranteed in poor network conditions?

Network- vs. Application-Centric Routing Metrics: In some existing work on routing for video communication, a set of paths is given *a priori*. With the network- (e.g. link delay) or application-centric metric (e.g., end-to-end video distortion) as the objective function, the best path is selected from the available path set via an exhaustive search. Such solutions may work in sparsely connected networks as the number of total combinatorial paths is small. However, in densely connected networks, these solutions fail owing to the exponential complexity of an exhaustive search in a large set of paths. As an example, the network in Figure 8.8(a) has only three paths from the source to the destination. However, by adding only another four links to the network, the resulting network shown in Figure 8.8(b) has a set of up to 11 paths, which may increase significantly the computational complexity of the exhaustive search to find the best path. It is worth noting that the emerging wireless mesh networks are generally densely connected networks, where a mixture of fixed and mobile nodes is interconnected via wireless links. Therefore, the routing approach of an exhaustive path search is not suitable for the emerging multi-hop wireless mesh networks.

To avoid an exhaustive search, much work uses classical graph-theoretical algorithms such as the *Dijkstra's algorithm* and the *Bellman-Ford algorithm* in many forms

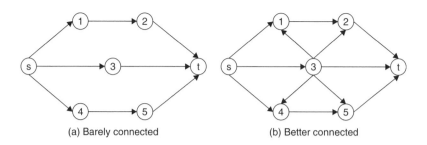

(a) Barely connected   (b) Better connected

Figure 8.8 Two examples of a multi-hop wireless network

in order to find the shortest path. In these routing algorithms, each link/arc is weighted by network-centric metrics such as the physical distance, the number of hops and the estimated transmission delay. These link metrics are additive. The goal of routing is to find the path with the minimum accumulated weight from the source to the destination. For example, the total weight of the path in Figure 8.9(a) from source s to destination t is $W = \sum_{h=1}^{H} w_h$, where the individual link weight w_h can be the physical distance, the number of hops ($w_h = 1$) and the measured transmission delay. However, traditional routing methods with network-centric routing metrics have their own shortcomings for multimedia communications. For example, the minimum hop-count protocol uses the minimum number of hops as the routing metric without considering load balance among nodes. If an intermediate node is chosen which is involved in multiple source-end pairs, heavy traffic and the resulting congestion might happen at the node. As another example, the minimum loss-ratio metric does not cope well with short-term channel variations because it uses mean packet loss ratios in making routing decisions. For instance, radio channels may have low average packet loss ratios, but with high variability, implying that the minimum loss-ratio metric will perform poorly because they do not adapt well to bursty loss conditions.

In multimedia communications, from the point of view of users, application-centric metrics is the most reasonable routing criteria. However, most application-centric metrics are non-additive. As an example, in Figure 8.9(b), the weight of the link ($h - 1 \rightarrow h$) for a packet is labeled by the resulting expected video distortion $E[D]_h$ of transmitting the packet from the node $h - 1$ to h. However, different from Figure 8.9(a) with network-centric routing metrics, the end-to-end expected distortion $E[D]$ of transmitting the packet along the holistic path from the source to the destination is not the sum of the distortion values on each individual link, i.e., $E[D] \neq \sum_{h=1}^{H} E[D]_h$. This means that both Dijkstra's algorithm and the Bellman-Ford algorithm cannot be directly used to solve the routing problems with application-centric metrics.

(a) Additive metrics   (b) Nonadditive metrics

Figure 8.9 Different metrics on an end-to-end path

The Application-Centric Routing in Existing Work: In literature, there has already been some work which uses video distortion as the criterion in selecting good paths, for example, selecting a path from a group of preset paths in overlay networks [13], distributed routing for multi-user video streaming [16] and multi-path routing algorithm video multi-path transport [14, 15]. However, the video distortion values used in this work are either pre-calculated [13, 16] or generated from video distortion-rate models [14, 15, 40] without considering the video coding process and the error concealment strategies. In fact, most existing work on video routing focuses on video streaming applications in which pre-encoded data is requested by the user. *Pre-encoded video streaming does not allow for real-time adaptation in finding optimal paths in the encoding process*. Therefore, to the best of our knowledge, little research has been done on integrating online video coding into network routing in a quality-driven sense. Furthermore, the existing research either performs an exhaustive search in order to obtain the optimal solution [13], or takes approaches such as heuristic analysis [14], relaxation techniques [15] or distributive analysis [16] and suboptimal solutions are the result.

In this section, we turn our attention to the problem of routing packets in a way that takes higher-layer video processing into account, finding a path that achieves the best delivered video quality. We develop a new routing metric utilizing the *expected end-to-end video distortion*. In the cross-layer framework of quality-driven routing, path determination and real-time video coding are optimized jointly so as to adapt to the time-varying network conditions while ensuring that the end-to-end delay constraint is satisfied. The motivation for integrating application-layer video processing into finding paths is based on the following observations in the sense of received-video quality maximization: 1) Various video processing options [38] lead to different pair of rate-distortion values. For different rate-distortion pairs, different paths may be be selected with the proposed routing metric. In other words, if the various options in video coding are not considered in routing, the path selected might not be the optimal path in the sense of maximizing received video quality under the given network conditions. 2) When network conditions are poor, video processing may need to be adapted in order to produce a lower rate video stream so as to find the optimal routing path. In contrast, in good network conditions video source processing may be allowed to perform finer source coding, outputting a higher rate data stream in order to take full advantage of network resource.

8.5.2 System Model

As shown in Figure 8.10, we model a multi-hop wireless network as a directed acyclic graph (DAG) $\mathcal{G} = (\mathcal{V}, \mathcal{E})$ (where $|\mathcal{V}| = n$), where $\mathcal{V}$ is the set of vertexes representing wireless nodes and $\mathcal{E}$ is the set of arcs representing directed wireless links. We characterize a link $(v, u) \in \mathcal{E}$ between nodes v and u with (i) $\gamma^{(v,u)}$: Signal-to-Interference-Noise-Ratio (SINR) of link (v, u), (ii) $P^{(v,u)}$: packet loss rate on the link (v, u), and (iii) $T^{(v,u)}$: packet delay on the link.

For a N-frame video sequence $\{f_1, f_2, \ldots, f_N\}$ to be encoded and transmitted from the source node s to the destination node t, we assume each frame f_n is coded and packetized into I_n packets. Each packet $\pi_{n,i}$ is decodable independently and represents a slice of the video. In the remainder of this section, the concept of packet and slice will be used alternatively unless otherwise specified. To provide a smooth video display

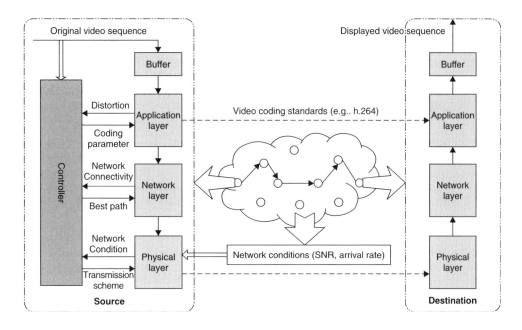

Figure 8.10   System model

experience to users, each frame f_n is associated with a frame delay deadline T_n^{budget}. For all packets $\{\pi_{n,1}, \pi_{n,2}, \ldots, \pi_{n,I_n}\}$ of frame f_n to be timely decoded at the receiver, the following delay constraint is derived:

$$\max\{T_{n,1}, T_{n,2}, \ldots, T_{n,I_n}\} \leq T_n^{\text{budget}} \quad n = 1, 2, \ldots, N. \tag{8.10}$$

where $T_{n,i}$ is the resulting delay for packet $\pi_{n,i}$ transmitted from s to t.

We design a system controller at the source node, which communicates with each layer in order to capture the updated network conditions, performs routing optimization with the end-to-end video distortion as the routing metric and determines dynamically the optimal values of control parameters residing in each layer so as to guarantee the entire system to perform efficiently. Table 8.2 summarizes the notations used in this section.

8.5.3   Routing Metric

To formulate the problem of quality-driven routing, we describe how to calculate the expected distortion as the routing metric with link statistics for a given path. That is, we need to first determine the routing metric used by the routing algorithm.

We assume that each intermediate node adopts some type of *link adaptation* scheme [41] in order to maximize its outgoing link throughput. It is a link adaptation scheme, meaning that the intermediate node can select adaptive modulation and coding schemes based on the detected link Signal-to-Interference-Noise-Ratio (*SINR*).

Packet Loss Rate: In wireless environment, the packet loss rate $p_{n,i}^{(u,v)}$ of packet $\pi_{n,i}$ transmitted over the link (u, v) is comprised mainly of two parts: packet dropping

Table 8.2 Notation

Symbol	Definition
$\mathcal{G} = (\mathcal{V}, \mathcal{E})$	Graph representation of the network
$\mathcal{V}$	Set of nodes
$\mathcal{E}$	Set of arcs
s	Source node
t	Destination node
(u, v)	A link from node u to v
f_n	The nth video frame
$f_{n,i}^j$	The original value of the jth pixel of the ith slice (packet) of frame f_n
$\hat{f}_{n,i}^j$	The reconstructed value at the encoder of the jth pixel of the ith slice (packet) of frame f_n
$\tilde{f}_{n,i}^j$	The reconstructed value at the decoder of the jth pixel of the ith slice (packet) of frame f_n
$\pi_{n,i}$	The ith slice/packet of video frame f_n
I_n	The slice/packet number of frame f_n
$p_{n,i}^u$	Packet dropping rate of packet $\pi_{n,i}$ at node u
$p_{n,i}^{(u,v)}$	Total packet loss rate of packet $\pi_{n,i}$ over the link (u, v)
$T_{n,i}^{(u,v)}$	Total delay of packet $\pi_{n,i}$ over the link (u, v)
$T_{n,i}$	End-to-end delay of packet $\pi_{n,i}$
T_n^{budget}	Delay deadline of f_n
$\mathcal{R}_{n,i}$	A path from s to t for packet $\pi_{n,i}$
$\mathcal{S}_{n,i}$	Source coding parameter of packet $\pi_{n,i}$

rate $p_{n,i}^u$ at the node u and packet error rate $p_{n,i}^{(u,v)}$ on the link (u, v). Owing to finite-length queuing at the node u, arriving packets could be dropped when the queue is full. To provide a timely packet dropping rate, we assume that $p_{n,i}^u$ is calculated at each node u. [42] provides all the details about how to calculate $p_{n,i}^u$ by using link adaptation schemes. Owing to the unreliable channel condition of link (u, v), wireless signals could be corrupted associated with the specific transmission scheme. A wireless link is assumed to be a memoryless packet erasure channel. The packet error rate $p_{n,i}^{(u,v)}$ for a packet of L bits can be approximated by the sigmoid function:

$$p_{n,i}^{(u,v)}(L) = \frac{1}{1 + e^{\zeta(SINR - \delta)}}, \qquad (8.11)$$

where ζ and δ are constants corresponding to the modulation and coding schemes for a given packet length L [16, 43]. Thus, the equivalent loss probability for packet $\pi_{n,i}$ over the link (u, v) is:

$$p_{n,i}^{(u,v)} = 1 - (1 - p_{n,i}^u)(1 - p_{n,i}^{(u,v)}(L)) \qquad (8.12)$$

Therefore, the loss probability for packet $\pi_{n,i}$ traversing a H-hop path $\mathcal{R}^{s,v}$ and reaching the node v is:

$$P_{n,i}^{s,v} = 1 - \prod_{h=1}^{H}(1 - p_{n,i}^{h}) \tag{8.13}$$

where $p_{n,i}^{h}$ is the packet loss rate over the hop h, which can be calculated by equation (8.12).

Packet Delay: The packet delay $T_{n,i}^{(v,u)}$ on the link (v, u) also includes mainly two parts: the waiting time $T_{n,i}^{u}$ in the queue of node u and the service time $T_{n,i}^{(u,v)}$ by the link (u, v), i.e.:

$$\begin{aligned} T_{n,i}^{(v,u)} &= T_{n,i}^{u} + T_{n,i}^{(u,v)} \\ &= \frac{\xi_{n,i}}{\overline{R}} + \frac{L_{n,i}}{R}. \end{aligned} \tag{8.14}$$

where $\xi_{n,i}$ is the queue state (the number of waiting packets before packet $\pi_{n,i}$) and can be easily measured by the intermediate node; $\overline{R}$ is the averaged service rate for all waiting packets before packet $\pi_{n,i}$; R is the instantaneous transmission rate when transmitting packet $\pi_{n,i}$. Therefore, the total delay for packet $\pi_{n,i}$ traversing a H-hop path $\mathcal{R}^{s,v}$ and reaching the node v is:

$$T_{n,i}^{s,v} = \sum_{h=1}^{H} T_{n,i}^{h} \tag{8.15}$$

where $T_{n,i}^{h}$ is the packet delay over the hop h, which can be calculated by equation (8.14).

Routing Metric: The routing metric is the expected video distortion of packets transmitting from the source to the target node. With the ROPE algorithm [10], the expected distortion for packet $\pi_{n,i}$ can be calculated by summing up the expected distortion of all the pixels whose coding bits form the payload of the packet. Therefore, for a J-pixel slice, the total expected distortion can be calculated as:

$$E\left[D_{n,i}\right] = \sum_{j=1}^{J} E\left[d_{n,i}^{j}\right] \tag{8.16}$$

where $E\left[d_{n,i}^{j}\right]$ is the expected distortion of the jth pixel in the ith slice of frame f_n at the receiver.

8.5.4 Problem Formulation

Let $\mathcal{R}_{n,i}$ and $\mathcal{S}_{n,i}$ be the transmission path and the source coding parameter of packet $\pi_{n,i}$. Both $E[D_{n,i}]$ and $T_{n,i}$ depend on the choices of $\mathcal{R}_{n,i}$ and $\mathcal{S}_{n,i}$. Our goal is to find the best transmission path and coding parameter values for all the slices of the holistic video sequence. Therefore, the quality-driven routing problem can be formulated as:

$$\min_{\{\mathcal{R}_{n,i}, \mathcal{S}_{n,i}\}} \sum_{n=1}^{N} \sum_{i=1}^{I_n} E[D_{n,i}]$$

$$s.t. : \max\{T_{n,1}, T_{n,2}, \ldots, T_{n,I_n}\} \le T_n^{\text{budget}} \quad n = 1, 2, \ldots, N. \tag{8.17}$$

The constraint in equation (8.17) guarantees that all packets $\{\pi_{n,1}, \pi_{n,2}, \ldots, \pi_{n,I_n}\}$ of frame f_n arrive in a timely fashion at the receiver for decoding within the delay deadline T_n^{budget}. *It is worth noting that the optimization is performed over all the slices of a N-frame video sequence at one time by considering the independencies of different slices which are introduced by predictions in coding and error concealment in decoding.* Therefore, N can be tuned depending on the computational capabilities of network nodes as well as the feedback speed of the network information required by the controller.

Within a quality-maximization framework, adaptable video coding is coupled into the above formulated routing problem in order to enhance the flexibility of routing. The motivation for coupling tunable video coding with routing is based on the fact that a network is in bad condition (for example, with a medium-to-heavy traffic load) and no path exists to deliver the video packets successfully within the corresponding frame delay deadlines. However, in such cases, if the video coding parameters are tunable and video frames are coded coarsely into a small number of packets with lower quality, it is most likely that there exists a path which can deliver these video packets successfully, leading to better video quality. In other words, the network load is alleviated significantly by putting a lesser amount of packets into the network through tuning video coding parameters. Similarly, when the network is in good condition, there are multiple good paths existing in the network. It will lead to better end-to-end video quality if the video coding parameters are tuned so as to obtain a finer video quality. Therefore, tuning video coding parameters can guarantee the existence of good paths (i.e., the feasibility of routing) in bad network conditions, and doing so can also make the network resource fully utilized in good network conditions. Thus, the task of routing turns into finding the best path in a quality-driven sense.

8.5.5 Problem Solution

In this section, we first calculate the best path for each possible individual packet created by each coding option. Owing to the dependencies between slices introduced by encoder prediction modes and decoder concealment strategies, we use dynamic programming in order to obtain the coding parameter values and path for each group of interdependent slices.

Best-Path Determination for Each Individual Packet Created by Each Coding Option: Any node v in the network can be abstracted into the model shown in Figure 8.11. Node v is connected to its $\mathcal{A}$ backward adjacent nodes $\{u_1, \cdots, u_a, \cdots, u_{\mathcal{A}}\}$ via $\mathcal{A}$ incoming links and connected to its $\mathcal{B}$ forwarding adjacent nodes $\{w_1, \cdots, w_{+b}, \cdots, w_{+\mathcal{B}}\}$ via $\mathcal{B}$ outgoing links.

Next, we will show how the controller calculates the optimal path for one packet based on the feedback information from all other nodes. We assume that the controller knows

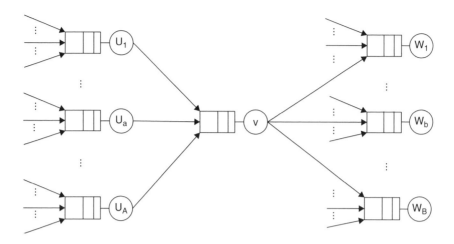

Figure 8.11   A nodal model in wireless multi-hop networks

the network topology. For the convenience of comparison, each node is labeled with the expected video distortion of the current packet along the path. Initially, no paths are known, so all nodes are labeled with infinity. As the algorithm proceeds and paths are found, the labels may be changed. Moreover, a label can be either tentative or permanent. Initially, all labels are tentative. When it is determined that a label is a part of the optimal possible path from the source, it will be made permanent.

Assuming that node v is made permanent, in the following we will show how the weights of its $\mathcal{B}$ outgoing links are calculated. v is made permanent which means that the best path from s to v has been determined. Let $\mathcal{R}_{n,i}^{(s,v)}$ denote the path, and $P_{n,i}^{(s,v)}$ denote the packet loss rate when packet $\pi_{n,i}$ is transmitted from s to v through the path $\mathcal{R}_{n,i}^{(s,v)}$. For intra-coded macro blocks, the first moment of $\tilde{f}_{n,i}^{j}$ is recursively calculated as

$$
E\left[\tilde{f}_{n,i}^{j}\right]_{\mathcal{R}_{n,i}^{(s,v)}} = \left(1 - P_{n,i}^{(s,v)}\right)\left(\hat{f}_{n,i}^{j}\right) + P_{n,i}^{(s,v)}\left(1 - P_{n,i-1}^{(s,t)}\right)E\left[\tilde{f}_{n-1,k}^{l}\right]
$$
$$
+ P_{n,i}^{(s,v)} P_{n,i-1}^{(s,t)} E\left[\tilde{f}_{n-1,i}^{j}\right] \tag{8.18}
$$

where the first term at right side is the case of $\pi_{n,i}$ being correctly transmitted to v, and we have $\tilde{f}_{n,i}^{j} = \hat{f}_{n,i}^{j}$. For the second term, if the current packet $\pi_{n,i}$ is lost and the previous packet is received, the concealment motion vector associates the jth pixel in the ith packet of the current frame f_n with the lth pixel in the kth packet $\pi_{n-1,k}$ of the previous frame f_{n-1}. We thus have $\tilde{f}_{n,i}^{j} = \tilde{f}_{n-1,k}^{l}$, and the probability of this event is $P_{n,i}^{(s,v)}(1 - P_{n,i-1}^{(s,t)})$. $P_{n,i-1}^{(s,t)}$ is the loss probability of packet $\pi_{n,i-1}$ being transmitted from s to t, which is known to the controller for each coding option. For the third term, if neither the current packet nor the previous is received, correctly the current pixel is concealed by the pixel with the same position in the previous frame. We will have $\tilde{f}_{n,i}^{j} = \tilde{f}_{n-1,i}^{j}$.

Similarly, the second moment is calculated as:

$$
E\left[\left(\tilde{f}_{n,i}^{j}\right)^{2}\right]_{\mathcal{R}_{n,i}^{(s,v)}} = \left(1 - P_{n,i}^{(s,v)}\right)\left(\hat{f}_{n,i}^{j}\right)^{2} + P_{n,i}^{(s,v)}\left(1 - P_{n,i-1}^{(s,t)}\right)E\left[\left(\tilde{f}_{n-1,k}^{l}\right)^{2}\right]
$$

$$
+ P_{n,i}^{(s,v)} P_{n,i-1}^{(s,t)} E\left[\left(\tilde{f}_{n-1,i}^{j}\right)^{2}\right] \tag{8.19}
$$

Therefore, by equation (8.16) and the ROPE algorithm, the total expected distortion of packet $\pi_{n,i}$ when transmitted to v can be written as:

$$
E\left[D_{n,i}\right]_{\mathcal{R}_{n,i}^{(s,v)}} = \sum_{j=1}^{J}\left\{\left(f_{n,i}^{j}\right)^{2} - 2f_{n,i}^{j}E\left[\tilde{f}_{n,i}^{j}\right]_{\mathcal{R}_{n,i}^{(s,v)}} + E\left[\left(\tilde{f}_{n,i}^{j}\right)^{2}\right]_{\mathcal{R}_{n,i}^{(s,v)}}\right\} \tag{8.20}
$$

Next, we consider that packet $\pi_{n,i}$ continues move forward one hop from v and reaches the node w_b. The resulting total packet loss rate is:

$$
P_{n,i}^{(s,w_b)} = 1 - (1 - P_{n,i}^{(s,v)})(1 - P_{n,i}^{(v,w_b)}) \tag{8.21}
$$

where $P_{n,i}^{(v,w_b)}$ is the packet loss rate for packet $\pi_{n,i}$ over the link (v, w_b). Similar to Eq. (8.20) the total expected distortion of packet $\pi_{n,i}$ when $\pi_{n,i}$ is transmitted to w_b can be written as:

$$
E\left[D_{n,i}\right]_{\mathcal{R}_{n,i}^{(s,w_b)}} = \sum_{j=1}^{J}\left\{\left(f_{n,i}^{j}\right)^{2} - 2f_{n,i}^{j}E\left[\tilde{f}_{n,i}^{j}\right]_{\mathcal{R}_{n,i}^{(s,w_b)}} + E\left[\left(\tilde{f}_{n,i}^{j}\right)^{2}\right]_{\mathcal{R}_{n,i}^{(s,w_b)}}\right\} \tag{8.22}
$$

where $E\left[\tilde{f}_{n,i}^{j}\right]_{\mathcal{R}_{n,i}^{(s,w_b)}}$ and $E\left[\left(\tilde{f}_{n,i}^{j}\right)^{2}\right]_{\mathcal{R}_{n,i}^{(s,w_b)}}$ can be calculated as in equation (8.18) and equation (8.19) with the packet loss rate $P_{n,i}^{(s,w_b)}$ in equation (8.21).

Figure 8.12 shows a flowchart for determining the optimal path from source s to destination t. The optimal path is $s \rightarrow a \rightarrow c \rightarrow d \rightarrow f \rightarrow t$. As shown in Figure 8.12, the distinctive differences between the proposed routing algorithm and the classical Dijkstra's shortest path algorithm [44] are: 1) the link weights in the proposed routing algorithm are calculated on the fly based on fedback network information; 2) the link weights are non-additive.

Optimal Solution: Owing to the dependencies between slices introduced by encoder prediction modes and decoder concealment strategies, we use dynamic programming to solve equation (8.17) in order to determine optimal the coding parameter values and paths for each group of interdependent slices.

Assume that the target video clip to be transmitted is compressed into W packets. Then, according to the packetization scheme in Section V-C, $W = I \times N$ we define that the current packet, the ith packet of the nth frame, is the wth packet of the entire video clip. The parameter vector $\Theta_{n,i} = \{\mathcal{R}_{n,i}, \mathcal{S}_{n,i}\}$ can be represented by $\Theta_w = \{\mathcal{R}_w, \mathcal{S}_w\}$. Most decoder concealment strategies introduce dependencies between packets. For example,

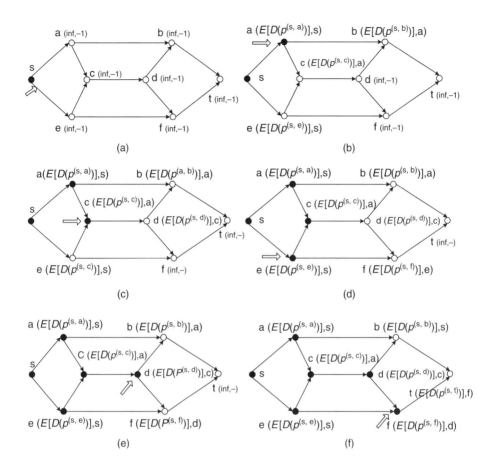

Figure 8.12   The flowchart of finding the optimal path $s \to a \to c \to d \to f \to t$

if the concealment algorithm uses the motion vector of the MB above to conceal the lost MB, then it would cause the calculation of the expected distortion of the current packet to depend on its previous packet. Without losing the generality, we assume that the current packet depends on its previous z packets ($z \geq 0$) by using the chosen concealment strategy. To solve the optimization problem, we define a cost function $\mathcal{J}_k(\Theta_{w-z}, \cdots, \Theta_w)$, which represents the minimum average distortion up to and including the wth packet, given that $\{\Theta_{w-z}, \ldots, \Theta_w\}$ are the decision vectors of packets $\{(w-z), \ldots, w\}$. Therefore, $\mathcal{J}_W(\Theta_{W-z}, \ldots, \Theta_W)$ represents the minimum total distortion for all the packets of the video clip. Clearly, solving (8.17) is equivalent to solving:

$$\min_{\Theta_{W-z}, \ldots, \Theta_W} \mathcal{J}_W(\Theta_{W-z}, \ldots, \Theta_W) \tag{8.23}$$

The key observation for deriving an efficient algorithm is the fact that given $z+1$ decision vectors $\{\Theta_{w-z-1}, \ldots, \Theta_{w-1}\}$ for the $(w-z-1)$th to $(w-1)$th packets and the cost function $\mathcal{J}_{w-1}(\Theta_{w-z-1}, \ldots, \Theta_{w-1})$, the selection of the next decision vector Θ_w is independent of the selection of the previous decision vectors $\{\Theta_1, \Theta_2, \ldots, \Theta_{w-z-2}\}$. This means that the cost function can be expressed recursively as:

$$\mathcal{J}_w (\Theta_{w-z}, \ldots, \Theta_w) = \min_{\Theta_{w-z-1}, \ldots, \Theta_{w-1}} \{\mathcal{J}_{w-1} (\Theta_{w-z-1}, \ldots, \Theta_{w-1}) + E[D_w]\}. \quad (8.24)$$

The recursive representation of the cost function above makes the future step of the optimization process independent from its past steps, which is the foundation of dynamic programming. The problem can be converted into a graph theory problem of finding the shortest path in a directed acyclic graph (DAG) [39].

The complete proposed optimization algorithm is summarized in Table 8.3.

8.5.6 Implementation Considerations

Timeliness Consideration of Network Feedback: In the quality-driven routing scheme, the controller at the source node calculates the optimal path based on the quality-based routing metric by using the feedback information from other nodes in the network. To guarantee obtaining the optimal path in real-time or near real-time, it is necessary to maintain an effective message passing mechanism so as to obtain the latest information.

From the standpoint of implementation, the quality-driven routing algorithm can be built on top. We propose to build our algorithm on top of any proactive routing protocols such as optimized link state routing (OLSR) [45]. Owing to its proactive nature, OLSR maintains up-to-date global network topology information in its link-state database. Once there are packets to be transmitted, the controller will retrieve the information and computation can be performed to find the best path.

Complexity and convergence: Lemma 1: Once a label is made permanent, it will be never changed.

Proof: Recall that the label with the minimal distortion among all tentative labels is the one which is determined to be made permanent. That is, all other tentative nodes have larger distortion values. For descriptive convenience, we name the node corresponding to the label as the just-made-permanent node. In subsequent procedures when other tentative nodes are examined, the distortion incurred when the packet travels to the just-made-permanent node from source s through any of other tentative nodes, is definitely larger than that of the just-made-permanent label. Therefore, the just-made-permanent label will be never changed.

Table 8.3 Proposed Optimization Algorithm

1.	For each slice
2.	For each coding option
3.	Find the path through which the compressed packet has minimized distortion by performing the routing algorithm described in V-E.1
4.	Perform the dynamic programming optimization described in V-E.2 to find the optimal coding parameter values and paths for all the slices

Dealing with link breakages and node breakdown: With the proposed routing algorithm, a set of paths can be discovered and maintained simultaneously. Once the optimal path breaks owing is to either link outage or node breakdown, the suboptimal path with the second least distortion will be selected immediately for transmission and no path rediscovery will be initiated.

8.5.7 Performance Evaluation

Experiments have been carried out to show the superior performance of joint optimization for video communication based on the quality-driven routing scheme. The experiments are designed using H.264/AVC JM 12.2. We encode the video sequences 'Foreman' and 'Glasgow' at 30 fps. The bandwidth of each link is assumed as 10^6 Hz.

We compare the quality-driven routing (QDR/min-dist) algoritm with the conventional 'minimum-hops' routing (MHR/min-hop) algorithm. In Figure 8.13(a), V–G we implement four video transmission schemes for the 'Foreman' sequence. For QP-fixed cases, the quantizer step size (for both intra-mode and inter-mode) is set equal to 10, 15, 20, 25, 30, 35, 40, 45, respectively. Based on the video display requirement at the application layer, the corresponding values of packet delay deadlines are also preset. For jointly optimized cases, the choice of QP (from 10, 15, 20, 25, 30, 35, 40, 45) and routing are performed jointly. The results in Figure 8.13(a) indicate that the quality-driven routing, with either QP fixed or jointly optimized, can provide significant gains in expected PSNR over the conventional MHR algorithm. Specifically, for both jointly optimized schemes, when the packet deadline is tight ($T_n^{budget} < 0.001s$), the quality-driven routing algorithm has a $1 - 2$ dB gain in PSNR over the minimum-hops routing algorithm. When the packet deadline ($T_n^{budget} \geq 0.002s$), the quality-driven routing algorithm provides up to 4-dB PSNR gain over the minimum-hops routing algorithm. In Figure 8.13(b), with the same experiment setting as in Figure 8.13(a), we encode the 'Glasgow' sequence so that we can test the advantages of the quality-driven

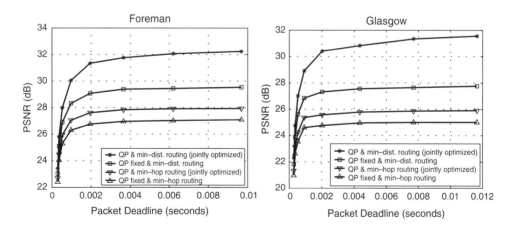

Figure 8.13 Average PSNR comparison of the quality-driven routing algorithm with the conventional minimum-hops routing algorithm, (a) Foreman (b) Glasgow

algorithm in the video sequences with different motion patterns. Compared with the 'Fore-man' sequence, the 'Glasgow' sequence takes on a feature of fast motion. We can observe that the quality-driven routing algorithm significantly outperforms the minimum-hops routing algorithm. More importantly, compared with Figure 8.13(a), for either tight or loose packet delay deadline, the performance of the quality-driven routing algorithm for fast-motion videos is better than that for slow and medium motion videos.

Figure 8.14 shows the frame PSNR comparison of the quality-driven routing algorithm with the conventional minimum-hops routing algorithm both for $T_n^{budget} = 0.005s$. We find that the quality-driven routing algorithm has a PSNR gain of 3-4dB over the minimum-hops routing algorithm for each frame of both the 'Foreman' and 'Glasgow' sequences. This further demonstrates the superior performance of the quality-driven routing algorithm.

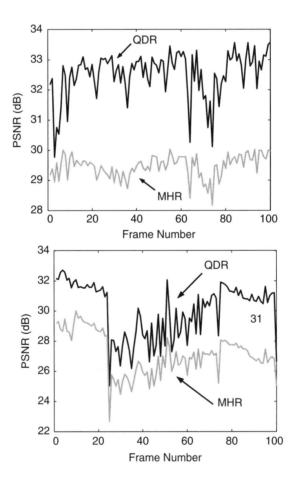

Figure 8.14 Frame PSNR comparison of the quality-driven routing algorithm with the conventional minimum-hops routing algorithm, (a) Foreman (b) Glasgow

8.6   Content-Aware Real-Time Video Streaming

8.6.1   Background

In wireless mesh networks, there exist multiple paths from the source to the destination. Thus, video delivery may take advantage of path diversity in order to achieve a better user perceived quality. Furthermore, path diversity also offers us flexibility in performing resource allocation for different video contents with different importance levels. Although a great deal of work on content-aware video transmissions can be found [46–48], much of it focuses either on the techniques of determining packet importance or on multi-user packet scheduling. For example, a content-aware resource allocation and packet scheduling scheme for video transmissions to multiple users was proposed in [18]. However, to the best of our knowledge, no work exists to jointly consider content-aware real-time video coding and content-aware network routing within a quality-driven framework.

In this section, we will discuss a unified content-aware quality-aware optimization framework for video communications over wireless mesh networks. To maximize user perceived video quality, we first use background subtraction techniques [49] in order to identify the region of interest (ROI), i.e., the foreground content. Then, considering the different video quality contributions of foreground and background packet losses, the video coding parameters (e.g., quantization step size, prediction modes or both) at the application layer and the path selection at the network layer for video packets corresponding to different types of contents are jointly optimized. It is expected that at the application layer, the foreground should be coded by a finer QP; at the network layer, the foreground packets be transmitted along good paths; and at the MAC layer, the foreground packets be given a prior scheduling in the queue. However, the task described above is not trivial owing to arbitrary foreground changing and video delay constraints. We formulate the problem to find the optimal video source coding, the transmission paths and the packet scheduling in order to achieve the best perceived video quality within the end-to-end video frame delay required by specific applications. In [17], we have shown a new design of a nodal controller residing on the control plane, which communicates with by the 'control knob' of each layer and determines dynamically the corresponding values of control knobs that guarantee the best perceived video quality.

8.6.2   Background

Application Layer: We consider a N-frame video sequence $\mathcal{C} = \{g_1, \cdots, g_N\}$. We assume that the content of each frame has been divided into a foreground part (ROI) and a background part. The foreground/background packets only consist of a group of foreground/background blocks (GOB). To provide a smooth video playback, we define the single-packet delay deadline as t_k^{max} which is associated with every packet k. We assume that t_k^{max} is always known to the controller. In the quality-driven routing, the controller always checks the delay t_k^v which packet k would have experienced if k came to intermediate node v. If t_k^v exceeds t_k^{max}, packet k should not get through the node v. Each packet k is characterized by: 1) source coding parameter S_k; 2) packet delay deadline t_k^{max}; 3) packet loss rate p_k; and 4) quality impact factor λ_k. The λ_k of foreground packets is largely different from those of background packets. Thus, the expected received video distortion $E[D_k]$ for packet k can be written as [16]:

$$E[D_k] = Q_k(S_k, t_k^{max}, p_k, \lambda_k) \tag{8.25}$$

Network Layer: We model a multi-hop wireless network as a directed acyclic graph (DAG) $\mathcal{G} = (\mathcal{V}, \mathcal{E})$ (where $|\mathcal{V}| = n$), where $\mathcal{V}$ is the set of vertexes representing wireless nodes and $\mathcal{E}$ the set of edges representing wireless links. We characterize a link $(v, u) \in \mathcal{E}$ between nodes v and u with (i) $\gamma^{(v,u)}$; Signal-to-Interference-Noise-Ratio (SINR) of link (v, u), (ii) $p^{(v,u)}$; packet loss rate on the link (v, u), and (iii) $t^{(v,u)}$; packet delay on the link (v, u). The network topology is assumed to be fixed over the duration of delivering a video clip. Each node runs a certain routing protocol such as the optimal link state routing protocol (OLSR), meaning that each node can send an update packet periodically to inform all the other nodes of its current status such as queue length and immediate link quality. Based on this information, the source node performs the distortion-driven routing, which will be discussed in more details in the following sections.

MAC Layer: We assume that each link enables packet-based retransmission. Let $\Pi_k^{(v,u)}$ be the maximum number of retransmissions for packet k over link (v, u). The optimal retransmission limit is determined jointly by packet delay constraint t_k^{max} and total delay t_k^v that the packet has experienced before it reaches the head of the queue at the node v, which will be discussed in more details in the following sections.

Considering the significant impact of foreground packets on perceived quality, when both foreground and background packets are queued at node v, the foreground packet will be scheduled first for transmission. If only foreground or background packets are present in the queue, the first-come first served (FCFS) scheduling rule is adopted. The packets whose delay constraint have been violated will be dropped from the queue.

8.6.3 Problem Formulation

Let $\mathcal{P}_k$ be the optimal transmission path for packet k. We assume that the video clip $\mathcal{C}$ is compressed into packet group $\{k_1^f, \cdots, k_I^f, k_{I+1}^b, \cdots, k_{I+J}^b\}$, which is comprised of I foreground packets and J background packets. Therefore, our goal is to jointly find the optimal coding parameter S_k and the optimal path $\mathcal{P}_k$ to maximize the expected received video quality. For notation simplicity, we define the following parameter vector for each packet k:

$$\text{CT}_k = \{S_k, \mathcal{P}_k\} \tag{8.26}$$

In reality, to provide timely decoding and playback, each video frame g_n is associated with a delay deadline T_n^{max}. All the packets k belonging to g_n have the same delay deadline, i.e., $t_k^{max} = T_n^{max}$. Therefore, the problem can be written as:

$$\mathbf{CT} := \{\text{CT}_k | k = 1 \ldots I + J\} = \arg\min \sum_{k=1}^{I+J} E[D_k]$$

$$s.t. : \max\{E[t_1], \cdots, E[t_{I+J}]\} \le T_n^{max} \tag{8.27}$$

where t_k is the end-to-end delay of packet k.

8.6.4 Routing Based on Priority Queuing

The quality-driven routing is inspired by the classical Dijkstra's shortest path algorithm [44]. The particular difference is that the routing metric is dynamically calculated at the source based on the feedback network information. In the source's routing table, each node is labeled with its expected distortion incurred along the shortest path from the source to the node. The shortest path is defined as the path leading to the minimized expected video distortion.

Computation of Gradual Distortion of Single-Hop: Any node v in the network can be abstracted into the model shown in Figure 8.15. We assume that node v is connected to its A backward adjacent nodes $\{v_1^-, \cdots, v_a^-, \cdots, v_A^-\}$ via A incoming links, and connected to its B forwarding adjacent nodes $\{v_1^+, \cdots, v_b^+, \cdots, v_B^+\}$ via B outgoing links. Let $\mathcal{P}_k^v$ represent the shortest path from the source to the node v for the packet k. Assume that the path $\mathcal{P}_k^v$ passes through the node v_a^-. Let D_k^v denote the resulting expected distortion after the packet k arrives at the node v from the source along the path $\mathcal{P}_k^v$. Then, the node v is labelled with (D_k^v, v_a^-). As in Dijkstra's algorithm [44], if the label is permanent, it means the optimal path will definitely pass through the node. Next, we consider the node v as the working node [44] and illustrate how to relabel each of the next-hop nodes adjacent to v.

We consider the next-hop from node v to v_b^+. Each packet will be retransmitted until it is either received successfully or dropped owing to the expiration of packet delay deadline. To obtain the packet dropping rate on the link (v, v_b^+), we need to first calculate the expected packet waiting time on that link. Let t_k^v be the current delay incurred by packet k as soon as it arrives at node v. The maximum retransmission limit for the packet k over the link (v, v_b^+) is determined as:

$$\Pi_k^{(v,v_b^+)} = \left\lfloor \frac{R_{k,goodput}^{(v,v_b^+)}(T_n^{max} - t_k^v)}{L_k} \right\rfloor - 1 \tag{8.28}$$

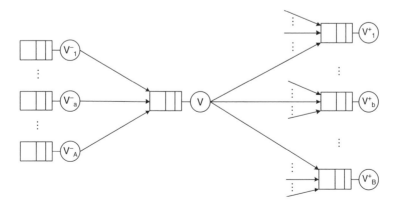

Figure 8.15 A nodal model in wireless multi-hop networks

where $\lfloor \cdot \rfloor$ is the floor operation; L_k is the packet size; and $R_{k,goodput}^{(v,v_b^+)}$ is the effective transmission rate which can be obtained by link quality and the used transmission scheme [41].

We assume that the arrival traffic in node v is a Poisson process. Let ϕ_g^v be the average arrival rate of the Poisson input traffic of queue at node v, where g is defined as:

$$
g = \begin{cases} 0, & \text{if packet } k \text{ is a foreground packet,} \\ 1, & \text{if packet } k \text{ is a background packet.} \end{cases}
$$

Let $W_g^{(v,v_b^+)}$ be the average waiting time of packet k transmitted on link (v, v_b^+). For a preemptive-priority M/G/1 queue, the priority queueing analysis gives the following result [50]:

$$
W_g^{(v,v_b^+)} = \frac{\sum_{g=0}^{1} \phi_g^v E\left[\left(Z_k^{(v,v_b^+)} \right)^2 \right]}{2\left(1 - g\phi_{g-1}^v E\left[Z_k^{(v,v_b^+)} \right] \right)\left(1 - \sum_{g=0}^{1} \phi_g^v E\left[Z_k^{(v,v_b^+)} \right] \right)}. \tag{8.29}
$$

where $E[Z_k^{(v,v_b^+)}]$ and $E[(Z_k^{(v,v_b^+)})^2]$ are the first and second moment of the packet service time, which can be calculated by formulating the packet service time as a geometric distribution [16].

We assume that the waiting time dominates the overall packet delay. Let $p_{k,drop}^{(v,v_b^+)}$ be the packet dropping rate due to the packet delay deadline expiration. Then $p_{k,drop}^{(v,v_b^+)}$ can be written as:

$$
p_{k,drop}^{(v,v_b^+)} = \text{Prob}\left(W_k^{(v,v_b^+)} > T_n^{max} - t_k^v \right). \tag{8.30}
$$

Let $P_k^{(v,v_b^+)}$ be the overall packet loss rate on link (v, v_b^+). Recall that $P_k^{(v,v_b^+)}$ consists of two parts: packet error rate $p_{k,corrupt}^{(v,v_b^+)}$ due to signal corruption and packet dropping rate $p_{k,drop}^{(v,v_b^+)}$ due to packet delay deadline expiration. Then, $P_k^{(v,v_b^+)}$ can be expressed as:

$$
P_k^{(v,v_b^+)} = p_{k,drop}^{(v,v_b^+)} + \left(1 - p_{k,drop}^{(v,v_b^+)} \right) p_{k,corrupt}^{(v,v_b^+)} \tag{8.31}
$$

Therefore, the expected distortion after packet k passes through the shortest path between the source and the node v and reaches node v_b^+ is

$$
D_k^{v_b^+} = D_k^v + \underbrace{\prod_{(x,y)\in\mathcal{P}_k^v} \left(1 - p_k^{(x,y)} \right) P_k^{(v,v_b^+)} E[D_k]}_{\Delta_k^{v_b^+}} \tag{8.32}
$$

where $E[D_k] = Q_k(S_k, t_k^{max}, p_k^{(v,v_b^+)}, \lambda_k)$; and $\Delta_k^{v_b^+}$ is the gradual distortion increase after packet k advances from node v to node v_b^+.

The Integrated Proposed Routing: Recall that $D_k^{(v,v_b^+)}$ is the distortion incurred after packet k passes through $\mathcal{P}_k^v$ and link (v, v_b^+) and reaches node v_b^+. Like Dijkstra's shortest path algorithm, we need to check whether $D_k^{(v,v_b^+)}$ is less than the existing label on node v_b^+. If it is, the node is relabeled by $D_k^{(v,v_b^+)}$. After all the forward nodes $\{v_1^+, \cdots, v_b^+, \cdots, v_B^+\}$ of the working node v have been inspected and the tentative labels [44] changed if possible, the entire graph is searched for the tentatively-labeled node with the smallest distortion value. This node is made permanent and becomes the working node. The distortion calculation is repeated, similar to the case of v being the working node as shown above.

As the algorithm proceeds and different nodes become working nodes, an optimal path will be obtained over which packet k is transmitted to the destination with the minimum incurred distortion. Different from the fixed weight value of each hop in the classical Dijkstra's shortest path routing, the calculation of the weight for the current hop in the proposed routing algorithm depends on the shortest sub-path that one packet has traversed before the current hop.

The running time of the above routing algorithm on a graph with edges $\mathcal{E}$ and vertices $\mathcal{V}$ can be expressed as a function of $|\mathcal{E}|$ and $|\mathcal{V}|$ using the Big-O notation. The simplest implementation of the above algorithm stores vertices of set Δ in an ordinary linked list or array, and operation Extract-Min(Δ) is simply a linear search through all vertices in Δ. In this case, the running time is $O(|\mathcal{V}|^2 + |\mathcal{E}|) = O(|\mathcal{V}|^2)$.

8.6.5 Problem Solution

We observe that, for packet k in (8.27), any selected parameter $CT_k = \{S_k, \mathcal{P}_k\}$ which results in a single-packet delay being larger than T_n^{max} cannot be chosen as part of the optimal parameter vector $CT_k^* = \{S_k^*, \mathcal{P}_k^*\}$. Therefore, to solve (8.27), we can make use of this fact by redefining the distortion as follows:

$$E[D_k'] = \begin{cases} \infty : & E[t_k] > T_n^{max} \\ E[D_k] : & E[t_k] \leq T_n^{max} \end{cases} \tag{8.33}$$

In other words, the average distortion for a given packet with a delay larger than the maximum permissible delay is set to infinity. This means that, given that a feasible solution exists, the parameter vector which minimizes the average total distortion, as defined in (8.27), will not result in any packet delay greater than T_n^{max}. That is to say, the minimum distortion problem, which is a constrained optimization problem, can be relaxed into an unconstrained dual optimization problem using the redefinition above of the single-packet delay.

Most decoder concealment strategies introduce dependencies between packets. Without losing the generality, we assume that owing to the concealment strategy, the current packet will depend on its previous j packets ($j \geq 0$). To solve the optimization problem, we define a cost function $G_k(CT_{k-j}, \cdots, CT_k)$, which represents the minimum average distortion up to and including the kth packet, given that $CT_{k-j}, \cdots, CT_k$ are the decision vectors of packets $\{(k-j), \cdots, k\}$ packets. For simplicity, let $U := I + J$ be the total packet number of the video clip. Therefore, $G_U(CT_{U-j}, \ldots, CT_U)$ represents the minimum total distortion for all the packets of the video clip. Clearly, to solve (8.27) is equivalent to solving:

$$\underset{CT_{U-j},\cdots,CT_U}{\text{minimize}} \; G_U(CT_{U-j}, \ldots, CT_U). \tag{8.34}$$

The key observation for deriving an efficient algorithm is the fact that given $j + 1$ decision vectors $CT_{k-j-1}, \cdots, CT_{k-1}$ for the $(k - j - 1)$st to $(k - 1)$st packets, and the cost function $G_{k-1}(CT_{k-j-1}, \cdots, CT_{k-1})$, the selection of the next decision vector CT_k is independent of the selection of the previous decision vectors $CT_1, CT_2, \cdots, CT_{k-j-2}$. This means that the cost function can be expressed recursively as:

$$G_k\left(CT_{k-j}, \cdots, CT_k\right) = \underset{CT_{k-j-1},\cdots,CT_{k-1}}{\text{minimize}} \left\{ G_{k-1}\left(CT_{k-j-1}, \cdots, CT_{k-1}\right) + E[D_k] \right\}.$$
$$\tag{8.35}$$

The recursive representation of the cost function above makes the future step of the optimization process independent from its past step, which is the foundation of dynamic programming. The problem can be converted into a graph theory problem of finding the shortest path in a directed acyclic graph (DAG) [39]. Let $|S|$ be the number of possible values of S_k, then computational complexity of the algorithm is $O(U \times |S|^{j+1})$, which depends directly on the value of j. For most cases, j is a small number, so the algorithm is much more efficient than an exhaustive search algorithm with exponential computational complexity.

8.6.6 Performance Evaluation

Experiments are designed using H.264/AVC JM 12.2 for the video clip called "Mother and Daughter". Before video encoding, the identification of the ROI's which commonly emerge as moving foregrounds, is performed by six stages: a) Background Subtraction, b) Split-and-Merge Segmentation, c) Region Growing, d) Morphological Operations, e) Geometric Correction and f) Contour Extraction. The results of the different stages of the operation above are shown in Figure 8.16.

Recall that the foreground packets are first scheduled when both the foreground and background packets are present simultaneously in the queue of a node. To compare the different quality of different received content, the peak signal-to-noise ratio (PSNR) for the foreground (region of interest), the background and the holistic video sequence case (corresponding to the case of no identification of regions of interest (NO-IRI)) are calculated separately. Figure 8.17 shows the PSNR performance comparison for the proposed framework with different single-packet delay deadlines. We consider three different deadline values ($5ms$, $10ms$, $15ms$). The ROI has a larger comparative PSNR improvement over NO-IRI and the background in the case of $5ms$ than the other two cases: $10ms$ and $15ms$. In other words, the more stringent the single-packet delay, the more comparative PSNR improvement of the ROI the proposed framework can offer. This shows that our

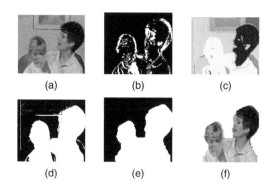

Figure 8.16   The identification of ROI's

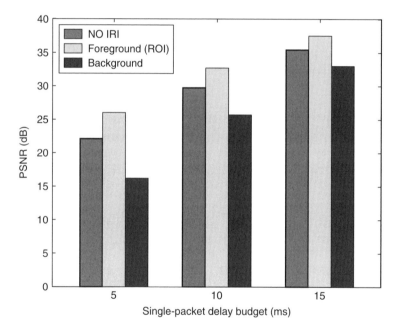

Figure 8.17   PSNR vs. different single-packet delay deadlines/budgets

cross-layer framework is extremely applicable to the content-aware video communications over delay-stringent or rate-limited wireless multi-hop networks.

To show the received video quality at the destination, we plot the 61th frame of the "Mother and Daughter" sequence obtained in different scenarios. From Figure 8.18(b) and 8.18(c), we can see that when the single-packet delay deadline becomes smaller, the quality of ROI does not deteriorate as much as the quality of the background. From Figure 8.18(c) and 8.18(d), we can see that with the same single-packet delay deadline, the ROI in the cross-layer framework has a better quality than in the case of NO-IRI.

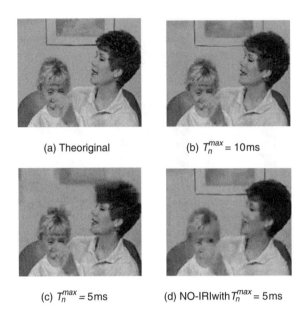

(a) Theoriginal   (b) $T_n^{max} = 10$ms

(c) $T_n^{max} = 5$ms (d) NO-IRIwith $T_n^{max} = 5$ms

Figure 8.18 Video frame quality

8.7 Cross-Layer Optimization for Video Summary Transmission

8.7.1 Background

In recent years, universal multimedia access (UMA) [51–53] has emerged as one of the most important components for the next generation of multimedia applications. The basic idea of UMA is universal or seamless access to multimedia content by automatic selection or adaptation of content following user interaction. As mobile phones have grown in popularity and capability, people have become enthusiastic about watching multimedia content using mobile devices and personalizing the content, for example, summarizing the video for real-time retrieval or for easy transmission. In general, the video summarization algorithm will generate a still-image storyboard, called a video summary, which is composed of a collection of salient images extracted from the underlying video sequence (as shown in Figure 8.19). Therefore, video summary is a special format of the video clip whose correlation between frames is not as high as with normal clips, but where losing consecutive or continuous summary frames might cause severe damage to an understanding of the summary content. The summary could be generated automatically or selected by user's interaction.

Although a great deal of work on video summarization can be found in the literature [54–56], the issue of transmission of video summary has gained little attention. [57] extends the work of [56] into the wireless video streaming domain, however packet loss factor owing to unsatisfactory wireless channel conditions has not been considered in the framework. In [58], packet loss is considered in video summary transmission, and the key frames that minimize the expected end-to-end distortion are selected as the summary frames. However, source coding has not been optimized in the optimization framework,

Figure 8.19   An example of a 10-frame video summary of the Stefan sequence

which might impact directly on the perceived quality of the results. In addition, the algorithm does not guarantee a good content coverage of the selected frames because the potential packet loss penalty biases heavily the selection process.

In wireless networks, packet loss is due mainly to the fading effect of time-varying wireless channels. Adaptive modulation and coding (AMC) has been studied extensively and advocated at the physical layer, in order to match transmission rates to time-varying channel conditions. For example, in order to achieve high reliability at the physical layer, one has to reduce the transmission rate using either small size constellations, or powerful but low-rate error-correcting codes [59–61]. An alternative way to decrease packet loss rate is to rely on the automatic repeat request (ARQ) protocol at the data link layer, which requests retransmissions for those packets received in error. Obviously, by allowing for a very large retransmission number, ARQ can guarantee a very low packet loss rate. However, in order to minimize delays and buffer sizes in practice, truncated ARQ protocols have been widely adopted so as to limit the maximum number of transmissions [62]. To sum up, the link adaptation technique formed by AMC and ARQ provides greater flexibility in delivering the summary frames.

In this section, within an quality-driven framework, we focus our study on the cross-layer optimization of the video summary transmission over lossy networks. We assume that a video summarization algorithm that can select frames based on some optimality criteria is available in the system. Therefore, a cross-layer approach is proposed optimize to jointly the AMC parameters at the physical layer, the ARQ parameters at the data link layer and the source coding parameters at the application layer in order to achieve the best quality of reconstructed video clips from received video summaries. Clearly, owing to the spectacular characteristics of video summary data, the general cross-layer optimization schemes proposed recently for normal video sequences in [63] and [64] do not cover automatically summary data transmission. As an example, the neighboring summary frames have typically less correlation in order to cover the content variation of the video clip, and thus the normal temporal-based error concealment algorithm considered in [63] and [64] would not be efficient in the current scenario. In [65], a cross-layer multi-objective optimized scheduler for video streaming over the 1xEV-DO system is presented. With the usage of decodability and semantic importance feedback from the application layer to the scheduler, [65] focuses on determining the best allocation of channel resources (time slots) across users. However, the joint optimization of source coding and transmission parameters has not been considered in the framework.

Clearly, to achieve the best delivered video quality while maintaining a good content coverage by providing tunable parameters so as to avoid consecutive summary frames being lost simultaneously is not a trivial task. The tradeoff among the selected parameters in these layers is mixed; for example, to maintain a reasonable delay, the source coding might choose a coarser parameter, or AMC chooses a larger size constellation or a higher rate FEC channel code, which will increase the vulnerability of coded frames and will result in unacceptable video quality. However, if source and channel coding use more bits, with the increase of packet length, the probability of packet loss increases, and ARQ might have to increase the number of retransmission trials in order to reduce the problem of quality owing to packet loss, which will then result in excessive delay. Therefore, the cross-layer optimization approach is a natural solution to for improving overall system performance.

8.7.2 Problem Formulation

We study a cross-layer framework that optimizes parameter selection in AMC at the physical layer, ARQ at the data link layer and source coding at the application layer in order to achieve the best quality of reconstructed video clip from the received video summary.

The following notation will be used. Let us denote by n the number of frames of a video clip $\{f_0, f_1, \ldots, f_{n-1}\}$, and m the number of frames of its video summary $\{g_0, g_1, \ldots, g_{m-1}\}$. Let S_i and B_i be the coding parameters and the resultant consumed bits of the ith ($i = 0, 1, \ldots, m - 1$) video summary frame in lossy source coding. The summarization with different coding parameters will produce summary frames with different frame lengths. Large size frames will be fragmented into multiple packets for transmission at lower layers. Let Q_i denote the number of fragmented packets of the ith summary frame. Let $N_{i,q}$ and $F_{i,q}$ be the number of transmissions and the packet size for the qth packet of the ith summary frame, respectively. To improve channel utilization, AMC is designed to update the transmission mode for every transmission and retransmission of each packet. Let $R_{i,q,n}(A_{i,q,n}, C_{i,q,n})$ be the rate (bits/symbol) of AMC mode used at the nth transmission attempt when transmitting the qth packet of the ith summary frame, where $A_{i,q,n}$ and $C_{i,q,n}$ are the corresponding modulation order and coding rate. We assume that the transmission rate of the physical layer channel is fixed, denoted by r (symbols/second). Clearly, the delay in transmitting the whole summary can be expressed by:

$$T = \sum_{i=0}^{m-1} \sum_{q=1}^{Q_i} \sum_{n=1}^{N_{i,q}} \left[\frac{F_{i,q}(S_i, B_i)}{R_{i,q,n}(A_{i,q,n}, C_{i,q,n}) * r} + T_{\text{RTT}} \right] \qquad (8.36)$$

where T_{RTT} is the maximum allowed RTT to get the acknowledgement packet via the feedback channel before a retransmission trial.

Let $\rho_i(S_i, B_i, N_{\max}, A_{i,q,n}, C_{i,q,n}, \gamma_{i,q,n})$ be the loss probability of the ith summary frame, where $N_{\max}$ is the maximum transmission number for one packet and $\gamma_{i,q,n}$ is the instantaneous channel SNR. Note that any summary frame may possibly get lost during video transmission. However, in order to simplify the problem of formulation, we assume that the first summary frame would be guaranteed to be received. Then the expected distortion of the video clip can be calculated by:

$$E[D] = \sum_{k=0}^{n-1} E[D(f_k, \tilde{f}_k)]$$

$$= \sum_{i=0}^{m-1} \sum_{j=l_i}^{l_{i+1}-1} \sum_{b=0}^{i} \left\{ (1 - \rho_{i-b})d[f_j, \tilde{g}_{i-b}(S_{i-b})] \cdot \prod_{a=0}^{b-1} \rho_{i-a} \right\} \qquad (8.37)$$

where $\tilde{f}_k$ is the reconstructed kth frame from the received summary at the receiver side, $\tilde{g}_k(S_k)$ is the reconstructed kth summary frame, l_i is the index of the summary frame g_i in the video clip and function $d()$ is the distortion between two frames. We use the mean squared error (MSE) between the two frames as the metric for calculating the distortion. The same distortion measure for the video summary result has been used in [56–58].

The problem at hand can be formulated as:

$$\text{Min } E[D], \quad \text{s.t.}: \quad T \leq T_{\max} \qquad (8.38)$$

where $T_{\max}$ is a given delay budget for delivering the whole video clip.

In this work, we consider the content coverage issue of the received summary. In other words, if a chunk of continuous summary frames is lost owing to channel error, then the coverage of the received summary for the original clip would be degraded significantly. To prevent such a problem but still keep the problem as general as possible, we define L so that the case of L or more than L consecutive summary frames being lost will never happen. For instance, if $L = 2$ then no neighboring summary frames can be lost together during transmission. So that the distortion can be calculated by:

$$E[D] = \sum_{k=0}^{n-1} E[D(f_k, \tilde{f}_k)]$$

$$= \sum_{i=0}^{m-1} \sum_{j=l_i}^{l_{i+1}-1} \sum_{b=0}^{\min(i,L-1)} \left\{ (1 - \rho_{i-b})d[f_j, \tilde{g}_{i-b}(S_{i-b})] \cdot \prod_{a=0}^{b-1} \rho_{i-a} \right\} \qquad (8.39)$$

It is important to realize that the value of L is a constant programmable by the system, and the introduction of L does not narrow down the original problem. As you may notice, when we set $L = m$, equation (8.39) is equal to equation (8.37).

If we set $L = 2$ in (8.39), it is very clear that for the ith summary frame, there are only two possibilities: either it is received or it is lost but its previous summary frame is received. Let us denote by G_i the chance of the ith summary frame being not lost, so $G_i = 1$ means it is guaranteed to be received, otherwise it is not guaranteed. Based on the constraint, we need $\max(G_i, G_{i-1}, \ldots, G_{i+1-L}) = 1$ for all $i \in [0, m-1]$. G_i can be guaranteed and derived by link adaptation, which will be discussed in a later section.

For delay issue, we hope that the total delay T of delivering all summary frames satisfies $T \leq T_{max}$. Therefore, the problem is:

$$\mathrm{Min}\,E[D], \quad \mathrm{s.t.:} \quad T \leq T_{max}, \quad \mathrm{and}$$

$$\max(G_i, G_{i-1}, \ldots, G_{i+1-L}) = 1, \quad i \in [0, m-1] \tag{8.40}$$

Once we work out problem (8.40), the optimal parameter combinations, i.e. source coding parameter S_i, AMC modulation order $A_{i,q,n}$ and channel coding rate $C_{i,q,n}$, and ARQ transmission number N_{max} are obtained to transmit the ith summary frame, which minimizes the whole clip distortion and satisfies certain predefined delay constraints.

8.7.3 System Model

The system model of the cross-layer framework is shown in Figure 8.20, which consists of a three-layer structure and a controller. At the application layer, summarization is performed on the target video clip and large size summary frames are fragmented into multiple packets for transmission at lower layers. At the data link layer, the ARQ protocol is adopted. If an error is detected in a packet, a retransmission request is generated by the receiver and is sent to the transmitter via a feedback channel. The transmitter arranges retransmission of the requested packet. If a packet is not received correctly after N_{max} transmission attempts, we will declare packet loss, then the summary frame to which the lost packet belongs is also regarded as lost. At the physical layer, we assume that multiple transmission modes are available as shown in Table 8.4, with each mode consisting of

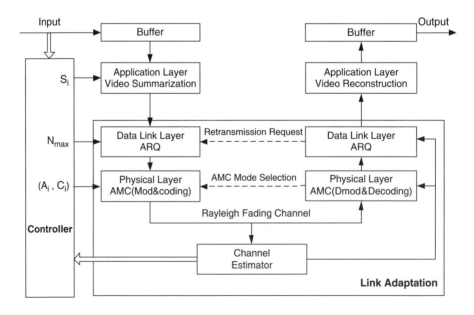

Figure 8.20 The system model of cross-layer optimization for video summary transmission

Table 8.4 AMC Modes at the Physical Layer

	Mode1	Mode2	Mode3	Mode4	Mode5	Mode6
Modulation	BPSK	QPSK	QPSK	16-QAM	16-QAM	64-QAM
Coding Rate C_m	1/2	1/2	3/4	9/16	3/4	3/4
R_m (bits/sym.)	0.50	1.00	1.50	2.25	3.00	4.50
a_m	1.1369	0.3351	0.2197	0.2081	0.1936	0.1887
b_m	7.5556	3.2543	1.5244	0.6250	0.3484	0.0871

a specific modulation and FEC code pair as in the 3GPP, HIPERLAN/2, IEEE 802.11a, and IEEE 802.16 standards [66–68]. Based on channel state information (CSI) from the channel estimator, the transmitter updates the AMC mode for the next packet transmission. Coherent demodulation and maximum-likelihood (ML) decoding are used at the receiver. The decoded bit streams are mapped to packets, which are pushed upwards to the data link layer. If all fragmented packets of the summary frame g_i are delivered correctly, summary frame g_i is saved into the buffer, and video clip frames f_{l_i} through $f_{l_{i+1}-1}$ are reconstructed with g_i. If g_i does not reach the receiver after some fixed time, the video clip frames f_{l_i} through $f_{l_{i+1}-1}$ will be reconstructed with the previously received summary frame. Obviously, the receiver only need a buffer that can contain one summary frame, i.e. the latest received summary frame.

From description above, it is obvious that AMC combined with ARQ performs a link adaptation in a joint approach. For a fixed video summary, say $\{g_0, g_1, \ldots, g_{m-1}\}$, the link adaptation can guarantee the constraint $\max(G_i, G_{i-1}, \ldots, G_{i+1-L}) = 1$ in problem (8.40), and produce the summary frame error rate (FER) ρ_i and transmission time T_i for each summary frame g_i. The detailed link adaptation and close-form expressions for (ρ_i, T_i) will be clarified in section 8.7.4.

The controller is the most important part of the system, which is equipped with all possible values of the key parameters of each layer. These parameters include the coding parameter S at the application layer, the allowed maximum transmission number $N_{\max}$ at the data link layer, and the available AMC modes with modulation order and FEC code rate pair (A, C). Note that here S, A, and C are parameter allocation vectors for $m - 1$ summary frames, for example, $S = \{S_1, S_2, \ldots, S_{m-1}\}$.

The following is a brief list of the performance flows of the cross-layer framework.

- When there is a video clip to transmit, based on the current average SNR $\bar{\gamma}$ from the channel estimator, from all possible values of parameter set $\{S, N_{\max}, A, C\}$, the controller first calculates all possible theoretical values of pair (ρ_i, T_i) for all possible summary frames by using the close-form expressions of link adaptation performance with the constraint $\max(G_i, G_{i-1}, \ldots, G_{i+1-L}) = 1$, $i \in [0, m-1]$.
- With the total delay budget $T_{\max}$, the controller use all possible (ρ_i, T_i)s for the whole summary to solve problem (8.40). The group values of $\{S, N_{\max}, A, C\}$ corresponding to the optimal solution of problem (8.40) are the optimal parameters to transmit the whole video summary.
- The obtained optimal parameters $\{S, N_{\max}, A, C\}$ are assigned to the corresponding layers, then the whole video summary is sent out frame by frame.
- Corresponding video clip frames are reconstructed with the newly received summary frame.

We next list the operating assumptions adopted in this work.

- The channel is frequency-flat, meaning that it remains time invariant during a packet but varies from packet to packet. Thus, AMC is adjusted on a packet-by-packet basis. In other words, AMC scheme is updated for every transmission and retransmission attempt. The channel quality is captured by a single parameter, namely the received SNR γ. we adopt Rayleigh channel model to describe γ statistically. The received SNR γ per packet is thus a random variable with a probability density function (pdf):

$$p_\gamma(\gamma) = \frac{1}{\overline{\gamma}} \exp\left(-\frac{\gamma}{\overline{\gamma}}\right) \tag{8.41}$$

where $\overline{\gamma} := E\{\gamma\}$ is the average received SNR.
- Perfect channel state information (CSI) is available at the receiver. The corresponding mode selection is fed back to the transmitter without error and latency. This assumption could be at least satisfied approximately by using a fast feedback channel with powerful error control information as adopted in IEEE 802.16 [68].
- Error detection based on CRC is perfect, provided that sufficiently reliable error detection CRC codes are used.

8.7.4 Link Adaptation for Good Content Coverage

In this section, we explain how link adaptation can guarantee constraint $\max(G_i, G_{i-1}, \ldots, G_{i+1-L}) = 1$, and derive the close-form expression of (ρ_i, T_i).

Actually, it is impossible to guarantee strictly constraint $\max(G_i, G_{i-1}, \ldots, G_{i+1-L}) = 1$, owing to the fading characteristics of wireless channels. Let P_L be the probability that L consecutive summary frames are lost simultaneously. We assume P_L to be a very small value, say 10^{-2}, to approximate constraint $\max(G_i, G_{i-1}, \ldots, G_{i+1-L}) = 1$ is satisfied. Then the goal of link adaptation becomes to guarantee P_L with the least total transmission delay.

Since the processing unit of link adaptation is packet, we need to transform P_L into target packet error rate P_{target} of lower layers. Let L_s, L_f and L_a be summary frame size, fragmentation packet size and the actual packet length of link adaptation, respectively, According to different summary frame sizes, there are two possible cases:

- The summary frame size is smaller than the fragmentation packet size. Since there is no need to do fragmentation, we have $P_{\text{target}} = P_L^{1/L}$ and $L_a = L_s$.
- The summary frame size is larger than the fragmentation packet size, where fragmentation is necessary. A summary frame of length L_s will be fragmented into $N_p = \lceil L_s/L_f \rceil$ packets. $\lceil \rceil$ is the smallest integer greater than or equal to a given real number. The actual packet size L_a of the first $\lceil L_s/L_f \rceil - 1$ packets equals to L_f, and the actual packet size L_a' of the final packet is $L_s - \lfloor L_s/L_f \rfloor \cdot L_f$. The target PER should be as follows:

$$P_{\text{target}} = 1 - (1 - P_L^{1/L})^{1/N_p} \tag{8.42}$$

In both of the cases above, P_{target} can be regarded as the required PER at the data link layer. Next we explain how to guarantee P_{target} with transmission packet size L_a by AMC and ARQ. Let us define a PER upper bound P_{AMC} such that the instantaneous PER is guaranteed to be no greater than P_{AMC} for each chosen AMC mode at the physical layer. Then the PER at the data link layer after N_{max} transmissions is no larger than $P_{AMC}^{N_{max}}$. To satisfy P_{target}, we need to impose:

$$P_{AMC}^{N_{max}} = P_{target}, \ i.e., \ P_{AMC} = P_{target}^{1/N_{max}} \tag{8.43}$$

We assume each bit inside the packet has the same bit error rate (BER) and bit-errors are uncorrelated, the PER can be related to the BER through:

$$PER = 1 - (1 - BER)^{L_a} \tag{8.44}$$

for a packet containing L_a bits. For any AMC mode, to guarantee the upper bound P_{AMC}, the required BER to achieve is:

$$BER_{AMC} = 1 - (1 - P_{AMC})^{1/L_a} \tag{8.45}$$

Since exact closed-form BERs for the AMC modes in Table 1 are not available, to simplify the AMC design, we adopt the following approximate BER expression:

$$BER_m(\gamma) = a_m exp(-b_m \gamma) \tag{8.46}$$

where m is the mode index and γ is the received SNR. Parameters a_m and b_m are obtained by fitting (8.46) to the exact BER. To guarantee P_{AMC} with the least delay when transmitting a packet, we set the mode switching threshold γ_m for the AMC mode m to be the minimum SNR required to achieve BER_{AMC}. By (8.46) γ_m can be expressed as:

$$\gamma_m = \frac{1}{b_m} \ln \left(\frac{a_m}{BER_{AMC}} \right), \quad m = 1, 2, \cdots, M$$

$$\gamma_{M+1} = +\infty \tag{8.47}$$

where M is the total number of AMC modes available ($M = 6$).

Since the instantaneous PER is upper-bounded by P_{AMC} in our AMC design, the average PER at the physical layer will be lower than P_{AMC}. Taking expectations over channel realizations, the average PER at the physical layer is:

$$\overline{P} = \frac{1}{P_T} \sum_{m=1}^{M} \int_{\gamma_m}^{\gamma_{m+1}} PER_m(\gamma) p_\gamma(\gamma) d\gamma$$

$$= \frac{1}{P_T} \sum_{m=1}^{M} \int_{\gamma_m}^{\gamma_{m+1}} [1 - (1 - a_m exp(-b_m \gamma))^{L_a}] \cdot p_\gamma(\gamma) d\gamma \tag{8.48}$$

where $P_T = \int_{\gamma_1}^{+\infty} p_\gamma(\gamma) d\gamma$ is the probability that channel has no deep fades and at least one AMC mode can be adopted. Similarly, the average delay for one transmission attempt at the physical layer can be expressed as:

$$\overline{T} = \frac{1}{P_T} \sum_{m=1}^{M} \int_{\gamma_m}^{\gamma_{m+1}} \left(\frac{L_a}{R_m \cdot r} + T_{\text{RTT}} \right) p_\gamma(\gamma) d\gamma \tag{8.49}$$

Then the average number of transmission attempts per packet can be found as [69]:

$$\overline{N} = 1 + \overline{P} + \overline{P}^2 + \cdots + \overline{P}^{N_{\max}-1}$$

$$= \frac{1 - \overline{P}^{N_{\max}}}{1 - \overline{P}} \tag{8.50}$$

Then the actual PER at the data link layer is:

$$P_{\text{actual}} = \overline{P}^{N_{\max}} \tag{8.51}$$

and the actual transmission delay for each packet at the data link layer is:

$$T_{\text{actual}} = \overline{T} \cdot \overline{N} \tag{8.52}$$

When to calculate the actual FER ρ_i and the actual delay T_i for transmitting the ith summary frame, two cases should be considered as mentioned above:

- If the summary frame size L_s is smaller than the fragmentation packet size L_f, we adopt $L_a = L_s$ to compute $\overline{P}(L_a)$ and $\overline{T}(L_a)$ and we can have $\rho_i = P_{\text{actual}}(L_a)$ and $T_i = T_{\text{actual}}(L_a)$ with [48–50].
- If the summary frame size L_s is larger than the fragmentation packet size L_f, we adopt $L_a = L_f$ and $L_a' = L_s - \lfloor L_s/L_f \rfloor \cdot L_f$ to compute $P_{\text{actual}}(L_a)$, $T_{\text{actual}}(L_a)$, $P_{\text{actual}}'(L_a')$ and $T_{\text{actual}}'(L_a')$. Then we can have:

$$\rho_i = 1 - (1 - P_{\text{actual}})^{N_p - 1} \cdot (1 - P_{\text{actual}}') \tag{8.53}$$

$$T_i = (N_p - 1) \cdot T_{\text{actual}} + T_{\text{actual}}' \tag{8.54}$$

The closed-form expressions above of ρ_i and T_i will be used by the controller to calculate all possible (ρ_i, T_i) to solve problem (8.40), which we will discuss in detail in section VII-E.

8.7.5 Problem Solution

Optimal Solution: Since problem (8.40) is a constrained minimization problem, it can be solved by Lagrangian relaxation. So the problem can be converted into:

$$\text{Min}\{E[D] + \lambda T\}, \quad \text{s.t. :}$$

$$\max(G_i, G_{i-1}, \ldots, G_{i+1-L}) = 1, i \in [0, m-1] \tag{8.55}$$

The target to be minimized can be derived as the following Lagrangian cost function:

$$J_\lambda = E[D] + \lambda T$$

$$= \sum_{i=0}^{m-1} \sum_{j=l_i}^{l_{i+1}-1} \sum_{b=0}^{\min(i,L-1)} \left\{ (1 - \rho_{i-b}) d[f_j, \widetilde{g}_{i-b}(S_{i-b})] \cdot \prod_{a=0}^{b-1} \rho_{i-a} + \lambda T_i \right\} \qquad (8.56)$$

Let us define a cost function $H_i(u_i)$ to represent the sum of distortion and delay for up to ith summary frame, where u_i represents the parameter vector $\{S_i, N_{\max}, A_{i,q,n}, C_{i,q,n}, \gamma_{i,q,n}\}$. Clearly it can be observed that:

$$H_i(u_i) = H_{i-1}(u_{i-1}) + \sum_{j=l_i}^{l_{i+1}-1} \sum_{b=0}^{\min(i,L-1)} \left\{ (1 - \rho_{i-b}) \right.$$

$$\left. \cdot d[f_j, \widetilde{g}_{i-b}(S_{i-b})] \prod_{a=0}^{b-1} \rho_{i-a} + \lambda T_i \right\} \qquad (8.57)$$

which means the process of choosing u_i for the ith summary frame is independent of $\{u_0, u_1, \ldots, u_{i-2}\}$, the parameters selected for the first $i-1$ summary frames. This is the fundamental of dynamic programming (DP). So the optimal solution can be found by a shortest path algorithm.

As a toy example, we assume there are three summary frames $\{g_0, g_1, g_2\}$ to be sent and assume $L = 2$. In addition, we suppose for each summary frame, there are k different source coding options. Then the path graph will be like Figure 8.21. In this figure, each node u_i^a corresponds to a cost value $H(u_i^a)$. The weight $h(u_i^a u_{i+1}^b)$ on each branch from node u_i^a to u_{i+1}^b corresponds the incremental cost value when transmitting the $(i+1)$th summary frame with the bth source coding option. $h(u_i^a u_{i+1}^b)$ can be computed by the second term of the right hand side of (8.57).

As discussed before, the solution to problem (8.40) is to minimize the average distortion D for a total delay budget $T_{\max}$ in transmitting a whole video summary. With a path graph

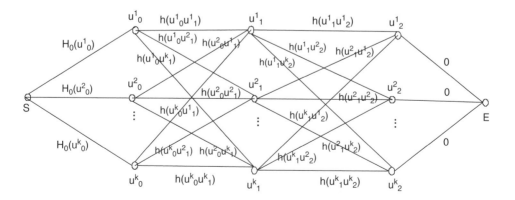

Figure 8.21 Path graph of a 3-frame toy video summary transmission

like above, the main function of the controller is to find the shortest path in the graph with the forward DP. The obtained shortest path has the minimal distortion D, and at the same time indicates the optimal choice of parameters $\{S_i, N_{max}, A_{i,q,n}C_{i,q,n}\}$ for source coding and transmitting the ith summary frame.

For J_λ in [56], it has been shown that if there is a λ^* such that:

$$\{S^*, N^*_{max}, A^*, C^*\} = \arg \quad \min J_{\lambda^*}(S, N_{max}, A, C) \qquad (8.58)$$

leads to $T(S, N_{max}, A, C) = T_{max}$, then $\{S^*, N^*_{max}, A^*, C^*\}$ is also an optimal solution to (8.40) [70]. It is well known that when λ sweeps from zero to infinity, the solution to problem (8.58) traces out the convex hull of the distortion delay curve, which is a non-increasing function. Hence λ^* can be obtained via a fast convex recursion in λ using the bisection algorithm.

Next we list the algorithm to find λ^*.

- Step 1: We judiciously choose two values of λ, λ_l and λ_u with $\lambda_l \le \lambda_u$ which satisfy the relation:

$$\sum_i T^*_i(\lambda_u) \le T_{max} \le \sum_i T^*_i(\lambda_l) \qquad (8.59)$$

where $\sum_i T^*_i(\lambda)$ is the total delay corresponding to the shortest path found by forward DP. A conservative choice for a solvable problem would be $\lambda_l = 0$ and $\lambda_u = \infty$;

- Step 2: $\lambda_{next} \longleftarrow \frac{\lambda_l + \lambda_u}{2}$.
- Step 3: Perform forward DP through the path graph for λ_{next};
  $\Longrightarrow$ if $\left\{\sum_i T^*_i(\lambda_{next}) = \sum_i T^*_i(\lambda_u)\right\}$, then stop, $\lambda^* = \lambda_u$;
  $\Longrightarrow$ else if $\left(\sum_i T^*_i(\lambda_{next})\rangle T_{max}\right)$, $\lambda_l \longleftarrow \lambda_{next}$, Go to Step 2,
  $\Longrightarrow$ else $\lambda_u \longleftarrow \lambda_{next}$, Go to step 2.

Thus $\sum_i T^*_i(\lambda)$ is made successively closer to T_{max} and finally we obtain the expected λ^*. With λ^*, we perform DP for the last time and obtain the optimal shortest path. The values of $\{S_i, N_{max}, A_{i,q,n}C_{i,q,n}\}$ corresponding to the shortest path are just the optimal parameter values for source coding and transmitting the ith summary frame.

Implementation Considerations: From the analysis above, we can say that problem (8.40) is converted into a graph theoretic problem of finding the shortest path in a directed acyclic graph (DAG) [39]. The computational complexity of the algorithm above is $O(N \times |U|^L)$, with $|U|$ denoting the cardinality of U, which depends on the number of the optional values of parameters $\{S, N_{max}, A, C\}$, but is still much more efficient than the exponential computational complexity of an exhaustive search algorithm. Clearly for cases with smaller L, the complexity is quite practical to perform the optimization. On the other hand, for larger L, the complexity can be limited by reducing the cardinality of U. The practical solution would be an engineering decision and trade off between the computational capability and optimality of the solution. For a storage issue, it is important to emphasize that the problem formulation and the proposed solution are quite generic and flexible for devices with various storage and computational capabilities. For the transmitter with a buffer size that only allows to store some portion of the video clip, the clip has to be divided into a number of segments and problem (8.5) is solved for each segment. In such cases, although the solution is not fully optimal for the video clip, the optimization would still bring sufficient gains compared to those without optimization.

8.7.6 Performance Evaluation

In this section, experiments are designed using H.264/AVC JM 10.2 for the video clip called "Glasgow", which is a typical test clip. For comparison, we summarize the first 300 frames into 30 and 60 summary frames, respectively. The case with more summary frames means a higher sampling rate thus less distortion. To simplify the problem, we compress the summary by choosing different QP (quantization step size), and we consider 10 possible QPs (5, 10, 15, 20, 25, 30, 35, 40, 45, 50). According to each QP, the frames have different rates and distortion values. The video summary is coded with an intra-coding mode for each summary frame due to the lesser correlation between neighboring frames. In addition, without loss of generality, we consider the case of $L = 2$, in other words, we impose the constraint $\max(G_i, G_{i-1}) = 1\tilde{}(i \in [0, m-1])$ which needs to be guaranteed by the link adaptation. Besides of the parameters to be optimized, we assume a fixed channel transmission rate $r = 6 * 10^6$ symbols/second and a fixed round trip time $T_{RTT} = 100$ milliseconds in our experiment.

Figure 8.22 and Figure 8.23 are comparisons between QP adaptation and No QP adaptation. In both figures, the average channel SNR $\overline{\gamma}$ is 25dB and we fix $P_L = 10^{-2}$ and $N_{max} = 3$. P_L is the target probability that $L = 2$ consecutive frames are being lost simultaneously, which should be small enough to approximately satisfy the constraint $\max(G_i, G_{i-1}) = 1$. In Figure 8.22, where the total summary frame number is 30, the

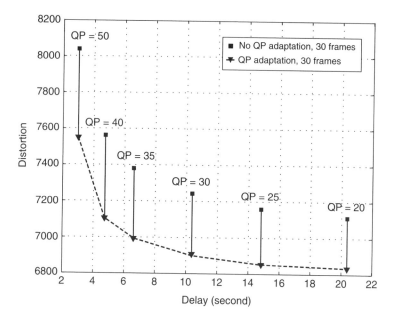

Figure 8.22 Distortion vs. delay comparison between QP adaptation and no QP adaptation with 30 summary frames

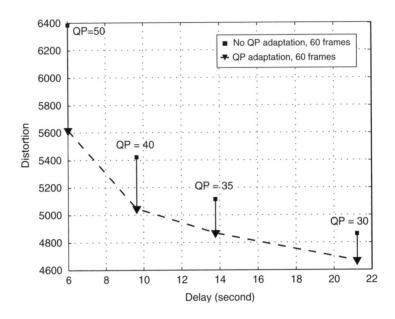

Figure 8.23 Distortion vs. delay comparison between QP adaptation and no QP adaptation with 60 summary frames

square nodes show the distortion-delay pairs when the video summary is source coded with the labeled QPs. The 'v' nodes refer to the distortion-delay budget pairs with QP adaptation when delay budget is set equal to the delay time that the corresponding labeled single QP takes to transmit the video summary. We can observe that QP adaptation, i.e., the cross-layer framework, has much distortion gain up to 6.2 % over fixed QP video transmission when the delay is small. In the case of summary frame number equal to 60 as in 23, much more significant distortion gain up to 12 % can be obtained in small delay regions.

Figure 8.24 shows the distortion vs. SNR comparisons between QP adaptation and QP = 50 with different prescribed maximum transmission number for ARQ. Owing to link adaptation performed by ARQ and AMC in a cross-layer fashion, both QP adaptation and QP = 50 have a stable distortion level along all SNR values. Of course here the delay difference of different schemes is not considered. We also notice that for either QP adaptation or QP = 50, $N_{max} = 3$ has better performance than $N_{max} = 1$. This is because the case with $N_{max} = 3$ can achieve lower actual PER than with $N_{max} = 1$ even though they both aim to guarantee $P_L = 10^{-2}$. The same conclusion goes for Figure 8.25 where the total summary frame number is 60.

Different distortion vs. delay budget with different P_L (L = 2 in this work) is shown in Figure 8.26. We observe that there is a large distortion-delay difference between $P_L = 10^{-1}$ and $P_L = 10^{-2}$. Once P_L achieves 10^{-2}, there is no big distortion vs. delay difference even though N_{max} is different. However, in the two cases with different P_L and same $N_{max} = 3$, the difference in distortion vs. delay is marginal. This is because with larger N_{max}, the actual PER is much lower than P_L. From this figure, we can conclude that the maximum transmission number impacts greatly on video transmission quality in

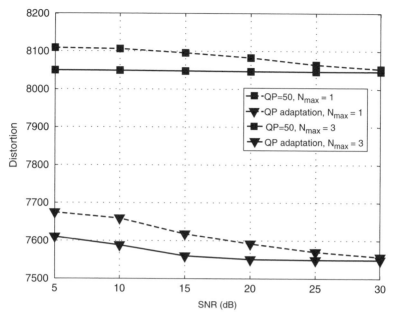

Figure 8.24 Distortion vs. average SNR comparison between QP adaptation and QP = 50 with different N_{max} for a 30-frame video summary

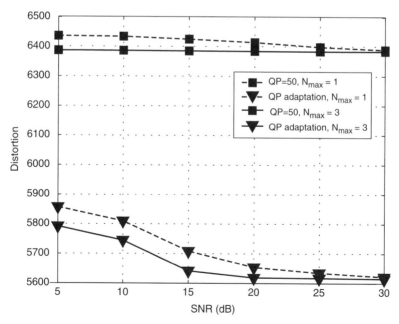

Figure 8.25 Distortion vs. average SNR comparison between QP adaptation and QP = 50 with different N_{max} for a 60-frame video summary

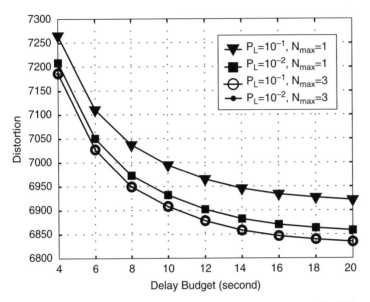

Figure 8.26   Comparisons of distortion vs. delay budget with different P_L

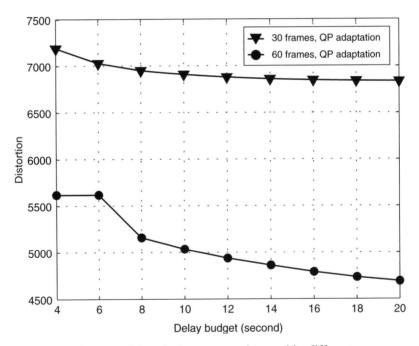

Figure 8.27   Distortion vs. delay budget comparison with different summary frame number

our cross-layer optimization framework. With the same delay budget, a larger allowed maximum transmission number leads to better video transmission quality.

Figure 8.27 shows the distortion vs. delay budget in our proposed framework when the summary frame number is 30 and 60 respectively both with $\overline{\gamma} = 25$dB. With the same delay budget, the case with 60 frames has better performance than the case with 30 frames. This is because with a higher sampling rate (that is, using 60 summary frames instead of 30), the similarity and correlation between neighboring summary frames have increased. Therefore, the distortion caused by losing one frame is reduced in this case because the lost frame would be concealed by its neighboring summary frame with a higher similarity.

8.8 Conclusions

In this chapter, we have discussed cross-layer optimized wireless multimedia communications and networking. A quality-driven cross-layer design framework has been discussed in details. Then, based on this quality-driven framework, different design scenarios, such as TFRC, content-aware delivery, routing and video summary, have been studied.

References

1. M. van der Schaar and S. Shankar, "Cross-layer wireless multimedia transmission: Challenges, principles, and new paradigms," *IEEE Wireless Commun. Mag.*, pp. 50–58, Aug. 2005.

2. D. W. S. Ci, H. Wang, "A theoretical framework for quality-aware cross-layer optimized wireless multimedia communications," *Advances in Multimedia*, 2008.

3. M. V. K. Ramchandran, "Best wavelet packet bases in a rate-distortion sense," *IEEE Trans. Image Process.*, Vol. 20, No. 2, pp. 160–175, Apr. 1993.

4. "Generic coding of moving pictures and associated audio information- part 2: Video," ITU-T and ISO/IEC JTC1, ITU-T Recommendation H.262–ISO/IEC 13 818-2 (MPEG-2), 1994.

5. "Video coding for low bitrate communication version 1," ITU-T, ITU-T Recommendation H.263, 1995.

6. "Coding of audio-visual objects part 2: Visual," ISO/IEC JTC1, ISO/IEC 14 496-2 (MPEG-4 visual version 1), 1999.

7. T. Wiegand, G. J. Sullivan, and A. Luthra, *Draft ITU-T Recommendation H.264 and Final Draft International Standard 14496-10 Advanced Video Coding*. Joint Video Team of ISO/IEC JTC1/SC29/WG11 and ITU-T SG16/Q.6, Doc. JVT-G050rl, Geneva, Switzerland, May 2003.

8. T. Wiegand, H. Schwarz, A. Joch, F. Kossentini, and G. Sullivan, "Rate-constrained coder control and comparison of video coding standards," *IEEE Trans. Circuits Syst. Video Technol.*, Vol. 13, No. 7, pp. 688–703, 2003.

9. R. Hinds, "Robust mode selection for block-motion-compensated video encoding," Ph.D. dissertation, Massachusetts Inst. Technol, Cambridge, MA, 1999.

10. R. Zhang, S. L. Regunathan, and K. Rose, "Video Coding with Optimal Inter/Intra-Mode Switching for Packet Loss Resilience," *IEEE J. Sel. Areas Commun.*, Vol. 18, No. 6, pp. 966–976, Jun. 2000.

11. Z. He, J. Cai, and C. W. Chen, "Joint source channel rate-distortion analysis for adaptive mode selection and rate control in wireless video coding," *IEEE Trans. Circuits Syst. Video Technol.*, Vol. 12, No. 6, pp. 511–523, June 2002.

12. D. Wu, T. Hou, B. Li, W. Zhu, Y.-Q. Zhang, and H. J. Chao, "An end-to-end approach for optimal mode selection in internet video communication: Theory and application," *IEEE J. Sel. Areas Commun.*, Vol. 18, No. 6, pp. 977–995, June 2000.

13. Y. Andreopoulos, N. Mastronade, and M. van der Schaar, "Cross-layer optimized video streaming over wireless multi-hop mesh networks," *IEEE J. Sel. Areas Commun.*, Vol. 24, No. 11, pp. 2104–2115, Nov. 2006.

14. S. Mao, Y. T. Hou, X. Cheng, and H. D. Sherali, "Multipath Routing for Multiple Description Video in Wireless Ad Hoc Networks," in *Proc. IEEE INFOCOM*, Miami, FL, Mar. 2005, pp. 740–750.

15. S. Kompella, S. Mao, Y. Hou, and H. Sherali, "Cross-layer optimized multipath routing for video communications in wireless networks," *IEEE J. Sel. Areas Commun.*, Vol. 25, No. 4, pp. 831–840, May 2007.

16. H. Shiang and M. van der Schaar, "Multi-user video streaming over multi-hop wireless networks: A distributed, cross-layer approach based on priority queuing," *IEEE J. Sel. Areas Commun.*, Vol. 25, No. 4, pp. 770–785, May 2007.

17. D. Wu, S. Ci, and H. Wang, "Cross-layer optimization for video summary transmission over wireless networks," *IEEE J. Sel. Areas Commun.*, Vol. 25, No. 4, pp. 841–850, May 2007.

18. P. Pahalawatta, R. Berry, T. Pappas, and A. Katsaggelos, "Content-aware resource allocation and packet scheduling for video transmission over wireless networks," *IEEE J. Sel. Areas Commun.*, Vol. 25, No. 4, pp. 749–759, May 2007.

19. T. Stockhammer, D. Kontopodis, and T. Wiegand, "Rate-distortion optimization for H.26L video coding in packet loss environment," in *Proc. Packet Video Workshop*, Pittsburgh,PA, 2002.

20. G. Cote, S. Shirani, and F. Kossentini, "Optimal mode selection and synchronization for robust video communications over error-prone networks," *IEEE J. Sel. Areas Commun.*, Vol. 18, No. 6, pp. 952–965, June 2000.

21. T. Wiegand, N. Farber, K. Stuhlmuller, and B. Girod, "Error-resilient video transmission using long-term memory motion-compensated prediction," *IEEE J. Sel. Areas Commun.*, Vol. 18, No. 6, pp. 1050–1062, June 2000.

22. C. Kim, D. Kang, and I. Kwang, "High-complexity mode decision for error prone environment," JVT-C101, May 2002.

23. Y. Wang, J. Ostermann, and Y. Q. Zhang, *Video Processing and Communications*. Prentice-Hall, Englewood Cliffs, NJ, 2002.

24. S. Mao, D. Bushmitch, S. Narayanan, and S. Panwar, "Mrtp: A multiflow real-time transport protocol for ad hoc networks," *IEEE Trans. Multimedia*, Vol. 8, No. 2, pp. 356–369, Apr. 2006.

25. H. Schulzrinne, S. Casner, R. Frederick, and V. Jacobson, "Rtp: A transport protocol for realtime applications," IETF Request For Comments 3550, Jul. 1999.

26. R. Stewart, K. Morneault, C. Sharp, H. Schwarzbauer, T. Taylor, I. Rytina, M. Kalla, L. Zhang, and V. Paxson, "Stream control transmission protocol," IETF RFC 2960, Oct. 2000.

27. H. Hsieh, K. Kim, Y. Zhu, and R. Sivakumar, "A receivercentric transport protocol for mobile hosts with heterogeneous wireless interfaces," in *Proc. ACM Mobicom*, San Diego, CA, Sep. 2003, pp. 1–15.

28. J. Yan, K. Katrinis, M. May, and B. Plattner, "Media- and TCP-Friendly Congestion Control for Scalable Video Streams," *IEEE Trans. Multimedia*, Vol. 8, pp. 196–206, Apr. 2006.

29. S. Floyd and K. Fall, "Promoting the use of end-to-end congestion control in the Internet," *IEEE/ACM Trans. Networking*, Vol. 7, pp. 458–472, Aug. 1999.

30. S. Floyd, M. Handley, J. Padhye, and J. Widmer, "Equation-Based Congestion Control for Unicast Applications," *Pro. ACM SIGCOMM*, Aug. 2000.

31. A. Argyriou, "A Joint Performance Model of TCP and TFRC with Mobility Management Protocols," *Wiley Wireless Communications and Mobile Computing (WCMC) Special Issue on Mobile IP*, Vol. 6, pp. 547–557, Aug. 2006.

32. M. Li, C. Lee, E. Agu, M. Claypool, and R. Kinicki, "Performance Enhancement of TFRC in Wireless Ad Hoc Networks," *Distributed Multimedia Systems (DMS)*, September 2004.

33. Z. Fu, X. Meng, and S. Lu, "A Transport Protocol For Supporting Multimedia Streaming in Mobile Ad Hoc Networks," *IEEE JSAC*, December 2004.

34. K. Chen and K. Nahrstedt, "Limitations of Equation-based Congestion Control in Mobile Ad Hoc Networks," *Workshop on Wireless Ad Hoc Networking (WWAN)*, March 2004.

35. M. Handley, S. Floyd, J. Pahdye, and J. Widmer, "Tcp friendly rate control (tfrc): protocol specification," RFC 3448, Jan. 2003.

36. X. Zhu and B. Girod, "Media-Aware Multi-User Rate Allocation over Wireless Mesh Networks," in *IEEE OpComm-06*, Sep. 2006, pp. 1–8.

37. _____, "Distributed Rate Allocation for Video Streaming over Wireless Networks with Heterogeneous Link Speeds," in *ISMW-07*, Aug. 2007, pp. 296–301.

38. T. Wiegand, G. J. Sullivan, G. Bjontegaard, and A. Luthra, "Overview of the H.264/AVC Video Coding Standard," *IEEE Trans. Circuits Syst. Video Technol.*, Vol. 13, No. 7, pp. 560–576, Jul. 2003.

39. G. M. Schuster and A. K. Katsaggelos, *Rate-Distortion Based Video Compression: Optimal Video Frame Compression and Object Boundary Encoding*. Norwell, MA: Kluwer, 1997.

40. S. Adlakha, X. Zhu, B. Girod, and A. Goldsmith, "Joint Capacity, Flow and Rate Allocation for Multiuser Video Streaming over Wireless Ad-Hoc Networks," in *IEEE ICC, Proc. Glasgow, Scotland*, June 2007, pp. 1747–1753.

41. D. Qiao, S. Choi, and K. G. Shin, "Goodput analysis and link adaptation for ieee 802.11a wireless lan," *IEEE Trans. Mobile Comput.*, Vol. 1, No. 4, 2002.

42. Q. Liu, S. Zhou, and G. B. Giannakis, "Queuing with Adaptive Modulation and Coding over Wireless Links: Cross-Layer Analysis and Design," *IEEE Trans. Wireless Commun.*, Vol. 4, No. 3, pp. 1142–1153, May 2005.

43. D. Krishnaswamy, "Network-Assisted Link Adaptation with Power Control and Channel Reassignment in Wireless Networks," in *Proc. 3G Wireless Conf.*, 2002, pp. 165–170.

44. A. S. Tanenbaum, *Computer Networks*. Upper Saddle River, NJ: Prentice Hall, 2003.

45. T. Clausen and P. Jacquet, "Optimized link state routing protocol," *IETF RFC 3626*, Oct. 2006.

46. R. Tupelly, J. Zhang, and E. Chong, "Opportunistic scheduling for streaming video in wireless networks," in *Proc. Conference on Information Sciences and Systems*, 2003.

47. G. Liebl, T. Stockhammer, C. Buchner, and A. Klein, "Radio link buffer management and scheduling for video streaming over wireless shared channels," in *Proc. Packet Video Workshop*, 2004.

48. G. Liebl, M. Kalman, and B. Girod, "Deadline-aware scheduling for wireless video streaming," in *Proc. IEEE ICME*, July 2005.

49. Stauffer, Chris, and Eric, "Learning patterns of activity using real-time tracking," *IEEE Transactions on Pattern Analysis and Machine Intelligence*, Vol. 22, No. 8, pp. 747–757, 2000.

50. D. Bertsekas and R. Gallager, Data Networks. Upper Saddle River, NJ: Prentice Hall, 1992.

51. P. V. Beek et al., "Metadata-driven multimedia access," *IEEE Signal Processing Magazine*, pp. 40–52, Mar. 2003.

52. A. Hanjalic and H. Zhang, "An integrated scheme for automatic video abstraction based on unsupervised cluster-validity analysis," *IEEE Trans. Circuits Syst. Video Technol.*, Vol. 9, Dec. 1999.

53. B. L. Tseng, C. Lin, and J. R. Smith, "Using MPEG-7 and MPEG-21 for personalizing videoMetadata-driven multimedia access," *IEEE Multimedia*, Vol. 11, pp. 40–53, Jan. 2004.

54. D. DeMenthon, V. Kobla, and D. Doermann, "Video summarization by curve simplification," in *Proc. ACM Multimedia*, Jul. 1998.

55. Y. Gong and X. Liu, "Video summarization using singular value decomposition," *Computer Vision and Pattern Recognition*, Vol. 2, pp. 13–15, Jun. 2000.

56. Z. Li et al., "Rate-distortion optimal video summary generation," *IEEE Trans. Image Process.*, Vol. 14, pp. 1550–1560, Oct. 2005.

57. Z. Li, F. Zhai, and A. K. Katsaggelosothers, "Video summarization for energy efficient wireless streaming," in *Proc. SPIE Visual Communication and Image Processing (VCIP)*, Beijing, China, 2005.

58. P. V. Pahalawatta et al., "Rate-distortion optimized video summary generation and transmission over packet lossy networks," in *Proc. SPIE Image/Video Communication and Processing (IVCP)*, San Jose, US, 2005.

59. D. L. Goeckel, "Adaptive coding for time-varying channels using outdated fading estimates," *IEEE Trans. Commun.*, Vol. 47, pp. 844–855, Jun. 1999.

60. A. J. Goldsmith and S. G. Chua, "Adaptive coded modulation for fading channels," *IEEE Trans. Commun.*, Vol. 46, pp. 595–602, May 1998.

61. M. Alouini and A. J. Goldsmith, "Adaptive modulation over nakagami fading channels," *Kluwer J. Wireless Communications*, Vol. 13, pp. 119–143, May 2000.

62. E. Malkamaki and H. Leib, "Performance of truncated type-ii hybrid arq schemes with noisy feedback over block fading channels," *IEEE Transactions on Communications*, Vol. 48, pp. 1477–1487, Sep. 2000.

63. F. Zhai et al., "Rate-distortion optimized hybrid error control for packetized video communications," *IEEE Trans. Image Process.*, Vol. 15, pp. 40–53, Jan. 2006.

64. H. Wang, F. Zhai, Y. Eisenburg, and A. K. Katsaggelosothers, "Cost-distortion optimal unequal error protection for object-based video communications," *IEEE Trans. Circuits Syst. Video Technol.*, Vol. 15, pp. 1505–1516, Dec. 2005.

65. T. Ozcelebi, F. D. Vito, M. O. Sunay, M. R. C. A. M. Tekalp, and J. D. Martin, "Cross-Layer Scheduling with Content and Packet Priorities for Optimal Video Streaming over 1xEV-DO," in *Proc. VLBV05*, Sardinia, Italy, Sep. 2005, pp. 76–83.

66. (2004) 3GPP TR 25.848 V4.0.0, Physical Layer Aspects of UTRA High Speed Downlink Packet Access (release 4).

67. A. Doufexi, S. Armour, M. Butler, A. Nix, D. Bull, J. McGeehan, and P. Karlsson, "A Comparison of the HIPERLAN/2 and IEEE 802.11a Wireless LAN Standards," *IEEE Communication Magazine*, Vol. 40, pp. 172–180, May 2002.

68. (2002) IEEE Standard 802.16 Working Group, IEEE Standard for Local and Metropolitan Area Networks Part 16: Air Interface for Fixed Broadband Wireless Access Systems.

69. Q. Liu, S. Zhou, and G. B. Giannakis, "Cross-layer combining of adaptive modulation and coding with truncated arq over wireless links," *IEEE Transactions on Wireless Communications*, Vol. 3, pp. 1746–1755, Sep. 2004.

70. K. Ramchandran and M. Vetterli, "Best wavelet packet bases in a rate-distortion sense," *IEEE Transaction on Signal Processing*, Vol. 2, pp. 160–175, Apr. 1993.

9

Content-based Video Communications

9.1 Network-Adaptive Video Object Encoding

Network-adaptive video encoding is a robust video compression approach that designs and optimizes the source encoder by considering transmission factors such as error control, packetization, packet scheduling and retransmission, routing and error concealment. Error resilience is gained by selecting optimally the encoding mode that enables the decoded video to reach the minimum expected distortion for the available resources.

In this framework, we assume that the encoder knows the transmission channel characteristics such as the bit error rate (BER) or the probability of packet loss (in packet-based networks, the packet could be dropped or lost by bit error or excessive delay). This can be either specified in the initial negotiations, or calculated adaptively from messages exchanged by the transmission protocol. In a wireless channel, when transmission delay is not a major issue, the probability of packet loss can be calculated easily from the channel BER and the length of the packet. For ease of discussion, we assume that there is a lossy channel in this section and in section 9.2. In this way, packets are either received error-free or lost. The issues related to channels with BER will be addressed in section 9.3.

In order to evaluate received video quality, we assume that the transmitter knows the background VOP on which the transmitted video object will be composed at the receiver, which is possible if the encoder has sufficient knowledge of the application or if there is a feedback channel from the decoder which provides to the encoder the information about the composition order of the video objects and the success or failure of the delivering status of the video packet. Otherwise, a default background VOP will be adopted. Therefore, the distortion is calculated as the total intensity error of the composed frame. Owing to packet loss in the transmission channel, the decoded video at the decoder is not deterministic, and the distortion is a random variable. Therefore, the expected distortion is used as our objective distortion metric. Clearly, the expected distortion for the ith slice can be

4G Wireless Video Communications   Haohong Wang, Lisimachos P. Kondi, Ajay Luthra and Song Ci
© 2009 John Wiley & Sons, Ltd

calculated by summing up the expected distortion of all the pixels in the slice:

$$E[D_i] = \sum_{j=iN}^{iN+N-1} E[d_j] \tag{9.1}$$

where $E[d_j]$ is the expected distortion at the receiver for the jth pixel in the VOP, and N is the total number of pixels in the slice. Let us denote by f_n^j the original value of pixel j in VOP n, and $\tilde{f}_n^j$ its reconstructed value at the decoder. By definition:

$$f_n^j = s_n^j t_n^j + (1 - s_n^j)g_n^j, \; \tilde{f}_n^j = \tilde{s}_n^j \tilde{t}_n^j + (1 - \tilde{s}_n^j)g_n^j \tag{9.2}$$

and

$$\begin{aligned} E[d_j] &= E[(f_n^j - \tilde{f}_n^j)^2] = (f_n^j)^2 - 2f_n^j E[\tilde{f}_n^j] + E[(\tilde{f}_n^j)^2] \\ &= (f_n^j)^2 - 2f_n^j E[\tilde{s}_n^j \tilde{t}_n^j] - 2f_n^j g_n^j + 2f_n^j g_n^j E[\tilde{s}_n^j] + E[(\tilde{s}_n^j \tilde{t}_n^j)^2] + (g_n^j)^2 \\ &\quad - 2g_n^j E[\tilde{s}_n^j] + (g_n^j)^2 E[(\tilde{s}_n^j)^2] + 2g_n^j E[\tilde{s}_n^j \tilde{t}_n^j] - 2g_n^j E[(\tilde{s}_n^j)^2 \tilde{t}_n^j] \end{aligned} \tag{9.3}$$

where $s_n^j (s_n^j = 0$ for transparent or 1 for opaque block; only binary shape is considered here) and t_n^j are the corresponding shape and texture component of f_n^j, $\tilde{s}_n^j$ and $\tilde{t}_n^j$ are the corresponding shape and texture component of $\tilde{f}_n^j$, and g_n^j is the background pixel value at the same position.

The further calculation of $E[d_j]$ in equation (9.3) is dependent on a concealment strategy at the decoder. Out of a number of possible strategies, we consider the following (an extension of ROPE [1]), which allows for the recursive calculation of $E[d_j]$ at the encoder. We assume that each row of macroblocks becomes a packet, and use ρ_i to denote the probability of loss for the ith packet. To recover the lost shape (texture) macroblocks in a packet, the decoder uses the shape (texture) motion vector of the neighboring macroblocks above as the concealment motion vector. If the concealment motion vector is not available, e.g., the macroblock above is also lost, then the decoder uses a zero motion vector for concealment. In calculating $E[d_j]$, the first and second moments of the reconstructed shape and texture intensity value for the jth pixel are needed. In the following paragraph, we demonstrate how the first moment can be calculated recursively in time. The second moment is computed in a similar fashion, but omitted here, owing to lack of space.

In object-based video, since texture and shape information can be independently encoded in the intra and inter modes, we consider the following four cases:

$$E[\tilde{s}_n^j \tilde{t}_n^j](I, I) = (1 - \rho_i)\hat{s}_n^j \hat{t}_n^j + \rho_i(1 - \rho_{i-1})E[\tilde{s}_{n-1}^{k_s} \tilde{t}_{n-1}^{k_t}] + \rho_{i-1}\rho_i E[\tilde{s}_{n-1}^j \tilde{t}_{n-1}^j] \tag{9.4}$$

$$\begin{aligned} E[\tilde{s}_n^j \tilde{t}_n^j](I, P) &= (1 - \rho_i)\hat{s}_n^j (\hat{e}_n^j + E[\tilde{t}_{n-1}^{m_t}]) + \rho_i(1 - \rho_{i-1})E[\tilde{s}_{n-1}^{k_s} \tilde{t}_{n-1}^{k_t}] \\ &\quad + \rho_{i-1}\rho_i E[\tilde{s}_{n-1}^j \tilde{t}_{n-1}^j] \end{aligned} \tag{9.5}$$

$$E[\tilde{s}_n^j \tilde{t}_n^j](P, I) = (1 - \rho_i)E[\tilde{s}_{n-1}^{m_s}]\hat{t}_n^j + \rho_i(1 - \rho_{i-1})E[\tilde{s}_{n-1}^{k_s} \tilde{t}_{n-1}^{k_t}] + \rho_{i-1}\rho_i E[\tilde{s}_{n-1}^j \tilde{t}_{n-1}^j] \tag{9.6}$$

$$E[\tilde{s}_n^j \tilde{t}_n^j](P, P) = (1 - \rho_i)(E[\tilde{s}_{n-1}^{m_s}]\hat{e}_n^j + E[\tilde{s}_{n-1}^{m_s} \tilde{t}_{n-1}^{m_t}]) + \rho_i(1 - \rho_{i-1})E[\tilde{s}_{n-1}^{k_s} \tilde{t}_{n-1}^{k_t}] \quad (9.7)$$

$$+ \rho_{i-1}\rho_i E[\tilde{s}_{n-1}^j \tilde{t}_{n-1}^j]$$

where shape is intra coded in (9.4) and (9.5), and inter coded in (9.6) and (9.7); texture is intra coded in (9.4) and (9.6), and inter coded in (9.5) and (9.7); $\hat{s}_n^j$ and $\hat{t}_n^j$ are the encoder reconstructed shape and texture of the jth pixel; pixel j in frame n is predicted by pixel m in frame $n-1$ if the motion vector is available, otherwise predicted by pixel k if the concealment motion vector is available. The subscript s and t of k_s, k_t, m_s, and m_t in (9.4)–(9.7) are used to distinguish shape from texture, because the motion vector or concealment motion vector of shape could be different from that of texture. Computing $E[\tilde{s}_n^j \tilde{t}_n^j]$ in (9.4)–(9.7) depend on the computing of $E[\tilde{s}_n^j]$ and $E[\tilde{t}_n^j]$, which are calculated recursively as follows:

$$E[\tilde{s}_n^j](I) = (1 - \rho_i)\hat{s}_n^j + \rho_i(1 - \rho_{i-1})E[\tilde{s}_{n-1}^k] + \rho_{i-1}\rho_i E[\tilde{s}_{n-1}^j] \quad (9.8)$$

$$E[\tilde{t}_n^j](I) = (1 - \rho_i)\hat{t}_n^j + \rho_i(1 - \rho_{i-1})E[\tilde{t}_{n-1}^k] + \rho_{i-1}\rho_i E[\tilde{t}_{n-1}^j] \quad (9.9)$$

$$E[\tilde{s}_n^j](P) = (1 - \rho_i)E[\tilde{s}_{n-1}^m] + \rho_i(1 - \rho_{i-1})E[\tilde{s}_{n-1}^k] + \rho_{i-1}\rho_i E[\tilde{s}_{n-1}^j], \quad (9.10)$$

$$E[\tilde{t}_n^j](P) = (1 - \rho_i)(\hat{e}_n^j + E[\tilde{t}_{n-1}^m]) + \rho_{i-1}\rho_i E[\tilde{t}_{n-1}^j] + \rho_i(1 - \rho_{i-1})E[\tilde{t}_{n-1}^k] \quad (9.11)$$

where shape and texture are intra coded in (9.8) and (9.9), and inter coded in (9.10) and (9.11). It is not hard to understand that $E[\tilde{s}_n^{k_s} \tilde{t}_n^{k_t}]$ and $E[\tilde{s}_n^{m_s} \tilde{t}_n^{m_t}]$ in (9.4)–(9.7) can be computed recursively in a similar manner as $E[\tilde{s}_n^j \tilde{t}_n^j]$. However, computing these inter-pixel cross-correlation terms recursively may require computing and storing all inter-pixel cross-correlation values for all frames in the video sequence. The amount of this computation and storage is not feasible even for moderate size frame. It is natural to use a model-based cross-correlation approximation method to estimate $E[st]$ in terms of $E[s]$, $E[t]$, $E[s^2]$, $E[t^2]$ and standard deviations σ_s and σ_t. The simplest idea is to assume that s and t are uncorrelated, then $E[st] = E[s]E[t]$. In general, the model is quite accurate to use and be proved with experiments.

Once the distortion estimation model is set up, the optimization problem can be represented by:

$$\text{Minimize } E[D], \text{ subject to } R \le R_{budget} \quad (9.12)$$

which is obviously a simple instance of the expected distortion. In the equation, R is the total bit rate, and R_{budget} is the bit budget for the frame. The optimization is over the source coding parameters and is restricted to the frame level. Clearly, the expected frame distortion E[D] can be accumulated by the N packets within the frame, like $E[D] = \sum_{i=1}^{N} E[D_i]$.

The problem can be easily solved by using Lagarangian relaxation and dynamic programming, as discussed in section 6.2.4. Figure 9.1 shows the resulted R-D curves of a simulation based on MPEG-4 VM 18.0, where the proposed network adaptive approach and the non network-adaptive encoding method are compared by transmitting video bitstreams over lossy channels with packet loss of 10% and 20%. Clearly, the network-adaptive method has constant gains of around $2\,dB$ for both channels.

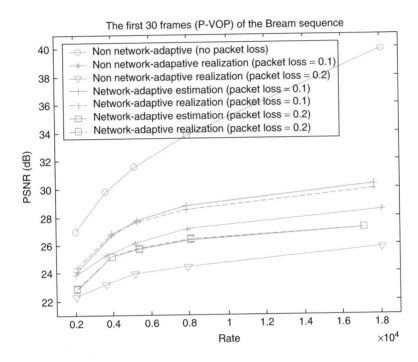

Figure 9.1   Comparison of proposed approach with non-network-adaptive approach

9.2   Joint Source Coding and Unequal Error Protection

In video communications, various video packets may contribute differently to overall video quality; unequal error protection (UEP) is a natural way of protecting transmitted video data. The idea is to allocate more resources to the parts of the video sequence that have a greater impact on video quality, while expending fewer resources on the parts that are less significant. In [2], a priority encoding transmission scheme is proposed so as to allow a user to set different priorities of error protection for different segments of the video stream. This scheme is suitable for MPEG video, in which there is a decreasing importance among I, P and B frames. In general, error protection can come from various sources such as forward error correction, retransmission and transmission power adaptation. In [3], a generic UEP forward error correction scheme is proposed, utilizing the knowledge that almost all video packet formats have more important data closer to the packet header. Thus, the algorithm requires that better protection be applied to the data closer to the packet header. In [4], an optimal unequal error protection scheme for layered video coding was proposed in order to provide an optimal bit allocation between source coding and channel coding. In [5, 6], the trade off between transmission energy consumption and video quality for wireless video communications is studied, where the goal is to minimize the energy needed to transmit a video sequence with an acceptable level of video quality and tolerable delay. By assuming that the transmitter knows the relationship between the transmission power and the probability of the packet loss, transmission power can be adjusted dynamically so as to control the level of protection provided for each packet [5, 6]. Similarly, in [7] and [8], the cost-distortion problem in a DiffServ network is studied, by assigning unequal protection (prices) to the different packets.

In object-based coding, video data is composed of shape and texture information, which have completely different stochastic characteristics and bit rate proportions. For example, in moderate or high bit rate applications, the shape data generally only occupies 0.5–20 % of the total bit rate [9], but can have a stronger impact on reconstructed video quality than texture. The unbalanced contribution of shape and texture information makes UEP a natural solution to protect this type of video sequence in lossy networks. In [3, 10, 11] and [12], UEP schemes have been explored and applied to the transmission of MPEG-4 compressed video bitstreams over error-prone wireless channels. In their work, the video is compressed using a standard block-based approach with rectangular shape, i.e., there is no shape information contained in the bitstream. Thus, the resulting bitstream is divided into partitions labeled header, motion and texture, which are assumed to have a decreasing order of importance. Therefore, the header and motion bits receive higher levels of protection and the texture bits receive the lowest level of protection. In [10], I-frame data is treated as having the same level of importance as header and motion data. In [3], I-frame DCT data is subdivided into DC coefficients and AC coefficients, where DC is considered to have a higher subjective importance than AC. In [10, 11] and [12], UEP is implemented using different channel coding rates, while in [12], it is implemented by separating the partitions into various streams and transmitting those streams over different carrier channels meeting different QoS levels. It is reported in [3, 10, 11] and [12] that the adoption of UEP results in a better performance than equal error protection (EEP) methods. However, this work has not considered video with arbitrarily shaped video objects. Furthermore, it is based on pre-encoded video. Thus it does not consider optimal source coding, nor does it incorporate the error concealment strategy used at the decoder in the UEP framework.

Recently, a rate-distortion optimal source-coding scheme was proposed for solving the bit allocation problem in object-based video coding [13]. There, the experimental results indicate that for some applications shape may have a stronger impact on reconstructed video quality than texture. This result motivates the unequal protection of the shape and texture components of the video objects in video transmission. In this section, a general cost-distortion optimal unequal error protection scheme for object-based video communications is demonstrated where source coding, packet classification and error concealment are considered jointly within the framework of cost-distortion optimization. The scheme is applicable for packet lossy networks, that is, there is an assumption that the packets are either received error-free or lost.

9.2.1 Problem Formulation

The problem at hand is to choose coding parameters for the shape and texture of a VOP, so as to minimize the total expected distortion, given a cost constraint and a transmission delay constraint in a lossy network environment. This objective can also be represented by:

$$\text{Minimize } E[D_{tot}], \text{ Subject to } C_{tot} \leq C_{\max} \text{ and } T_{tot} \leq T_{\max} \tag{9.13}$$

where $E[D_{tot}]$ is the expected total distortion for the frame, C_{tot} is the total cost for a frame, T_{tot} is the total transmission delay for the frame, C_{max} is the maximum allowable cost for the frame and T_{max} is the maximum amount of time which can be used to transmit the entire frame. We assume that there exists a higher-level controller which assigns a bit budget and a cost budget to each frame in order to meet any of the delay constraints imposed by the application. Therefore, the value of C_{max} and T_{max} can vary from frame to frame, but are known constants in (9.13).

9.2.1.1   System Model

We consider an MPEG-4 compliant object-based video application, where the video is encoded using different algorithms for shape and texture. As mentioned in [9], compared to texture data, shape data requires relatively fewer bits to encode but has a very strong impact on video quality. Therefore, it is natural to imagine that the unequal error protection scheme for shape and texture may provide improved performance over an equal error protection scheme. However, implementing an unequal error protection scheme is not straightforward because in the MPEG-4 video packet syntax, the shape and texture data are placed in the same packet (using a *combined packetization* scheme). If data partitioning is enabled, a motion marker is placed between the shape and texture data for resynchronization. One way to enable unequal error protection is to use a *separated packetization* scheme, where shape and texture are packed into separate packets. In a similar way as is proposed in [10] and [11], we insert an adaptation layer between the MPEG-4 video application and the network, which can reorganize the MPEG-4 compressed bitstream into corresponding shape packets and texture packets. In addition, the adaptation layer can add some forward error protection optimally to those packets. Figure 9.2 depicts the architecture of the proposed video transmission system. For a wireless network using an H.223

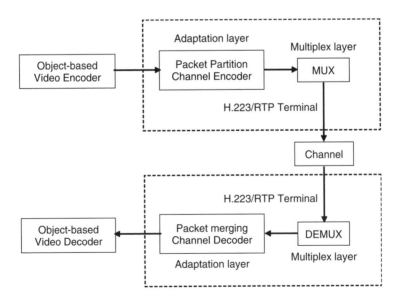

Figure 9.2   System block diagram

MUX [11], we simply replace the standard adaptation layer in the H.223 multiplexing protocol with our new layer. At the receiver side, the adaptation layer merges the shape and texture packets and makes the output bitstream compatible with the MPEG-4 syntax.

In the following text, the *separated packetization* scheme is used as the default packetization scheme. The coded video frame is divided into 16×16 macroblocks, which are numbered in scan line order and divided into groups called slices. For each slice, there is a shape packet and a corresponding texture packet. Let I be the number of slices in the given frame and i the slice index. For each macroblock, both coding parameters for shape and texture are specified. We use μ_{S_i} and μ_{T_i}, respectively, to denote the coding parameters for all macroblocks in the ith shape and texture packets, and use $B_{S_i}(\mu_{S_i})$ and $B_{T_i}(\mu_{T_i})$, respectively, to denote the corresponding encoding bit rates of these packets. It is important to point out that each packet is decodable independently in our system; that is, each packet has enough information for decoding and is independent of other packets in the same frame. This guarantees that a lost packet will not affect the decoding of other packets in the same frame. Of course, errors may propagate from one frame to the next owing to motion compensation.

9.2.1.2   Channel Model

In addition to the source coding parameters, we assume that the transmission parameters may also be adapted per packet. By adapting the transmission parameters, we can control the effective channel characteristics, such as the probability of loss. Another way to view this is that each encoded packet can be sent over a set of possible transmission channels. Each transmission channel is classified using a set of parameters, e.g., the probability of packet loss and transmission rate. Let us denote by π_{S_i} and π_{T_i}, the selected service classes for the ith shape and texture packet, respectively. Similarly, let $\rho(\pi_{S_i})$ and $\rho(\pi_{T_i})$ denote the corresponding probability of packet loss, and $R(\pi_{S_i})$ and $R(\pi_{T_i})$ the corresponding transmission rate. Then the total transmission time per fame is represented by:

$$T_{tot} = \sum_{i=1}^{I} \left[\frac{B_{S_i}(\mu_{S_i})}{R(\pi_{S_i})} + \frac{B_{T_i}(\mu_{T_i})}{R(\pi_{T_i})} \right] \tag{9.14}$$

Let $C(\pi_{S_i})$ and $C(\pi_{T_i})$ denote respectively the transmission cost per bit for the ith shape and texture packets. The total cost used to transmit all the packets in a frame is therefore:

$$C_{tot} = \sum_{i=1}^{I} [B_{S_i}(\mu_{S_i})C(\pi_{S_i}) + B_{T_i}(\mu_{T_i})C(\pi_{T_i})] \tag{9.15}$$

In a DiffServ network, the cost represents the price for each QoS channel, e.g., cents per Kbyte. We assume that the service level can be pre-specified in the service level agreement (SLA) between the Internet service provider (ISP) and the users [14]. Typically a set of parameters is used to describe the state of each service class, including the transmission rate bound and probability of packet loss. In this setting, a cost is associated with each service class as specified in the SLA. By adjusting the prices for each service class, the network can influence the class a user selects. The sender classifies each packet according to its importance in order to better utilize the available network resources.

In a wireless network, the cost we consider in this work is represented by energy per bit, i.e.:

$$C(\pi_{S_i}) = \frac{P(\pi_{S_i})}{R(\pi_{S_i})}, \text{ and } C(\pi_{T_i}) = \frac{P(\pi_{T_i})}{R(\pi_{T_i})} \tag{9.16}$$

where $P(\pi_{S_i})$ and $P(\pi_{T_i})$ are the corresponding transmission power for the ith shape and texture packet, respectively. The exact relationship between transmission power, transmission rate and the probability of packet loss varies for different channel models.

9.2.1.3 Expected Distortion

We assume that the transmitter only knows the probability with which a packet has arrived at the receiver. Thus, the distortion at the receiver is a random variable. Let $E[D_i]$ represent the expected distortion at the receiver for the ith slice. In this case:

$$E[D_i] = [1 - \rho(\pi_{S_i})][1 - \rho(\pi_{T_i})]E[D_{R,i}] + [1 - \rho(\pi_{S_i})]\rho(\pi_{T_i})E[D_{LT,i}]$$

$$+ \rho(\pi_{S_i})[1 - \rho(\pi_{T_i})]E[D_{LS,i}] + \rho(\pi_{S_i})\rho(\pi_{T_i})E[D_{L,i}] \tag{9.17}$$

where $E[D_{R,i}]$ is the expected distortion for the ith slice if both the shape and texture packets are received correctly at the decoder, $E[D_{LT,i}]$ is the expected distortion if the texture packet is lost, $E[D_{LS,i}]$ is the expected distortion if the shape packet is lost and $E[D_{L,i}]$ is the expected distortion if both the shape and texture packets are lost. Clearly, $E[D_{R,i}]$ depends only on the source coding parameters for the ith packet, while $E[D_{LT,i}]$, $E[D_{LS,i}]$ and $E[D_{L,i}]$ depend on the concealment strategy used at the decoder.

Note that the problem formulation and solution approach presented in this chapter are general. Therefore, the techniques developed here are applicable to various concealment strategies used by the decoder. The only assumption we make is that the concealment strategy is also known at the encoder. A common concealment strategy is to conceal the missing macroblock by using the motion information of its neighboring macroblocks. When the neighboring macroblock information is not available, the lost macroblock is typically replaced with the macroblock from the previous frame at the same location. It is also important to note that the formulation presented here is applicable to various distortion metrics. In our experimental results we use the expected mean squared error (MSE), as is commonly done in the literature [1, 5, 6, 8].

9.2.1.4 Optimization Formulation

The optimization problem (9.13) can be rewritten as:

$$\underset{\{\mu_{S_i}, \mu_{T_i}, \pi_{S_i}, \pi_{T_i}\}}{Minimize} \quad E[D_{tot}]$$

$$\text{subject to:} \qquad \sum_{i=1}^{I} [B_{S_i}(\mu_{S_i})C(\pi_{S_i}) + B_{T_i}(\mu_{T_i})C(\pi_{T_i})] \leq C_{\max} \qquad (9.18)$$

$$\text{and} \qquad \sum_{i=1}^{I} \left[\frac{B_{S_i}(\mu_{S_i})}{R(\pi_{S_i})} + \frac{B_{T_i}(\mu_{T_i})}{R(\pi_{T_i})} \right] \leq T_{\max}$$

We assume that the processing and propagation delays are constant and can therefore be ignored in this formulation. The only delay we are concerned with is the transmission delay.

9.2.2 Solution and Implementation Details

9.2.2.1 Packetization and Error Concealment

The packetization scheme, i.e., the number of macroblocks per packet, is not standardized within the MPEG-4 standard. Some applications pack each macroblock into a packet. This approach provides significant error resilience and encoding flexibility, but suffers from the large transmission overhead required for each packet header. In addition, since each packet must be decodable independently, this approach does not benefit from differential encoding between macroblocks. In other applications, each frame is packed into a separate packet. In this case, the transmission overhead is very small, but the resilience to channel errors is poor, e.g., an uncorrectable local error may cause the entire frame to be discarded. In order to balance error resilience and coding efficiency, we consider packing each row of macroblocks into a packet. In our simulations, we consider both the *combined* packetization scheme and the *separated* packetization scheme. It is important to note that four extra bits per MB, indicating the 8×8 shape block transparency, must be added to the texture packets in the separated packetization scheme, in order to make each packet decodable independently.

The error concealment strategy in the simulations is identical for both shape and texture packets. To recover the lost shape (texture) information, the decoder uses the shape (texture) motion vector of the neighboring macroblock above as the concealment motion vector. If the concealment motion vector is not available, e.g., because the above macroblock is also lost, then the decoder uses zero motion vector concealment.

9.2.2.2 Expected Distortion

For separated packetization, since the shape data and texture data are transmitted and decoded independently:

$$E[s_n^j t_n^j] = E[s_n^j]E[t_n^j], \text{ and } E[\tilde{s}_n^j \tilde{t}_n^j] = E[\tilde{s}_n^j]E[\tilde{t}_n^j] \qquad (9.19)$$

This observation enables us to calculate the expected distortion efficiently (9.3) by calculating recursively the necessary moments for shape and texture independently.

9.2.2.3 Optimal Solution

In this section, we present an optimal solution for problem (9.18). We use the Lagrange multiplier method in order to relax the cost and delay constraints. The relaxed problem can then be solved using a shortest path algorithm.

The Lagrangian relaxation method leads to a convex hull approximation for the constrained problem (9.18). Let U be the set of all possible decision vectors u_i for the ith slice ($i = 1, 2, \ldots, I$), where $u_i = (\mu_{S_i}, \mu_{T_i}, \pi_{S_i}, \pi_{T_i})$. We first define a Lagrangian cost function:

$$J_{\lambda_1,\lambda_2}(u) = E[D_{tot}] + \lambda_1 C_{tot} + \lambda_2 T_{tot} \tag{9.20}$$

$$= \sum_{i=1}^{I} \left\{ E[D_i] + \lambda_1 [B_{S_i}(\mu_{S_i})C(\pi_{S_i}) + B_{T_i}(\mu_{T_i})C(\pi_{T_i})] \right.$$

$$\left. + \lambda_2 \left[\frac{B_{S_i}(\mu_{S_i})}{R(\pi_{S_i})} + \frac{B_{T_i}(\mu_{T_i})}{R(\pi_{T_i})} \right] \right\}$$

where λ_1 and λ_2 are the Lagrange multipliers. It can easily be derived from [15] and [16] that if there exists a pair λ_1^* and λ_2^* such that $u^* = \arg[\min_u J_{\lambda_1^*,\lambda_2^*}(u)]$, which leads to $C_{tot} = C_{\max}$ and $T_{tot} = T_{\max}$, then u^* is also an optimal solution to (9.18). Therefore, the task of solving (9.18) is converted into an easier one, which is to find the optimal solution to the unconstrained problem:

$$\min \sum_{i=1}^{I} \left\{ E[D_i] + \lambda_1 [B_{S_i}(\mu_{S_i})C(\pi_{S_i}) + B_{T_i}(\mu_{T_i})C(\pi_{T_i})] + \lambda_2 \left[\frac{B_{S_i}(\mu_{S_i})}{R(\pi_{S_i})} + \frac{B_{T_i}(\mu_{T_i})}{R(\pi_{T_i})} \right] \right\} \tag{9.21}$$

Most decoder concealment strategies introduce dependencies between slices. For example, if the concealment algorithm uses the motion vector of the macroblock above to conceal the lost macroblock, then it would cause the calculation of the expected distortion of the current slice to depend on its previous slice. Without loss of the generality, we assume that the concealment strategy will cause the current slice to depend on its previous a slices ($a \geq 0$). To implement the algorithm for solving optimization problem (9.18), we define a cost function $G_k(u_{k-a}, \ldots, u_k)$, which represents the minimum total cost, delay and distortion up to and including the kth slice, given that $u_{k-a}, \ldots, u_k$ are decision vectors for the $(k-a)$th to kth slices. Therefore, $G_I(u_{I-a}, \ldots, u_I)$ represents the minimum total cost, delay and distortion for all the slices of the frame, and thus:

$$\min_{u} J_{\lambda_1,\lambda_2}(u) = \min_{u_{I-a}, \ldots, u_I} G_I(u_{I-a}, \ldots, u_I) \tag{9.22}$$

The key observation for deriving an efficient algorithm is the fact that given $a+1$ decision vectors $u_{k-a-1}, \ldots, u_{k-1}$ for the $(k-a-1)$th to $(k-1)$th slices, and the cost function $G_{k-1}(u_{k-a-1}, \ldots, u_{k-1})$, the selection of the next decision vector u_k is independent of the selection of the previous decision vectors $u_1, u_2, \ldots, u_{k-a-2}$. This is true since the cost function can be expressed recursively as:

$$
\begin{aligned}
G_k(u_{k-a}, \ldots, u_k) = \min_{u_{k-a-1}, \ldots, u_{k-1}} & \left\{ G_{k-1}(u_{k-a-1}, \ldots, u_{k-1}) + E[D_k] \right. \\
& + \lambda_1 \cdot [B_{S_k}(\mu_{S_k}) C(\pi_{S_k}) + B_{T_k}(\mu_{T_k}) C(\pi_{T_k})] \\
& + \left. \lambda_2 \left[\frac{B_{S_k}(\mu_{S_k})}{R(\pi_{S_k})} + \frac{B_{T_k}(\mu_{T_k})}{R(\pi_{T_k})} \right] \right\}
\end{aligned}
\tag{9.23}
$$

The recursive representation of the cost function above makes the future step of the optimization process independent from its past step, which is the foundation of dynamic programming.

The problem can be converted into a graph theory problem of finding the shortest path in a directed acyclic graph (DAG) [16]. The computational complexity of the algorithm is $O(I \times |U|^{a+1})$ ($|U|$ is the cardinality of U), which depends directly on the value of a. For most cases, a is a small number, so the algorithm is much more efficient than an exhaustive search algorithm which has exponential computational complexity.

9.2.3 Application on Energy-Efficient Wireless Network

In this section, we consider an application of video transmission over a narrow-band block-fading wireless channel with additive white Gaussian noise. The main objective is to compare three error protection schemes: 1) UEP-UST, an unequal error protection scheme using the *separated* packetization scheme, where the shape and texture data are placed in separate packets and therefore can be transmitted over different service channels; 2) UEP-EST, an unequal error protection scheme using *combined* packetization (i.e., the shape and texture are placed in the same packet) where the packets can be transmitted over different service channels; 3) EEP, an equal error protection scheme using combined packetization, where all the packets are transmitted over the same service channel.

9.2.3.1   Channel Model

In our simulation for video transmission over wireless channels, we assume that each packet is sent over a narrow-band block-fading channel with additive white Gaussian noise. We further assume the channel fading for each packet is independent, and can be modeled by a random variable H. Thus the received signal $y(t)$ can be represented by:

$$
y(t) = \sqrt{H} \, x(t) + n(t)
\tag{9.24}
$$

where $x(t)$ is the transmitted signal, and $n(t)$ is an additive white Gaussian noise process with power spectral density N_0. We assume that H stays fixed during the transmission of a packet, and varies randomly between packets. Each realization h of H is chosen according to the *a priori* distribution $f_H(h \,|\, \theta)$, where θ is the channel state information (CSI) parameters known by the transmitter. Here we assume $\sqrt{H}$ is Rayleigh distributed, and assume $\theta = E[H]$, thus:

$$f_H(h \mid \theta) = \frac{1}{\theta} e^{-h/\theta}, \quad h \geq 0 \tag{9.25}$$

In our implementation, we assume that a packet is dropped if the capacity of the channel realization during that block is less than or equal to the information rate. For the ith shape packet, the capacity of the channel over which this packet is sent is:

$$\delta(\pi_{S_i}) = \frac{1}{W} \log_2 \left(1 + \frac{h_{S_i} P(\pi_{S_i})}{N_0 W} \right) \tag{9.26}$$

where W is the channel bandwidth. Therefore, the probability of loss for the ith shape packet is:

$$\rho(\pi_{S_i}) = \Pr\{R(\pi_{S_i}) \geq \delta(\pi_{S_i})\} = \Pr\left\{ R(\pi_{S_i}) \geq \frac{1}{W} \log_2 \left(1 + \frac{h_{S_i} P(\pi_{S_i})}{N_0 W} \right) \right\} \tag{9.27}$$

$$= \Pr\left\{ h_{S_i} \leq \frac{N_0 W}{P(\pi_{S_i})} (2^{R(\pi_{S_i})/W} - 1) \right\}$$

$$= 1 - e^{-\frac{N_0 W (2^{R(\pi_{S_i})/W} - 1)}{\theta \cdot P(\pi_{S_i})}}$$

Inversely, the power can be represented by:

$$P(\pi_{S_i}) = -\frac{N_0 W \cdot (2^{R(\pi_{S_i})/W} - 1)}{\theta \cdot \ln \rho(\pi_{S_i})} \tag{9.28}$$

Similar results can be derived for the ith texture packet, and the ith packet in the combined packetization.

9.2.3.2 Experimental Results

We encode the QCIF 'Children' sequence at 10fps, and set $\frac{N_0 W}{\theta} = 6W$, $W = 5MHz$, and $R = 200\,kbits/s$. We use six classes of service channels with powers equal to 1, 2, 3, 4, 6, and 10 Watts.

We compare the UEP-UST and UEP-EST schemes with an optimal EEP reference system. Figure 9.3 shows the cost-distortion (C-D) curves for these settings. Each point on the C-D curve of the optimal EEP system is obtained by trying all the fixed power levels and choosing the one that achieves the best quality for each cost constraint. The results in the figure indicate that adapting jointly the source coding parameters along with the selection of the transmission channel can provide significant gains in expected PSNR over equal error protection methods. Typically, in the UEP approaches, those packets vulnerable to packet loss (hard to be recovered by the concealment strategy employed) but robust to compression (acceptable distortion from quantization or other approximation processing) are sent through the better-protected channel. That is, a higher compression ratio might be used in the higher-cost channels than the lower-cost channels in order to

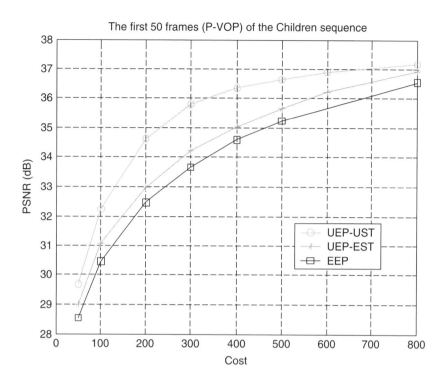

Figure 9.3 Comparison of Cost-Distortion curves for the UEP-UST, UEP-EST, and EEP schemes

distribute the total cost among the various packets efficiently. Furthermore, the UEP-EST approach outperforms the UEP-EST scheme because the UEP-EST approach has increased flexibility in providing unequal protection for shape and texture packets.

Figure 9.4 shows the distribution of the shape and texture packets among the six transmission channels for the UEP-UST approach when the cost constraint C_{max} equals 100, 200, 300 and 800. The results indicate that the shape packets are better protected than texture packets. During the increasing of C_{max} from 100 to 300, shape packets are selected more frequently than texture packets for transmission through the higher-cost channels. This is because shape packets have a lower bit consumption but a strong impact on the video quality. As shown in Figure 6.4, when $C_{max} = 300$ at least 80% of shape packets are transmitted over the most expensive channel, while over 60% of texture packets are transmitted over the two least expensive channels. In other words, the optimization process chooses to allocate more protection to the shape, because it impacts the end-to-end distortion more than the texture.

9.2.4 Application on Differentiated Services Networks

In this section, we consider video transmission over a differentiated services network. We simulate a simplified DiffServ network as an independent time-invariant packet erasure channel. Packet loss in the network is modeled as a Bernoulli random process. In addition, a packet is considered lost if it does not arrive at the decoder on time.

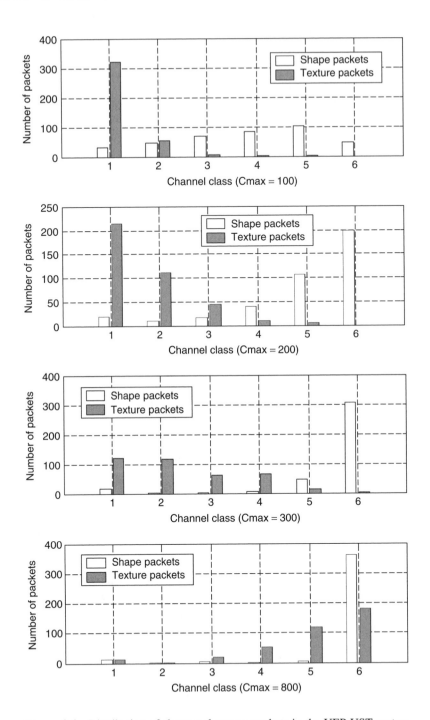

Figure 9.4   Distribution of shape and texture packets in the UEP-UST system

Table 9.1 Parameters of four service classes

Class	1	2	3	4
Probability of packet loss	0.2	0.1	0.05	0.01
Transmission rate (Kbps)	315	420	525	630
Cost (microcents/Kilobits)	10	30	60	100

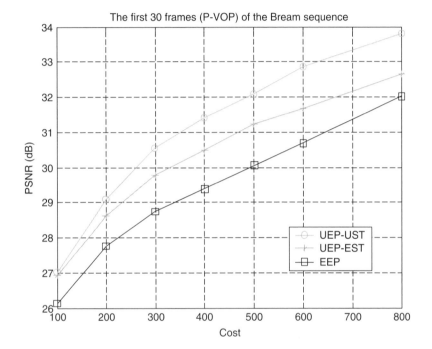

Figure 9.5 Comparison of UEP-UST with UEP-EST and EEP

There are four QoS channels available, whose parameters are defined in Table 9.1. The costs for each class is set proportionally to the average throughput of the class, which takes into account the transmission rate and probability of packet loss. In this experiment, a compressed RTP header (5 Bytes) has been added to each packet [17].

We encode the QCIF 'Bream' sequence at 30 fps and transmitted it over the simulated DiffServ network. Figure 9.5 shows the C-D curves for all the schemes. As expected, UEP-UST outperforms UEP-EST and both approaches outperform the EEP system.

9.3 Joint Source-Channel Coding with Utilization of Data Hiding

Data hiding [18] has been widely used in many applications. In copyright protection applications [19–21], specific signature information, called the watermark, is hidden invisibly inside a host image or video data source, by the owners before distributing their products. Later, the owner can retrieve the hidden signature from a distributed image or video product in order to prove its authenticity. In secure transmission applications [22], the secure data (for example, control information) are embedded inside the regular data to be transmitted in a standardized and open form, or over an insecure or readily available medium, such as the Internet, allowing only those authorized at the receiver side to retrieve the additional hidden informtion. An outstanding feature of data hiding scheme is that it provides the opportunity of retrival of hidden information even without the availability of the original host.

Using data hiding in error resilient video encoding and decoding has attracted much attention recently, for example, it can be used for increasing the error detection rate in video communication applications [23]. It has been proven that embedding redundant information, such as edge information and motion vectors, in encoding is very helpful for future error concealment. In [24] redundant information was embedded into the motion vectors of the next frame for protecting the motion vectors and coding modes of the current frame. Although this data embedded coding scheme degenerated the coding performance of half-pixel motion compensation back to integer-pixel motion compensation, the embedded information can effectively help the decoder to recover the motion vectors of the corrupted GOB (group of block) and conceal the missing blocks. The limited embedding capacity restricted the effectiveness of the proposed approach to cases of at most one corrupted GOB in each frame. In [25] the block type and major edge direction of content blocks of an image was embedded into the DCT coefficients of another block. In [26] and [3] one or several copies of an approximation version of the original frame was embedded inside the frequency coefficients. In [27] the FEC coded parity bits of the upper layer (the most important layer) were embedded in the least significant layer of a JPEG2000 code stream. In [28] the DCT coefficients of the segmented audio signal and the coarsest resolution chrominance components of the video frames were embedded in the luminance components. In [29], the data embedding scheme was used for error concealment for H.264 video encoding and transmission, where the edge information is embedded in I-frames, and the motion information, the reference frame, the code modes and the error concealment scheme are embedded in P-frames.

The general data hiding scheme is shown in Figure 9.6. At the sender side, a message m, encrypted optionally using a key K, is embedded in the host signal sequence x to form a signal sequence y, which is transmitted over a noisy channel. Signal y could be corrupted and become $\hat{y}$ at the receiver. The decoder extracts the estimated embedded signal $\hat{m}$. This process may require signal x, and key K. In error recovery and similar applications, we should note that a) key K may not be necessary, b) m can be correlated

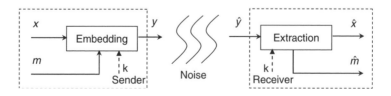

Figure 9.6 General data hiding scheme

to, or a subset of x, and c) at the receiver side the signal $\hat{x}$ is reconstructed from signals $\hat{y}$ and $\hat{m}$.

There are two popular data embedding and extracting approaches, the spread spectrum method [19, 30] and the odd-even method [31]. We describe both methods in the following, and for simplicity, we assume that the host data x is the DCT coefficients of the frame.

In the spread spectrum method, the data to be embedded is spread over the host data so that the embedded energy for each host frequency bin is negligible, which makes the embedded data imperceptible. First, an algorithm is used to generate a pseudo-noise signal sequence w, which has values in the range of $[-1,1]$ and is of zero mean. Then each embedding signal m_i is spread into M different locations $\{i, 2i, 3i, \ldots, Mi\}$ inside the host sequence as:

$$m_i^j = m_i w_{i \times j}, \; (j = 1, 2, \ldots, M) \tag{9.29}$$

The embedding process is:

$$y_{i \times j} = x_{i \times j} + \alpha m_i^j \tag{9.30}$$

where α is a scaling factor that determines the strength of the embedded information, that is, as α increases, the embedded data becomes more robust, but the difference between x and y increases. At the receiver side, the seed for generating the random sequence is known and thus w can be generated. Then an estimation of the spread version of the embedded signal is performed as follows:

$$\hat{m}_i = \frac{1}{M} \sum_{j=1}^{M} \hat{y}_{i \times j} w_{i \times j} \tag{9.31}$$

In the odd-even embedding method, the data is embedded in the non-zero quantized AC coefficients. If the bit to be embedded is '0', the AC coefficient will be forced to be an even number, otherwise, the AC coefficient will be forced to be an odd number. The scheme can be represented by:

$$Embed(p, b) = \begin{cases} p + 1 & \text{if } p > 0 \text{ and } (p - b) \bmod 2 \neq 0 \\ p - 1 & \text{if } p < 0 \text{ and } (p - b) \bmod 2 \neq 0 \\ p & \text{otherwise} \end{cases} \tag{9.32}$$

where p is the AC coefficient and b is the embedded bit.

The data extraction is quite straightforward, and can be represented by:

$$Extract(p') = p' \bmod 2 \tag{9.33}$$

where p' represents a received AC coefficient.

Clearly, in both methods, $x = \hat{x}$ cannot be guaranteed. However, if we assume that the embedded bitstream is either lost (dropped because of excessive delay or an uncorrectable bit error is detected) or received correctly ($y = \hat{y}$), then for the received bitstream, the odd-even embedding method guarantees that $m = \hat{m}$. This cannot be guaranteed by the spread spectrum method.

9.3.1 Hiding Shape in Texture

In MPEG-4, the data partitioned packetization scheme is applied so as to increase the error resilience. In this scheme, the shape and texture data are packed in the same packet (see Figure 9.7) but separated by a motion marker. In this way, when the partition containing texture data is corrupted, but shape and motion data are received, the obtained motion vector can be used to conceal the corrupted texture. However, since the decoding of the texture partition relies on information stored in the shape partition, such as texture motion vector, texture coding mode and shape BAB type, the entire packet will be discarded when the shape data is corrupted, even if the texture data is uncorrupted.

We consider the embedding of information of shape and motion partition into the texture data in order to make it self-decodable. Thus, texture data can be used even if the shape partition is corrupted. In addition, the embedded shape and motion data could also help to partially recover the lost shape and motion paritition. As shown in Table 9.1, the four types of data, shape BAB type, COD, texture coding mode and CBPC, are critical for the decoding of texture. We could embed the integer motion vectors in the range [−16, 16], and also a lossy version of the shape (using a lower-resolution of a 4×4 bitmap to represent the original 16×16 BAB).

It is very important to decide the amount of embedded information, because over-embedding could bring intolerable texture distortion, and the number of DCT coefficients limits the embedding capacity. Here, the odd-even method is used, which can guarantee the correct extraction of the embedded information. Then, the embedding capacity is further reduced to the number of non-zero DCT coefficients. To allocate the embedded data within limited spots efficiently, we propose five levels of embedding modes, that is, 0) No embedding; 1) Embed critical information (these data necessary for texture decoding as shown in Table 9.2) only; 2) Embed critical information, and motion vectors; 3) Embed critical information and lossy shape; 4) Embed critical information, motion vector and lossy shape. The embedding mode itself is also embedded inside the texture data.

The advantage of the scheme of 'hiding shape in texture' is demonstrated with a simple example shown in Figures 9.8 and 9.9. We encode the 'Children sequence' and transmit it over a noisy wireless channel. During transmission, the shape partition of the 3rd packet (3rd row of macroblocks) of the 16th frame (as shown in Figure 9.8(b)) is corrupted; thus the whole packet is discarded and a common concealment strategy, which uses the motion vectors of above macroblocks to get predictive data from the previous frame (as shown in Figure 9.8(a)), is called to recover the lost packet (as shown in Figure 9.8(c)). In Figure 9.9, the proposed data embedding scheme is used when the embedding level equals

Bab_type	MVDs	CR	ST	BAC	COD	MCBPC	MV	MVD2-4
Motion Marker		AC_pred_flag		CBPY	DC data	AC data		

Figure 9.7 MPEG-4 Data partitioned motion shape texture syntax for P-VOPs

Table 9.2 Information to be embedded into texture data

	Bits embedded	Necessary for texture decoding
Shape BAB type	Transparent: 1 Opaque: 2 Boundary: 2+4	Y
COD	1	Y
Texture mode	1	Y
CBPC	2	Y
Texture Motion vector	10	N
Lossy shape	4–16	N

(a) PSNR = 39.91 dB (b) PSNR = 37.31 dB, Rate = 5758 (c) PSNR = 23.08 dB

Figure 9.8 Example of reconstructed frame of 'Children' sequence: *a* corresponds to the 15th frame, *b* and *c* correspond to the 16th frame; in c, the shape partition of the 3rd packet is corrupted and then concealed

2, 3 and 4, respectively. Clearly, the embedding of shape and motion information increases the bit rate slightly and causes a very slight reduction in texture quality. However, the embedded information is very helpful in improving the concealed image quality. After the embedded lossy shape is extracted and used to recover the lost shape, the PSNR of the reconstructed image is increased by up to *2.5 dB*.

9.3.2 Joint Source-Channel Coding

According to Shannon's *Separation principle* [32], the design of source and channel coding can be separated without any loss in optimality as long as the source coding produces a bit rate which can be carried by the channel. The separation principle is powerful in that it divides a single complex problem into two simpler problems, where the designing of a communication system is divided into separate and independent source coding and channel coding subsystems. However, the principle relies on several assumptions that might break down in practice. For example, it assumes that the source and channel codes can be of an arbitrary large length, which is not practical because it could result in infinite delay and intolerable computation complexity. In addition, it assumes that both the source coder and the channel coder are optimal, that is, the source coder is designed assuming that the channel code will correct all errors introduced by the channel, and the

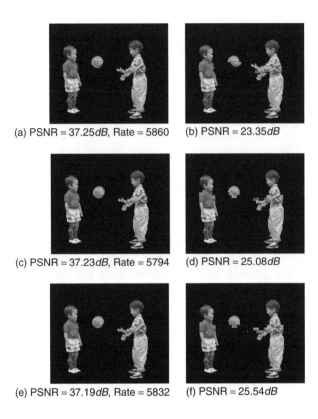

(a) PSNR = 37.25dB, Rate = 5860   (b) PSNR = 23.35dB

(c) PSNR = 37.23dB, Rate = 5794   (d) PSNR = 25.08dB

(e) PSNR = 37.19dB, Rate = 5832   (f) PSNR = 25.54dB

Figure 9.9 Example of reconstructing the 16th frame of 'Children' sequence using data hiding: a and b, c and d, e and f correspond to the case of using data embedding level 2, 3, and 4 respectively; in a, c, and e, no packet corruption took place, but in b, d, and f, the shape partition of the 3rd packet is corrupted and then concealed

channel codes are designed assuming that all bits created by the source code are equally important. Clearly, this assumption ignores the imperfections observed in real communications systems. Furthermore, the principle only applies to point-to-point communication system with a known source and channel distortion at code design time, which is thus not fit for the modern communications system with multi-user and time-varying transmission channels. Therefore, it is natural to consider joint source-channel coding. A practical state-of-the-art solution is to keep the source and channel coder separate but optimize their parameters jointly. Joint source-channel coding has become an active research area (see a recent review in [33]). It involves many facets of signal processing, communications, and information theory. In [4], a framework for the optimal selection of the source and channel coding rates over all scalable layers is proposed in order to minimize the overall distortion during video compression and transmission. The distortion caused by channel errors and the subsequent inter-frame propagation are analyzed theoretically, and an analytic solution for adaptive intra mode selection and joint source-channel rate control under time-varying wireless channel conditions is proposed. In [34], the joint source-channel coding is considered by controlling transmission power allocation at the physical layer for trading off video quality and resource allocation for energy efficient wireless video communications. Here, source-channel coding will be considered jointly with data hiding, which so far has not been addressed in the literature on object-based video communications.

9.3.3 Joint Source-Channel Coding and Data Hiding

Recent research into data hiding has considered data protection from the standpoint of network issues. In [35], distortion caused by data embedding, quantization and packet loss during transmission have been considered jointly in optimizing the source coding parameters. In [36], a hybrid method for lossy video communications was proposed which combines data hiding and unequal error protection. First, a second level wavelet approximation coefficient of the frame is embedded inside the DCT coefficients and the spread spectrum method is used to hide the marker for future concealment purposes. Then, the embedded video data is compressed and protected by channel coding. Packet protection levels are optimized in order to minimize the received video distortion. However, the research above only optimizes either the source coding or the channel coding, while none of it considers a joint optimization framework and adaptive data hiding.

Here, we consider jointly source coding, channel coding, data embedding and error concealment within a rate-distortion optimization framework. By selecting source coding and channel coding parameters and the level of data embedding, our goal is to minimize the total expected distortion given the frame bit budget. This is represented by:

$$\text{Minimize } E[D_{tot}], \text{ Subject to } R \leq R_{budget} \tag{9.34}$$

where $E[D_{tot}]$ is the expected total distortion for the frame, R is the total bit rate (including source and channel coding rate) for a frame, and R_{budget} is the bit budget for the frame. Clearly, almost the same problem formulation has been discussed in section 9.2.2.

Compared with the unequal error protection scheme on shape and texture described earlier, channel coding is used to provide error protection for the source data. In this way, the separate packetization of shape and texture data is not necessary, and the MPEG-4 data partitioning packet structure as shown in Figure 9.7 is used.

9.3.3.1 System Model

We consider an MPEG-4 compliant object-based video application, where the data hiding of the shape and motion information into the texture data, as mentioned earlier, can be applied here. In the encoding, the VOP is divided into 16×16 macroblocks, which are numbered in scan line order and divided into groups (rows) called packets. Each packet is decodable independently; that is, each packet has enough information for decoding and is independent of other packets and their related information. This guarantees that a single bit error only affects the decoding of a single packet. Let I be the number of packets in the given frame and i be packet index. For each macroblock, both shape coding parameters and texture coding parameters are specified. We use μ_{Si} and μ_{Ti} to denote the shape and texture coding parameters for all the macroblocks in the ith packet, and $B_{Si}(\mu_{Si})$ and $B_{Ti}(\mu_{Ti})$ the corresponding total number of bits used to encode these partitions. Let us denote by θ_i the level of shape embedding for the ith packet, and the total number of bits used to encode the texture partition in the ith packet is denoted by $B'_{T_i}(\mu_{T_i}, \theta_{i_i})$. Clearly, as mentioned earlier, θ_i is selected from $\{0, 1, 2, 3, 4\}$, and if $\theta_i = 0$, then $B'_{T_i}(\mu_{T_i}, \theta_{i_i}) \approx B_{T_i}(\mu_{T_i})$, because only one bit is embedded in the original packet to specify that there is no further information is embedded.

9.3.3.2 Channel Model

In wireless channels, harsh conditions often result in very high BERs (on the order of 10^{-1} to 10^{-3} BERs) [11], and thus channel coding is needed to bring the aggregate BER down to a level so that the error resilient tools at the decoder can be effective and provide reconstructed video with acceptable quality. Here, we apply channel coding separately on shape and texture partitions. As mentioned in Chapter 2, there are a number of channel coding approaches available. Without loss of generality, let us denote by r_{S_i} and r_{T_i}, respectively, the channel code rate for shape and texture of the ith packet, thus the total bit rate for the frame can be represented by:

$$R = \sum_{i=1}^{I} \left[\frac{B_{S_i}(\mu_{S_i})}{r_{S_i}} + \frac{B'_{T_i}(\mu_{T_i}, \theta_i)}{r_{T_i}} \right] \tag{9.35}$$

If this works, we assume that the burst errors can be converted into random errors with pre-interleaving [37], thus bit errors are considered. Let us denote by ρ_{Si} the probability of corruption of the ith shape data, and ρ_{Ti} the probability of corruption of the ith texture packet. Clearly,

$$\rho_{S_i} = 1 - (1 - p_e)^{\frac{B_{S_i}(\mu_{S_i})}{r_{S_i}}}, \quad \text{and } \rho_{T_i} = 1 - (1 - p_e)^{\frac{B'_{T_i}(\mu_{T_i}, \theta_i)}{r_{T_i}}} \tag{9.36}$$

where p_e is the BER at the decoder, which is usually smaller than the channel bit error rate.

9.3.3.3 Expected Distortion

We assume that the transmitter knows the channel condition. Since the distortion at the receiver is a random variable, let $E[D_i]$ represents the expected distortion at the receiver for the ith packet. It is equal to:

$$E[D_i] = (1 - \rho_{Si})(1 - \rho_{Ti})E[D_{R,i}] + (1 - \rho_{Si})\rho_{Ti}E[D_{LT,i}]$$
$$+ \rho_{Si}(1 - \rho_{Ti})E[D_{LS,i}] + \rho_{Si}\rho_{Ti}E[D_{L,i}] \tag{9.37}$$

where $E[D_{R,i}]$ is the expected distortion for the ith packet if both the shape and texture partitions are received correctly at the decoder, $E[D_{LT,i}]$ is the expected distortion if the texture partition is corrupted, $E[D_{LS,i}]$ is the expected distortion if the shape partition is corrupted, and $E[D_{L,i}]$ is the expected distortion if both partitions are corrupted. Clearly, $E[D_{R,i}]$ depends only on the source coding parameters for the packet, while $E[D_{LT,i}]$,

$E[D_{LS,i}]$ and $E[D_{L,i}]$ may also depend on the embedding level and the concealment strategy used at the decoder.

Note that the problem formulation and solution approach presented are general. Therefore, the techniques developed here are applicable to various concealment strategies used by the decoder. The only assumption we make is that the concealment strategy is also known at the encoder. A common concealment strategy is to conceal the missing macroblock by using the motion information of its neighboring macroblocks. When the neighboring macroblock information is not available, the lost macroblock is typically replaced with the macroblock from the previous frame at the same location. It is also important to note that the formulation presented here is applicable to various distortion metrics. In our experimental results we use the expected mean squared error (MSE), as is commonly done in the literature [1, 5, 6, 8].

Therefore, the optimization problem can be rewritten as:

$$\underset{\{\mu_{S_i}, \mu_{T_i}, r_{S_i}, r_{T_i}, \theta_i\}}{Minimize} \quad E[D_{tot}], \text{ subject to} : \sum_{i=1}^{I} \left[\frac{B_{S_i}(\mu_{S_i})}{r_{S_i}} + \frac{B'_{T_i}(\mu_{T_i}, \theta_i)}{r_{T_i}} \right] \leq R_{budget} \quad (9.38)$$

Similarly, the problem can be solved by Lagrangian relaxation and dynamic programming, as mentioned in previous sections.

9.3.3.4 Implementation Details

As the calculation of the expected distortion is closely related to the error concealment strategy, here we give an example and show how the E[D] can be derived afterwards. A typical error concealment strategy is as follows:

- If both shape and texture partitions are corrupted, then the decoder uses the shape (texture) motion vector of the neighboring macroblock above as the concealment motion vector. If the concealment motion vector is not available, e.g., because the above macroblock is also lost, then the decoder uses zero motion vector concealment.
- If only the shape partition is corrupted, then the decoder checks if enough shape and motion information is embedded inside the texture. If so, both shape and texture can be recovered; otherwise, if not enough shape information is available, the shape has to be concealed using the motion vector of its neighboring macroblock above; if the embedded motion information is not available and the texture is inter-coded, then the texture has to be concealed using zero motion vector concealment.
- If only the texture partition is corrupted, then the texture is recovered by using the motion vector stored in the shape partition.

The internal terms for calculating the expected distortion can be calculated as the follows:

$$E[\tilde{s}_n^{js} \tilde{t}_n^{jt}](I, I) = (1 - \rho_{S_i})(1 - \rho_{T_i})\hat{s}_n^{js}\hat{t}_n^{jt} + (1 - \rho_{S_i})\rho_{T_i}\hat{s}_n^{js} E[\tilde{t}_{n-1}^{m_t}]$$
$$+ \rho_{S_i}(1 - \rho_{T_i})\xi_n + \rho_{S_i}\rho_{T_i}(1 - \rho_{S_{i-1}})E[\tilde{s}_{n-1}^{m_s}\tilde{t}_{n-1}^{m_t}]$$
$$+ \rho_{S_i}\rho_{T_i}\rho_{S_{i-1}} E[\tilde{s}_{n-1}^{js}\tilde{t}_{n-1}^{jt}] \qquad (9.39)$$

$$E[\tilde{s}_n^{js}\tilde{t}_n^{jt}](I, P) = (1 - \rho_{S_i})(1 - \rho_{T_i})\hat{s}_n^{js}(\hat{e}_n^{jt} + E[\tilde{t}_{n-1}^{m_t}]) + (1 - \rho_{S_i})\rho_{T_i}\hat{s}_n^{js}E[\tilde{t}_{n-1}^{m_t}]$$

$$+ \rho_{S_i}(1 - \rho_{T_i})\xi_n + \rho_{S_i}\rho_{T_i}(1 - \rho_{S_{i-1}})E[\tilde{s}_{n-1}^{k_s}\tilde{t}_{n-1}^{k_t}]$$

$$+ \rho_{S_i}\rho_{T_i}\rho_{S_{i-1}}E[\tilde{s}_{n-1}^{js}\tilde{t}_{n-1}^{jt}] \tag{9.40}$$

$$E[\tilde{s}_n^{js}\tilde{t}_n^{jt}](P, I) = (1 - \rho_{S_i})(1 - \rho_{T_i})E[\tilde{s}_{n-1}^{m_s}]\hat{t}_n^{jt} + (1 - \rho_{S_i})\rho_{T_i}E[\tilde{s}_{n-1}^{m_s}\tilde{t}_{n-1}^{m_t}]$$

$$+ \rho_{S_i}(1 - \rho_{T_i})\xi_n + \rho_{S_i}\rho_{T_i}(1 - \rho_{S_{i-1}})E[\tilde{s}_{n-1}^{k_s}\tilde{t}_{n-1}^{k_t}]$$

$$+ \rho_{S_i}\rho_{T_i}\rho_{S_{i-1}}E[\tilde{s}_{n-1}^{js}\tilde{t}_{n-1}^{jt}] \tag{9.41}$$

$$E[\tilde{s}_n^{js}\tilde{t}_n^{jt}](P, P) = (1 - \rho_{S_i})(1 - \rho_{T_i})(\hat{e}_n^{jt}E[\tilde{s}_{n-1}^{m_s}] + E[\tilde{s}_{n-1}^{m_s}\tilde{t}_{n-1}^{m_t}])$$

$$+ (1 - \rho_{S_i})\rho_{T_i}E[\tilde{s}_{n-1}^{m_s}\tilde{t}_{n-1}^{m_t}] + \rho_{S_i}(1 - \rho_{T_i})\xi_n$$

$$+ \rho_{S_i}\rho_{T_i}(1 - \rho_{S_{i-1}})E[\tilde{s}_{n-1}^{k_s}\tilde{t}_{n-1}^{k_t}] + \rho_{S_i}\rho_{T_i}\rho_{S_{i-1}}E[\tilde{s}_{n-1}^{js}\tilde{t}_{n-1}^{jt}] \tag{9.42}$$

where

$$\xi_n = \begin{cases} (1 - \rho_{S_{i-1}})E[\tilde{s}_{n-1}^{k_s}\tilde{t}_{n-1}^{k_t}] + \rho_{S_{i-1}}E[\tilde{s}_{n-1}^{js}\tilde{t}_{n-1}^{jt}] & embedding\ level = 0 \\ \{(1 - \rho_{S_{i-1}})E[\tilde{s}_{n-1}^{k_s}] + \rho_{S_{i-1}}E[\tilde{s}_{n-1}^{js}]\}\hat{t}_n^{jt} & embedding\ level = 1\ or\ 2, \\ & intra\ texture \\ (1 - \rho_{S_{i-1}})E[\tilde{s}_{n-1}^{k_s}\tilde{t}_{n-1}^{k_t}] + \rho_{S_{i-1}}E[\tilde{s}_{n-1}^{js}\tilde{t}_{n-1}^{jt}] & embedding\ level = 1, inter\ texture \\ (1 - \rho_{S_{i-1}})E[\tilde{s}_{n-1}^{k_s}\tilde{t}_{n-1}^{m_t}] + \rho_{S_{i-1}}E[\tilde{s}_{n-1}^{js}\tilde{t}_{n-1}^{m_t}] & embedding\ level = 2, inter\ texture \\ \hat{s}_n^{\prime js}\hat{t}_n^{jt} & embedding\ level \geq 3, intra\ texture \\ \{(1 - \rho_{S_{i-1}})E[\tilde{t}_{n-1}^{k_t}] + \rho_{S_{i-1}}E[\tilde{t}_{n-1}^{k_t}]\}\hat{s}_n^{\prime js} & embedding\ level = 3, inter\ texture \\ \{(1 - \rho_{S_{i-1}})E[\tilde{t}_{n-1}^{m_t}] + \rho_{S_{i-1}}E[\tilde{t}_{n-1}^{m_t}]\}\hat{s}_n^{\prime js} & embedding\ level = 4, inter\ texture \end{cases} \tag{9.43}$$

shape is intra coded in (9.39) and (9.39), and inter coded in (9.40) and (9.41); texture is intra coded in (9.39) and (9.41), and inter coded in (9.40) and (9.42); $\hat{s}_n^j$ and $\hat{t}_n^j$ are the encoder reconstructed shape and texture of the jth pixel, and $\hat{s}_n^{\prime j}$ is the encoder reconstructed shape from embedded information in texture; pixel j in frame n is predicted by pixel m in frame $n-1$ if the motion vector is available, otherwise predicted by pixel k if the concealment motion vector is available; The subscript s and t of $k_s, k_t, m_s,$ and m_t in (9.40)–(9.42) are used to distinguish shape from texture, because the motion vector or concealment motion vector of shape could be different from that of texture. Computing $E[\tilde{s}_n^j\tilde{t}_n^j]$ in (9.40)–(9.42) depend on the computing of $E[\tilde{s}_n^j]$ and $E[\tilde{t}_n^j]$, which are calculated recursively as follows:

$$E[\tilde{s}_n^j](I) = (1 - \rho_{S_i})\hat{s}_n^j + \rho_{S_i}(1 - \rho_{T_i})\gamma_n + \rho_{S_i}\rho_{T_i}(1 - \rho_{S_{i-1}})E[\tilde{s}_{n-1}^k]$$

$$+ \rho_{S_i}\rho_{T_i}\rho_{S_{i-1}}E[\tilde{s}_{n-1}^j] \tag{9.44}$$

$$E[\tilde{t}_n^j](I) = (1 - \rho_{t_i})\hat{t}_n^j + \rho_{T_i}(1 - \rho_{S_{i-1}})E[\tilde{t}_{n-1}^k] + \rho_{T_i}\rho_{S_{i-1}}E[\tilde{t}_{n-1}^j] \tag{9.45}$$

$$E[\tilde{s}_n^j](P) = (1 - \rho_{S_i})E[\tilde{s}_{n-1}^m] + \rho_{S_i}(1 - \rho_{T_i})\gamma_n + \rho_{S_i}\rho_{T_i}(1 - \rho_{S_{i-1}})E[\tilde{s}_{n-1}^k]$$

$$+ \rho_{S_i}\rho_{T_i}\rho_{S_{i-1}}E[\tilde{s}_{n-1}^j] \tag{9.46}$$

$$E[\tilde{t}_n^j](P) = (1 - \rho_{t_i})(\hat{e}_n^j + E[\tilde{t}_{n-1}^m]) + \rho_{T_i}(1 - \rho_{S_{i-1}})E[\tilde{t}_{n-1}^k] + \rho_{T_i}\rho_{S_{i-1}}E[\tilde{t}_{n-1}^j] \tag{9.47}$$

where

$$\gamma_n = \begin{cases} (1 - \rho_{S_{i-1}})E[\tilde{s}_{n-1}^{k_s}] + \rho_{S_{i-1}}E[\tilde{s}_{n-1}^{j_s}] & \text{shape is not embedded} \\ \hat{s}_n^{\prime j_s} & \text{shape is embedded} \end{cases} \tag{9.48}$$

where shape and texture are intra coded in (9.44) and (9.45), and inter coded in (9.46) and (9.47).

9.3.4 Experimental Results

Our simulations are based on MPEG-4 VM18.0. The available Intra mode quantizers are of step sizes 2, 4, 6, 8, 10, 14, 18, 24, 30, and the available Inter mode quantizers are of step sizes 3, 5, 7, 11, 15, 19 and 25. The texture component of each macroblock can be coded as INTRA or INTER mode. The shape can be coded as transparent, opaque or boundary mode. For each boundary BAB, the scan type and resolution (conversion ratio of 1, 1/2 or 1/4) are also selected. As discussed earlier, Inter-mode shape coding has not been considered here because it violates the assumption that each packet is decodable independently [38]. We assume that the first frame in the sequence is coded as Intra mode with a quantization step size of 6 and that enough protection is used so that it arrives correctly at the decoder. This assumption makes the initial conditions of all the experiments identical.

In the first set of experiments, we consider the advantage of using an adaptive data hiding scheme in the source coding. We encode the first 80 frames of the 'Children' sequence and transmit them (without channel coding) over the wireless channels with BER $= 10^{-3}$ and BER $= 10^{-4}$, respectively. Two approaches are compared in Figure 9.10 with the adaptive data hiding method proposed applied in one of them while not in the other. In both approaches, however, the source coding is optimized. As expected, the method with adaptive data hiding outperformed the other. In the figures, the method using adaptive data hiding starts to gain after the bit rate is larger than a certain number, for example when rate is greater than *2400* bits in Figure 9.10(a). In other words, when the bit rate is very low, the two approaches have the same performance. This is reasonable because the texture data are coded in a very coarse visual quality, which makes the number of non-zero DCT coefficients fall below even the requirement for the first embedding level. When the bit rate increases, there is more room in the DCT coefficients for embedding, thus showing the benefits of using data hiding. Moreover, the figures indicate that the gain in using adaptive data hiding become smaller when a better channel with a smaller BER is used. It is reasonable because the gain of using data hiding comes mainly from the ability to recover shape and make use of the texture when the shape partition becomes corrupted. However, when the BER of the channel decreases, the probability that shape partition becomes corrupted also decreases, thus the gain decreases. In Figure 9.11, the

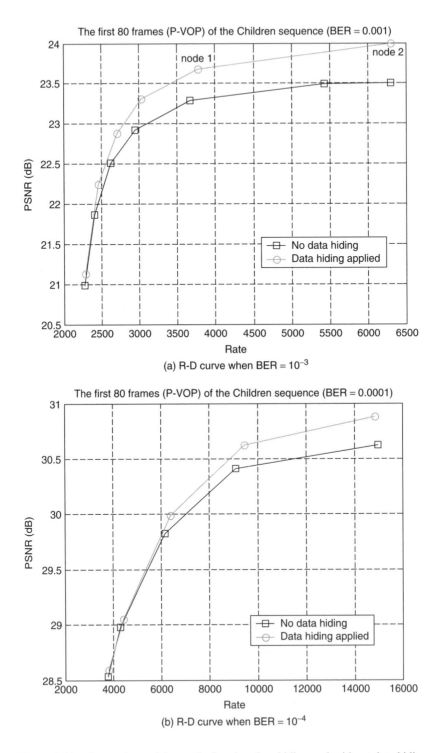

(a) R-D curve when BER $= 10^{-3}$

(b) R-D curve when BER $= 10^{-4}$

Figure 9.10   Comparison of the methods using data hiding and without data hiding

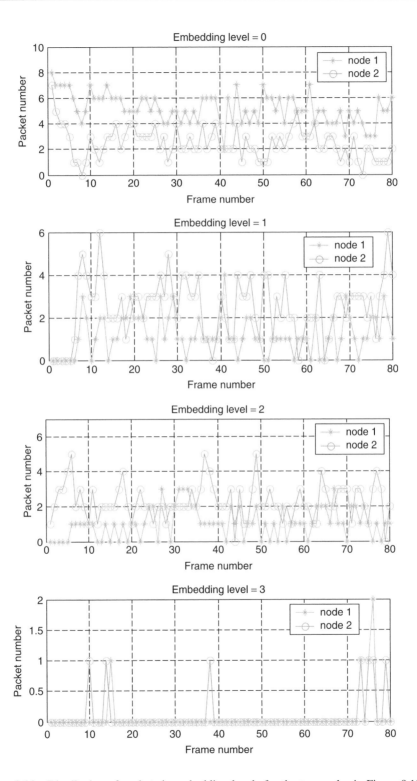

Figure 9.11   Distribution of packets in embedding levels for the two nodes in Figure 9.10(a)

distribution of packets of the frames in each embedding level for the two selected nodes in Figure 9.10(a) is shown. We find that the node corresponding to the lower bit rate (node 1) allocates more packets in embedding level 0 (which also means no embedding) than the other node (node 2), while allocates less packets in embedding level 1 and level 2. This indicates that when bit budget permits, data embedding is another effective way to help improve the reconstructed video quality other than source coding modes selection. However, the this phenomenon should not be interpreted as showing that more embedding is always better, because more embedding could cause higher distortion for texture data. As we pointed out previously, the relationship between source bit rate, distortion and data embedding is quite complicated. The distortion is caused by compression and transmission. When the source bit rate increases, the distortion caused by compression decreases, however, the probability of partition error increases, which results in an increase of transmission error. In addition, the increase of source bit rate could provide more room for data embedding, which generally reduces transmission error but increases compression error. This complicated relationship confirms indirectly the importance of our proposed optimal scheme. By adjusting dynamically the parameters in coding and embedding, the optimization procedure can find the best solution for minimizing the target function (or distortion).

In the second set of experiments, we consider the advantage of joint source-channel coding. We compare three approaches: 1) EEP method, where the shape and texture partition within a packet are protected equally (by channel coding) but different packets are allowed to be protected unequally; 2) UEP method, where the shape and texture partitions are protected unequally by channel coding; 3) Hybrid method, where the shape and texture partitions are protected unequally by channel coding, and the shape and motion data are allowed to be embedded into the texture data with various levels. Data hiding is not used in EEP and UEP methods, but source coding parameters are optimized for all of these methods.

In the simulation, we use an RCPC channel code with generator polynomials $(133,171)$, mother code rate $1/2$, and puncturing rate $P = 4$. This mother rate is punctured to achieve the $4/7, 2/3$ and $4/5$ rate codes. At the receiver, soft Viterbi decoding is used in conjunction with BPSK demodulation. We present experiments on Rayleigh fading channels, and the channel parameter is defined as $SNR = \alpha \dfrac{E_b}{N_0}$. The bit error rates for the Rayleigh fading with the assumption of ideal interleaving were obtained experimentally using simulations, as shown in Table 9.3.

We encode the first 80 frames of the 'Children' sequence and transmit them over the simulated wireless channels with SNR $= 6$ dB and SNR $= 10$ dB, respectively. The experimental results for the three approaches are shown in Figure 9.12. When SNR $= 6$ dB

Table 9.3 Performance of RCPC (in BER) over a Rayleigh fading channel with interleaving

Channel SNR (dB)	6	10	18
Channel rate $= 4/7$	5.3×10^{-4}	4.1×10^{-5}	3.8×10^{-6}
Channel rate $= 2/3$	7.4×10^{-3}	1.7×10^{-4}	1.2×10^{-5}
Channel rate $= 4/5$	4.0×10^{-2}	6.6×10^{-4}	3.6×10^{-5}

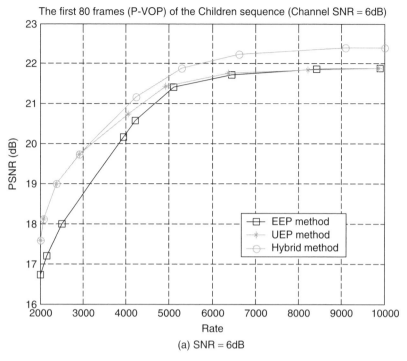

(a) SNR = 6dB

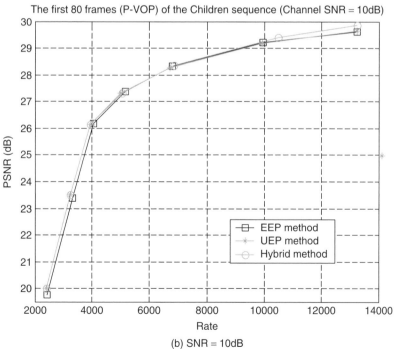

(b) SNR = 10dB

Figure 9.12   R-D curves for the 'Children' sequence

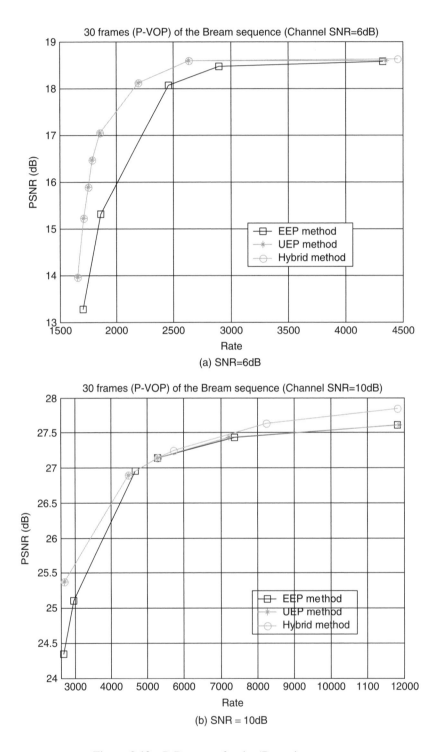

Figure 9.13   R-D curves for the 'Bream' sequence

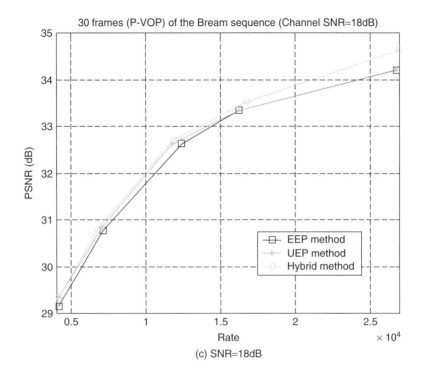

Figure 9.13   (*continued*)

(as shown in Figure 9.12(a)) the UEP method outperformed EEP at lower bit rates (when the rate was lower than 6000 bits) because it could use more channel bits to protect shape data, which have a stronger impact on decoded video quality. However, when the bit rate goes up, the probability of a corrupted partition becomes larger (recall that the probability of a corrupted partition is related to the partition length and BER of the channel). In order to control the error, both UEP and EEP methods are forced to choose the channel rate = 4/7 for channel coding, which corresponds to the smallest bit error rate. An interesting observation is that data hiding works well when the bit rate is high enough so that there are enough DCT coefficients available for embedding shape and motion information. Therefore, the hybrid method inherits the virtues of both UEP and data hiding, and thus performs well for all ranges of the bit rate.

However, the benefits of using UEP and hybrid methods decrease when the channel condition improves. The case for channel SNR = 10 dB is shown in Figure 9.12(b), where the hybrid method only has a maximum gain of 0.3 dB over EEP. The result is reasonable, because the differences between the BERs corresponding to various channel code rates become smaller, which directly affects the gains of UEP over EEP. On the other hand, better channel condition reduces the probability of the shape being corrupted, which directly affects the gains from using data hiding.

In another experiment, we tested the advantages of the proposed approach to sequences with higher motion. We encode the 'Bream' sequence from the 100th frame to the 130th frame, which corresponds to the flipping action of the Bream, and transmit them over wireless channels with SNR $=6$ dB, SNR $=10$ dB, and SNR $=18$ dB, respectively. The experimental results are shown in Figure 9.13. When channel SNR $=6$ dB (as shown in Figure 9.13(a)), the UEP performs better. It is understandable that data hiding has no contribution because the bit rate is too low to find enough spots to embed the lowest level shape/motion information. When channel SNR $=10$ dB (as shown in Figure 9.13(b)), the hybrid approach performs well at lower bit rates owing to UEP and at higher bit rates due to data hiding. When SNR $=18$ dB (as shown in Figure 9.9–9.13(c)), the proposed hybrid approach has a gain of 0.2–0.5 dB over EEP. The results of this experiment indicate that the proposed approach works better for video sequences containing higher motion and morphing activities. This is reasonable because the efficient encoding of those sequences relies heavily on shape and motion information, which makes the error resilience of such data very important and beneficial.

For readers interested to explore more relevant works, we would like to recommendation the following literatures: [21, 39–59].

References

1. R. Zhang, S.L. Regunathan, K. Rose, "Video Coding with Optimal Inter/Intra-Mode Switching for Packet Loss Resilience", *IEEE J. on Selected Areas in Communications*, Vol. 18, No. 6, pp. 966–976, June 2000.
2. A. Albanese, J. Blomer, J. Edmonds, M. Luby, and M. Sudan, "Priority encoding transmission", *IEEE. Trans. Information Theory*, Vol. 42, No. 6, pp. 1737–1744, Nov. 1996.
3. C. S. Lu, "Wireless multimedia error resilience via a data hiding technique", in *Proc. Int. Workshop Multimedia & Expo*, St. Thomas, USA, Dec. 2002.
4. L. P. Kondi, F. Ishtiaq, and A. K. Katsaggelos, "Joint Source/Channel Coding for SNR Scalable Video Processing", *IEEE Transactions on Image Processing*, Vol. 11, No. 9, pp. 1043–1052, Sept. 2002.
5. Y. Eisenberg, C. E. Luna, T. N. Pappas, R. Berry, and A. K. Katsaggelos, "Joint source coding and transmission power management for energy efficient wireless video communications", *IEEE Trans. Circuits and Systems for Video Technology*, Vol. 12, No. 6, pp. 441–424, June 2002.
6. Y. Eisenberg, C. E. Luna, T. N. Pappas, R. Berry, A. K. Katsaggelos, "Optimal source coding and transmission power management using a min-max expected distortion approach," in *Proc. IEEE International Conference on Image Processing, Rochester*, New York, Vol. I, pp. 537–540, September 2002.
7. A. Sehgal amd P. A. Chou, "Cost-distortion optimized streaming media over diffserv networks", in *Proc. IEEE International Conference on Multimedia and Expo*, Lausanne, August 2002.
8. F. Zhai, C. E. Luna, Y. Eisengberg, T. N. Pappas, R. Berry, and A. K. Katsaggelos, "Joint source coding and packet classification for real-time video transmission over differentiated service networks", *IEEE Trans. Multimedia*, 2004.
9. H. Wang, G. M. Schuster, A. K. Katsaggelos, "MINMAX optimal shape coding using skeleton decomposition", in *Proc. Int. Conf. Multimedia & Expo*, Baltimore, USA, July 2003.
10. J. Cai, Q. Zhang, W. Zhu, and C. W. Chen, "An FEC-based error control scheme for wireless MPEG-4 video transmission", in *Proc. IEEE WCNC'2000*, pp. 1243–1247, Chicago, Sept. 2000.
11. W. R. Heinzelman, M. Budagavi, and R. Talluri, "Unequal error protection of MPEG-4 compressed video", in *Proc Proceedings of the International Conference on Image Processing*, pp. 530–534, Oct. 1999.
12. S. Worrall, S. Fabri, A. H. Sadka, and A. M. Kondoz, "Prioritization of data partitioned MPEG-4 video over mobile networks", *European Transactions on Telecommunications, Special Issue on Packet Video*, Vol. 12, Issue No. 3, pp. 169–174, May/June 2001.
13. H. Wang, G. M. Schuster, A. K. Katsaggelos, "Operational rate-distortion optimal bit allocation between shape and texture for MPEG-4 video coding", in *Proc. Int. Conf. Multimedia & Expo*, Baltimore, USA, July 2003.
14. B. E. Carpenter and K. Nichols, "Differentiated service in the Internet", *Proceedings of the IEEE*, Vol. 90, No. 9, pp. 1479–1494, Sept. 2002.

15. H. Everett, "Generalized Lagrange multiplier method for solving problems of optimum allocation of resources", *Oper. Res.,* Vol. 11, pp. 399–417, 1963.

16. G. M. Schuster and A. K. Katsaggelos, *Rate-Distortion based video compression: optimal video frame compression and object boundary encoding*, Kluwer Academic Publishers, 1997.

17. S. Casner and V. Jacobson, "RFC 2508 - Compressing IP/UDP/RTP Headers for Low-Speed Serial Links", http://www.faqs.org/rfcs/rfc2508.html, The Internet Society, 1999.

18. F. A. Petitcolas, R. J. Anderson, and M. G. Kuhn, "Information hiding – a survey", *Proc. IEEE*, Vol. 87, No. 7, July 1999. pp. 1062–1078.

19. I. J. Cox, J. Killian, T. Leighton, and T. Shamoon, "Secure spread spectrum watermarking for multimedia", *IEEE Trans. Image Processing*, Vol. 6, pp. 1673–1687, Dec. 1997.

20. S. Craver, N. Memon, B. Yeo, and M. Yeung, "Can invisible watermarks resolve right ownership?", in *Proc. SPIE, Storage and Retrieval for Image and Video Database V*, Vol. 3022, 1997. pp. 310–321.

21. F. Hartung and B. Girod, "Digital watermarking of raw and compressed video", *Syst. Video Commun.,* pp. 205–213, Oct. 1996.

22. D. Mukherjee, J. J. Chae, S. K. Mitra, and B. S. Manjunath, "A source and channel-coding framework for vector-based data hiding in video", *IEEE Trans. Circuits Systems for Video Technology*, Vol. 10, No. 4, June 2000. pp. 630–645.

23. F. Bartolini, A. Manetti, A. Piva, and M. Barni, "A data hiding approach for correcting errors in H.263 video transmitted over a noisy channel", in *Proc. IEEE Multimedia Signal Processing Workshop*, 2001. pp. 65–70.

24. J. Song and K. J. R. Liu, "A data embedded video coding scheme for error-prone channels", *IEEE Trans. Multimedia,* Vol. 3, No. 4, Dec. 2001. pp. 415–423.

25. P. Yin, B. Liu, and H. H. Yu, "Error concealment using data hiding", in *Proc. Of IEEE Int. Conf. on Acoustics, Speech, and Signal Processing*, 2002. pp. 729–732.

26. M. Carli, D. Bailey, M. Farias, and S. K. Mitra, "Error control and concealment for video transmission using data hiding", in *Proc. Wireless Personal Multimedia Comm.*, Oct. 2002. pp. 812–815.

27. M. Kurosaki, K. Munadi, and H. Kiya, "Error correction using data hiding technique for JPEG2000 images", in *Proc. Int. Conf. Image Processing*, Barcelona, Spain, Sept. 2003. Vol. III, pp. 473–476.

28. A. Giannoula and D. Hatzinakos, "Compressive data hiding for video signals", in *Proc. Int. Conf. Image Processing*, Barcelona, Spain, Sept. 2003. Vol. I, pp. 529–532.

29. L. Kang and J. Leou, "An error resilient coding scheme for H.264 video transmission based on data embedding", In *Proc. IEEE Int. Conf. Acoustics, Speech, and Signal Processing*, Montreal, Canada, May 2004.

30. P. Campisi, M. Carli, G. Giunta, and A. Neri, "Tracing watermarking for multimedia communication quality assessment", in *Proc. IEEE Int. Conf. Communications*, New York, April 2002. pp. 1154–1158.

31. M. Wu, H. Yu and A. Gelman, "Multi-level data hiding for digital image and video", *SPIE*, Vol. 3854, 1999.

32. T. M. Cover and J.A. Thomas, *"Elements of information theory"*, John Wiley & Sons, New York, 1991.

33. R. E. V. Dyck, and D. J. Miller, "Transport of wireless video using separate, concatenated, and joint source-channel coding", *Proc. IEEE*, Vol. 87, No. 10, Oct. 1999. pp. 1734–1750.

34. F. Zhai, Y. Eisenberg, T. N. Pappas, R. Berry, and A. K. Katsaggelos, "Joint source-channel coding and power allocation for energy efficient wireless video communications," *Proc. 41st Allerton Conf. Communication, Control, and Computing*, Oct. 2003.

35. L. W. Kang, J. J. Leou, "A new error resilient coding scheme for H.263 video transmission", in *Proc. IEEE Pacific-Rim Conference on Multimedia*, Hsinchu, Taiwan, 2002. pp. 814–822.

36. C.B. Adsumilli, M. Carli, M.C.Q. de Farias, S.K. Mitra, "A hybrid constrained unequal error protection and data hiding scheme for packet video transmission", in *Proc. IEEE International Conference on Acoustics, Speech, and Signal Processing*, April 2003, Hong Kong.

37. B. Girod and N. Farber, "Wireless video", in *Compressed video over networks*, M.-T. Sun and A. R. Reibman, Eds. New York, Marcel Dekker Inc., 2000.

38. N. Brady, "MPEG-4 standardized methods for the compression of arbitrarily shaped video objects", *IEEE Trans. Circuits and System for Video Technology*, Vol. 9, No. 8, pp. 1170–1189, Dec. 1999.

39. M. R. Frater, W. S. Lee, and J. F. Arnold, "Error concealment for arbitrary shaped video objects", in *Proc. Int. Conf. Image Processing*, Vol. 3, Chicago, Oct. 1998. pp. 507–511.

40. U. Horn, K. Stuhlmuller, M. Link and B. Girod, "Robust Internet video transmission based on scalable coding and unequal error protection", *Image Communcation*, Special Issue on Real-time Video over the Internet, pp. 77–94. Sept. 1999.

41. A. K. Katsaggelos, L. Kondi, F. W. Meier, J. Ostermann, and G. M. Schuster, "MPEG-4 and rate distortion based shape coding techniques", *Proc. IEEE*, pp. 1126–1154, June 1998.

42. A. Li, J. Fahlen, T. Tian, L. Bononi, S. Kim, J. Park, and J. Villasenor, "Generic uneven level protection algorithm for multimedia data transmission over packet-switched network", in *Proc. ICCCN'01*, Oct. 2001.

43. G. M. Schuster and A. K. Katsaggelos, "Fast and efficient mode and quantizer selection in the rate distortion sense for H.263", in *SPIE Proc. Conf. Visual Communications and Image Processing*, pp. 784–795, March 1996.

44. S. Shirani, B. Erol, and F. Kossentini, "Concealment method for shape information in MPEG-4 coded video sequences", *IEEE Trans. Multimedia*, Vol. 2, No. 3, Sept. 2000, pp. 185–190.

45. K. Stuhlmuller, M. Link and B. Girod, "Scalable Internet video streaming with unequal error protection", *Packet video workshop' 99*, New York, April 1999.

46. M. D. Swanson, B. Zhu, and A. H. Tewfik, *"Data hiding for video-in-video"*, in *Proc. IEEE Int. Conf. Image Processing*, Vol. 2, Santa Barbara, CA, Oct. 1997, pp. 676–679.

47. H. Wang, G. M. Schuster, A. K. Katsaggelos, "Object-based video compression Scheme with Optimal Bit Allocation Among Shape, Motion and Texture", in *Proc. Int. Conf. Image Processing*, Barcelona, Spain, Sept. 2003.

48. H. Wang, A. K. Katsaggelos, "Robust network-adaptive object-based video encoding", in *Proc. IEEE Int. Conf. Acoustics, Speech, and Signal Processing*, Montreal, Canada, May 2004.

49. H. Wang, G. M. Schuster, A. K. Katsaggelos, "Rate-distortion optimal bit allocation scheme for object-based video coding", *IEEE Trans. Circuits Syst. Video Technol*, Vol. 15, No. 9, pp. 1113–1123, Sept. 2005.

50. H. Wang, F. Zhai, Y. Eisenburg, A. K. Katsaggelos, "Cost-distortion optimal unequal error protection for object-based video communications", *IEEE Trans. Circuits Syst. Video Technol*, Vol. 15, No. 12, pp. 1505–1516, Dec. 2005.

51. H. Wang, Y. Eisenburg, F. Zhai, A. K. Katsaggelos, "Joint Object-based Video Encoding, Transmission and Power Management for video streaming over Wireless Channels", in *Proc. IEEE International Conference on Image Processing*, Singapore, Sept. 2004.

52. H. Wang, F. Zhai, Y. Eisenburg, A. K. Katsaggelos, "Optimal Object-based Video Streaming over DiffServ Networks", in *Proc. IEEE International Conference on Image Processing*, Singapore, Sept. 2004.

53. H. Wang and A. K. Katsaggelos, "A hybrid source-channel coding scheme for object-based wireless video communications", in *Proc. Thirteenth International Conference on Computer Communications and Networks*, Chicago, Oct. 2004.

54. H. Wang and A. K. Katsaggelos, "Joint source-channel coding for wireless object-based video communications utilizing data hiding", *IEEE Trans. Image Processing*, August 2006.

55. D. Wu, Y. T. Hou, and Y. – Q. Zhang, "Transporting real-time video over the Internet: challenges and approaches", *Proc. IEEE*, Vol. 88, No. 12, Dec. 2000. pp. 1–19.

56. F. Zhai, C. E. Luna, Y. Eisenberg, T. N. Pappas, R. Berry, and A. K. Katsaggelos, "A novel cost-distortion optimization framework for video streaming over differentiated services networks," *Proc. IEEE International Conf. Image Processing (ICIP'03)*, Barcelona, Spain, Sept. 2003.

57. F. Zhai, Y. Eisenberg, T. N. Pappas, R. Berry, and A. K. Katsaggelos, "Rate-distortion optimized hybrid error control for packetized video communications," *IEEE Trans. Image Processing*, Vol. 15, pp. 40– 53, Jan. 2004.

58. A. K. Katsaggelos, Y. Eisenberg, F. Zhai, R. Berry, and T. N. Pappas, "Advances in Efficient Resource Allocation for Packet-Based Real-Time Video Transmission", *IEEE Proceedings*, Vol. 93, No. 1, pp. 135–147, Jan. 2005.

59. F. Zhai and A. K. Katsaggelos, "Joint Source-Channel Video Transmission", Synthesis Lectures on Image, Video, & Multimedia Processing, Morgan & Claypool Publishers, 2006.

10

AVC/H.264
Application – Digital TV

10.1 Introduction

Economically, digital TV is by far the most successful application of the digital video compression standards. It is also one of the main applications targeted by 4G Wireless systems. Since the mid 1990s, when TV began the transition from analog to digital, the number of channels over which digital TV services are distributed has increased significantly. Today those channels include Satellite, Cable, Terrestrial and Telephone networks as well as storage media such as DVD. The 4G wireless network is the latest addition to that list. The success of 4G networks may depend largely on the success they have in delivering digital video services such as digital TV. An end-to-end digital TV system uses multiple devices, such as encoders, decoders and receivers, designed, manufactured and deployed by several different companies. The ability to interoperate among those multiple devices is critical. Therefore, the standards for digital audio and video compression and the carriage of compressed bitstreams and related data play an important role in a digital TV system. They satisfy the critical need for interoperability. However, at the same time, there is also a great need and desire to have flexibility so as to be able to adapt to the varying requirements of different digital TV services – such as Video on Demand (VoD), free but commercial supported TV, broadcast or multicast TV and unicast based IPTV – and to adapt to the different characteristics – such as error rates, Quality of Service (QoS), two way connectivity and available channel capacity per program – of various delivery channels. Standards such as MPEG-2 [1–3] and AVC/H.264 [4] are designed to provide both interoperability and the flexibility so as to adapt the encoders as desired. It is important to understand the impact of the parameters, left at the discretion of the encoding side in a digital TV system, on key operations such as: random access, bitstream splicing and the implementation of trick modes. Therefore, it is not sufficient to have digital audio and video compression standards for the successful operation of a digital TV system. Recommended practices or the constraints on the flexibilities which those standards provide must also be specified. Various application layer standards bodies

4G Wireless Video Communications Haohong Wang, Lisimachos P. Kondi, Ajay Luthra and Song Ci
© 2009 John Wiley & Sons, Ltd

and industry consortia provide some of those constraints and recommended practices. Examples of such groups include: Digital Video Broadcasting (DVB) [5], Society of Cable Telecommunications Engineers (SCTE) [6], Advanced Television Systems Committee (ATSC) [7], 3rd Generation Partnership Project (3GPP) [8], 3rd Generation Partnership Project 2 (3GPP2) [9], Internet Streaming Media Alliance (ISMA) [10], DVD Forum [11], Blu-ray Disc Association [12] and Digital Living Network Alliance (DLNA) [13] among many.

Chapter 5 described the AVC/H.264 video coding standard, which is expected to play a key role in the success of 4G Wireless networks owing to its high coding efficiency. This chapter focuses on some of the encoder side flexibilities offered by that standard and their impact on key operations performed in a digital TV system.

In addition, it is equally important to specify in a digital TV system how the compressed bitstreams and other relevant information necessary for the proper working of the television system are packetized and carried, how the clocks on the sending and receiving side are synchronized and how video and audio are synchronized, etc. TheMPEG-2 Systems standard [1] is the most commonly used and recommended standard for this purpose. This chapter also provides a brief overview of the MPEG-2 Systems standard in section 10.5.

10.1.1   Encoder Flexibility

As explained in Chapter 5, in order to achieve both interoperability and flexibility, standards such as MPEG-2 and AVC/H.264 specify the requirements imposed on bitstream syntax and decoders and provide a great deal of flexibility for selecting various parameters on the encoding side. The rigid specification of the decoders provides for interoperability across various devices. As far as digital video decoding is concerned, the only flexibility allowed for at the decoding end is the selection of Profile and Level. On the other hand, a great deal of flexibility is left on the encoding side. That flexibility includes frequencies and arrangements of I, P, B, B_R pictures in the bitstream, number of reference pictures (only maximum number is specified), buffer size, coding and decoding delay, picture rates and rate control algorithms. The flexibility for selecting various parameters on the encoding side allows one to achieve desired user-experiences related to channel changing time, random accessing of a stored program and fast forwarding or reversing the play back, how much delay is added by encoding and decoding processes, visual quality for a given channel capacity and how often the battery in a portable player needs to replaced or charged. These parameters also influence coding efficiency and operations such as commercial insertions and multi-channel multiplexing performed in a head-end or a digital video server in a digital TV system. Therefore, proper selection and recommended practices for the use of those parameters are important in order for the deployed digital television system to behave in a way desired by the users.

10.2   Random Access

Random access in a coded bitstream is a critical requirement for a Digital TV system. It is used for changing channels as well as for bitstream splicing, as described in Section 10.3 and trick modes, such as fast forwarding or reversing the play back of a stored program, as described in Section 10.4. In order to achieve those capabilities with desired characteristics, application providers introduce one more hierarchy between the Sequence layer and

Picture layer described in section 5.2.8. It is called Group of Pictures (GOP). As the name suggests, it consists of a group of pictures. GOP was first defined formally in MPEG-1. In MPEG-1, GOP starts with an Intra coded picture in the compressed bitstream; and the last picture in the display order in a GOP is an I picture or a P picture. However, there is no definition of GOP in the AVC/H.264 standard. Therefore, even though the term GOP in relation to this standard is widely used by the user community, it remains a loosely defined term with the basic concept close to the one provided in MPEG-1. The number of pictures in a GOP is also not a fixed and universally accepted number. Although it is not a general practice, a GOP may contain more than one Intra coded picture. An Intra coded picture can either be an IDR or I picture. There can be an arbitrary number of inter coded (predicted) pictures – P, B or B_R – in a GOP (so far, the Extended Profile is not used in Digital TV applications, so SP and SI pictures are not used). The hierarchy of the video data organization now becomes:

Vidio sequence {GOP {picture { slices [MBs(sub-MBs (blocks (pixels)))]}}}.

Note the GOP layer between Sequence and Picture coding layers.

10.2.1 GOP Bazaar

As explained below, depending upon the number and the frequency of I, P, B and B_R pictures, GOPs can have various structures associated with them. These structures have a significant impact upon the performance of a digital TV system and the end user experiences mentioned above.

10.2.1.1 MPEG-2 Like, 2B, GOP Structure

The GOP structure most commonly used in MPEG-2 based Digital TV is shown in Figure 10.1.

The sequence of pictures in Figure 10.1 is shown in display order (it is sometimes also called capture order). The picture numbers (1 through 19) are shown in the row below the row showing the picture types. In this structure every 15th picture (picture numbers 1, 16, 31 and so on) is encoded as an Intra picture. The sub-structure between the two Intra coded pictures consists of two B pictures followed by one P picture. The arrows indicate the pictures used as references for motion estimation. For example, in order to compress picture number 19 as a P picture, picture number 16 (which in Figure 10.1 is shown to be compressed as an I picture) is used as a reference picture. As shown in this figure, B pictures use adjacent (one in future and one in past) I or P pictures as references and P

Figure 10.1 Pictures in capture or display order and temporal references

pictures use previous I or P pictures as references. In order to compress a B picture in this scheme, one has to wait for the reference P picture to arrive and be compressed before the B picture. This therefore causes a corresponding delay in an encoder and change in the order in which the pictures are compressed. The encoding order, which is also the order in which pictures appear in a bitstream, for the GOP of Figure 10.1 is shown in Figure 10.2 (note that in AVC/H.264 a video sequence stream has to start with an IDR picture, hence picture 1, the first picture in a sequence, is shown explicitly as an IDR picture). As the encoding order is not the same as the order in which the pictures are displayed, a decoder needs to re-order the pictures after decoding and before displaying them.

On the encoding side, B picture number 17 cannot be encoded until P picture number 19 arrives and is encoded. This causes delay in the encoding process. On the receiver side, P picture number 19 can not be displayed as soon as it is decoded. It has to be delayed and displayed after the B pictures numbered 17 and 18 are decoded and displayed. In order to avoid gaps in the displayed sequence, the entire video sequence needs to be delayed accordingly before displaying it. Therefore, the use of B pictures as shown in Figures 10.1 and 10.2 causes encoding and decoding (displaying) delays in this structure. However, the use of B pictures improves coding efficiency [14]. Therefore, from the standpoint of coding efficiency improvement, it is desirable to use B pictures. As a compromise between delay and the coding efficiency, two B pictures are most commonly used when B pictures are not used as references (see section 10.2.1.2).

When a channel is changed, one enters the bitstream at an arbitrary location. In this case, the history (past pictures) of the new bitstream is not available. Therefore, one cannot decode properly the P and B pictures which use those pictures as references which are before the entry point in the new bitstream. For example, if one enters the bitstream just before picture number 19, as shown in Figure 10.2(b), then reference picture number 16 is not available and picture number 19 can not be decoded properly. Consequently, the

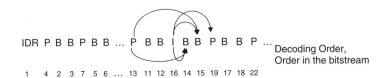

Figure 10.2(a) Pictures in encoding or decoding order corresponding to the structure in Figure 10.1

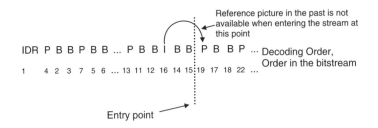

Figure 10.2(b) Random access in a bitstream corresponding to the structure in Figure 10.1

next pictures which use picture number 19 as a reference can also not be decoded properly and this chain reaction continues until the next I picture. As a result, the decoding process has to wait for the next Intra coded picture before the pictures can be decoded properly. Therefore, Intra coded pictures must be sent periodically and the spacing between the two Intra coded pictures is a key factor that determines the time it takes to change a channel. Shorter spacing between I pictures would provide quicker channel change. But, as I pictures contain a significantly higher number of bits in comparison to P and B pictures, shorter spacing between them will cause loss in coding efficiency. Therefore, as a compromise between quicker channel change time and coding efficiency, typically the spacing between the two Intra coded pictures is selected to be $\frac{1}{2}$ or 1 sec in broadcast bitstreams. The $\frac{1}{2}$ sec spacing corresponds to a spacing of 15 frames for a 30 frames/sec system used in North America. As quick channel change is not an issue in Video on Demand services or DVD playback, a longer spacing between I pictures is more commonly used.

Closed and Open GOPs

As explained above, in the case of random access, the decoder needs to wait until the Intra coded (I or IDR) picture arrives before being able to decode the pictures properly. However, even in that situation not all of the pictures appearing after the I pictures in the bitstream can be decoded properly in the GOP structure shown in Figure 10.1. For example, assume that one enters the bitstream after the Intra coded picture number – let us say just before picture number 7. Now, no pictures until I picture number 16 can be decoded properly. Consider the two B pictures (picture numbers 14 and 15) shown in Figure 10.2(a). They need to be displayed immediately before I picture number 16 but appear in the bitstream after that I picture as they use I picture number 16 as a reference for prediction. However, in order to be able to decode them properly, P picture number 13 is also needed as it is also used as a reference by those B pictures. However, P picture number 13 cannot be decoded properly because we entered the bitstream at picture number 7. Therefore B pictures numbered 14 and 15 can not be decoded and displayed properly even though they appear after the Intra coded picture in the bitstream. This type of GOP structure is called Open GOP. In this type of GOP, pictures in one GOP use pictures in other GOP as a reference and each GOP cannot be decoded fully and displayed as an independent entity. The GOP structure where each GOP can be decoded as an independent entity is called closed GOP. One way to re-structure and have closed GOPs is to end the GOPs with P pictures. Another way is to use only an I picture as a reference for those B pictures.

Closed, Partially Open and Fully Open GOPs in AVC

As described in Chapter 5, in AVC/H.264, both P and B pictures can use multiple pictures as references. So the temporal referencing in an AVC/H.264 bitstream corresponding to the GOP structure of Figure 10.1 can look like that shown in Figure 10.3. Note that P picture number 19 also uses P picture number 13 as a reference. Now, even if one enters in the bitstream of Figure 10.3 at I picture number 16, the decoder cannot decode B pictures numbered 14 and 15 properly, as well as P picture number 19. Pictures after P picture 19 use that picture as a reference and consequently they also cannot be decoded! Therefore, the impact of entering in the middle of the stream is much more profound than it was for the structure and temporal referencing shown in Figure 10.1. This type of GOP structure is called here 'Fully Open' GOP.

Figure 10.3   Pictures in capture or display order and temporal references

In this GOP structure, the entire GOP, not just the B pictures immediately following I pictures, becomes distorted when one enters in the middle and does not start decoding from the beginning of the stream. This is not a very practical situation in a digital TV system. This problem does not arise in MPEG-2 as P pictures can use only one picture in the immediate past as a reference. One possible way to avoid this problem is to place a restriction on the P pictures so that they cannot use the pictures before the I-picture as references. This implies that only one picture in the past can be used as reference by the P picture that is immediately after the I picture. The second P picture after the I picture can use up to two pictures in the past as references, the third P picture up to three references in the past and so on. This creates a random access related user experience that is similar to the open GOP situation explained in section 10.2.1.1 where few B pictures become distorted when one enters in the middle of the stream, and the rest of the pictures can be decoded and displayed properly. A reduction in the number of reference pictures will cause small loss in coding efficiency [14] for those P pictures that are close to the I pictures and this needs to be taken in to account when designing a rate control algorithm. This type of GOP structure is called here 'Partially Open' GOP in order to distinguish it from Fully Open GOP.

Another possible structure is called here 'closed' GOP. In this structure, neither P nor B pictures that arrive after the I picture in the bitstream use pictures before that I picture as references. This can be achieved in many different ways. One way is for B pictures, numbered 14 and 15 in Figure 10.3, to not use an I picture as a reference and use P pictures, numbered 13 and 10, as references (note that, unlike MPEG-2, in AVC both of the references for a B picture can be in the past). In this case, those B pictures will arrive before the I-picture. In addition, a restriction has to be placed for all the pictures appearing after the I-picture so that they do not use pictures before I pictures as references, i.e. pictures 17, 18, 19 and so on, do not use before 16 as references. AVC has defined such an I picture, where pictures appearing after it in the bitstream do not use pictures appearing before it as references, as an IDR (Instantaneous Decoding Refresh) picture.

In order to distinguish between the Intra coded pictures in these GOP structures, I-pictures in Partially Open GOP are sometimes also called 'Acquisition' I-pictures, so as to signify that unlike the case of Fully Open GOP, one can acquire the bitstream (with only a few distorted pictures in the beginning) after entering in the middle.

10.2.1.2   Reference B and Hierarchical GOP structures

Unlike MPEG-2, AVC allows for B pictures to be used references. Therefore, some applications use the GOP structure shown in Figure 10.4. In this figure, the pictures are shown in capture order and B_R denotes the B pictures that are used as references.

Figure 10.4 GOP with use of B-pictures as references. B_R denotes the B pictures used as references

An advantage of this structure is that the B pictures (pictures 2 and 4) have reference pictures that are closer than the traditional GOP structure with two B pictures as shown in Figure 10.1. This provides a better estimate of the motion in those pictures. However, the P pictures now are farther apart, which may require a wider motion search range for those pictures for a good estimate of the motion for the sequences with high motion. In addition, as B_R pictures are generally only used as references locally (by neighboring B pictures), they can be quantized more than the P pictures. This GOP structure can provide better coding efficiency than that in Figure 10.1 for sequences which do not have very large motion or for the encoders with a large motion search range. Another structure that uses B pictures as references is shown in Figure 10.5. It adds significant delay but can potentially provide higher coding efficiency. However, the B reference in the middle is quite far from the anchor I-pictures and may require a significantly wider motion search area in order to obtain a good estimate of motion. This type of B picture in a GOP is sometimes also called Hierarchical B pictures and the GOP structure is also sometimes called a Hierarchical GOP structure.

10.2.1.3 Low Delay Structure

In many applications, such as video phone, or video conferencing or conversational video programs, it is desirable to have low delay during the encoding and decoding processes. In those applications typically a IPP... GOP structure is used. As B pictures in AVC can have both of the reference pictures in the past, they may also be used for low delay applications with both references in the past.

10.2.1.4 Editable Structure

In studio and content creation environments, the ability to access each picture as well as the ability to enter the stream at any picture are key requirements. If long GOP structures,

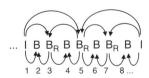

Figure 10.5 A Hierarchical GOP structure

as shown in Figures 10.1 through 10.5, are used, it becomes very difficult to decode any arbitrary picture. In order to ease that process short GOP structures, such as I-only or IPIPIP... or IBIBIB ..., structures are used. If P and B pictures are used then typically only one or two pictures, respectively, are used as references. These structures have a lower coding efficiency in comparison to the longer GOPs owing to I pictures appearing too frequently. Therefore, they are not used to send video to the home in a digital TV system.

In studio and content creation environments, higher video fidelity than that in consumer applications is also generally desired. Higher fidelity includes more than 8 bits per luma and chroma samples, denser chroma sampling (like 4:2:2 or 4:4:4) and higher compressed bit rates. In order to support those applications, as described in Chapter 5, specific profiles, such as High 10, High 4:2:2, Intra- only etc., are provided. Those profiles, except Intra-only profiles, also support arbitrary GOP structures and one can chose the one most suitable for any given situation.

10.2.1.5   Others

It should be clear by now that a virtually endless number of GOP structures can be created by using various combinations of the number of B pictures between P pictures, the number of B_R pictures between P or I pictures, the number of P pictures between I pictures, the number of reference pictures, the location (past or future) of reference pictures, etc. It is very hard, if not impossible, to tabulate all of those combinations. In a simple application where one only decodes and displays a stream, this does not give rise to too much complexity and hardship. However, in a real application environment such as digital TV, one also performs bitstream splicing, random access, special effects (fast forward, reverse etc.), transcoding, etc. As discussed in sections 10.3 and 10.4, the complexity of those operations increases significantly as the number of possible GOP variations increases. Therefore, various application related standards bodies, such as SCTE, DVB, etc., impose further constraints on this flexibility.

10.2.2   Buffers, Before and After

10.2.2.1   Coded Picture Buffer

In majority of the applications, digital video compression produces an unequal number of bits per picture. Intra pictures contain a significantly larger number of bits per picture than P and B pictures. Depending upon the content type, Intra pictures can contain ten or more times the number of bits used in P pictures. P pictures may contain two to four, or more, times the number of bits used in B pictures. These compressed bits are sent through two types of channels – Constant Bit Rate (CBR) or Variable Bit Rate (VBR). In CBR, as the name indicates, the bits are sent at constant bit rates. Therefore, the pictures do not arrive periodically. Even in VBR channels, the bit rates generally do not vary to such an extent as to allow one to send each and every picture during the same time period. Therefore, in order to match a codec to the channel, the bits produced at the encoder and received at the decoder are stored in buffers – after encoding on the encoding side and before decoding in a decoder. The buffer where the incoming bits are stored at the decoder is called the

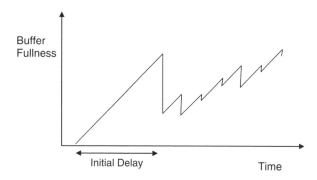

Figure 10.6   Buffer fullness at a decoder

Coded Picture Buffer (CPB) in AVC. An example of how this buffer's fullness may vary with time in the case of CBR is shown in Figure 10.6.

In the case of CBR, the buffer is filled at some constant bit rate corresponding to the channel rate. When bits corresponding to a picture are removed, the buffer is emptied by that amount. In the decoding model, it is assumed that all of the bits corresponding to a picture are taken out instantaneously and also that the pictures are decoded instantaneously. In a real implementation, a decoder needs to take into account how far it is from this model and make appropriate design modifications so that its buffer does not overflow or underflow (or if it underflows so that this condition does not cause distortion in the video presentation).

Initial CPB delay signals a decoder when to start taking the bits out of the input buffer. Longer delay allows the buffer to be filled up at a higher level before the bits are taken out for decoding. At the initial CPB delay time, the first picture (IDR) is removed from the buffer and decoded. As IDR pictures contain large number of bits in comparison to P and B pictures, the buffer occupancy drops by a larger amount when an IDR (or I) pictures is removed. Assuming that the GOP is as shown in Figure 10.2, a P picture is removed at a time equal to inverse of the frame rate (e.g. 1/30 sec in a 525 line system). After this P picture, two B pictures are removed at their corresponding decoding time. Notice that it takes more than one frame time to receive the I pictures. Therefore, an encoder has to compensate for that by making B pictures, and many times P pictures, small enough so that 'on average' the frames are received at the frame rate of the video. If CPB delay is large then one has a greater number of pictures to average over and to distribute the bits over so that the buffer does not underflow or overflow.

In order to allow for interoperability among various encoders and decoders, the minimum CPB buffer size in a decoder is also specified as part of the Level specification in the standard. The CPB sizes specified in the standard for Baseline, Extended, Main and High profiles for VCL are shown in Table 5.6 in Chapter 5. See the AVC standard specification for the buffer sizes corresponding to other profiles.

Encoders are required to produce the bitstreams so that they do not require a decoder to have an input buffer larger than that specified in the standard. An encoder does not need to, and a typical encoder does not, use the fully available CPB size. How much of a CPB buffer is used is controlled via the rate control algorithm in an encoder. A larger buffer size allows for greater variation in the bits from one part of a sequence to another. This allows one to maintain a closer to equal quality across the video sequence by allocating a larger number of bits for harder to compress complex scenes while allocating a smaller number of bits for simpler scenes. In other words, larger buffer size allows one to have a longer time window over which to maintain the bit rate corresponding to the CBR channel. However, a larger buffer size also means longer delay through the system as one has to wait longer for the buffer to fill in before starting to decode the pictures. This also implies a longer channel change time. The encoding algorithms choose the targeted buffer size in order to provide a good compromise between the delay and the video quality. For applications that are insensitive to delay, such as Video on Demand (VOD), larger buffer sizes can be used. In the AVC standard, the maximum buffer sizes were selected to be big enough to allow these applications to have better video quality by using a larger buffer size. As an example, the typical range of a compressed bitstream for SDTV (Level 3) is between 1 to 3 Mbps. Therefore, at 1 Mbps, the buffer size is big enough to allow storage for 10 second long duration. For applications such as video conferencing requiring low delay, a smaller buffer size is used. TV broadcasting, where both shorter channel change time as well as good video quality are desired, buffer sizes used are in between those used for VOD and video conferencing.

10.2.2.2 Decoded Picture Buffer (DPB)

Unlike in MPEG-2, an AVC encoder may choose multiple numbers of pictures as references while compressing P or B pictures. Therefore, a decoder needs to have a picture buffer which at least stores those decoded pictures that are used as references. To allow for interoperability among various encoders and decoders, the standard specifies the largest decoded picture buffer (DPB) size that an encoder can use while generating the bitstreams. An encoder is free to use a smaller than the maximum number specified in the standard. However, a decoder needs to have a decoded picture buffer (DPB) that is at least as big as that specified in the standard. Table 5.6 in Chapter 5 shows the maximum DPB size specified in the standard for 4:2:0 resolution. When the pictures are no longer needed for the reference, the decoder removes them from DPB. In digital TV applications the FIFO (First In First Out) principle – the first reference picture to arrive in a decoder is removed first – is commonly used. However, it is easy to conceive of scenarios where that may not be the case. One example is that in a video sequence with low motion, it may be more desirable to use an I picture, that is further away in the past, as a reference than the P or B_R picture that is closer in time (P or B_R pictures may contain more compression noise that an I picture). In this case the FIFO rule will need to be overridden and the decoder needs to be signaled as to which pictures to discard from the DPB buffer in order to make room for the new pictures. This signaling and DPB buffer management is done via Memory Management COmmands (MMCOs) specified in the standard. When a channel is changed, the pictures in the DPB need to be cleared out of the buffer to make room for the new reference pictures. Some of those reference pictures may not have been displayed at that time and the system may decide to display them before discarding them from the DPB.

10.3   Bitstream Splicing

In applications such as commercial insertion, two compressed bitstreams – the bitstream corresponding to the main program and another bitstream corresponding to a commercial – need to be spliced together. Various scenarios related to the commercial insertion exist. A commercial may be appended before or at the end of the main video stream. Alternatively, commercial may need to be inserted inside the main bitstream. In that case, the commercial may either be replacing other commercial already present at that spot and a new commercial may need to be inserted inside the bitstream or the main bitstream may need to be cut apart to make room for the commercial. Consider the case where the main bitstream needs to be cut apart and a new commercial be inserted. Generally, the main program stream is cut at IDR locations. The new bitstream created after the commercial insertion needs to satisfy the various constraints specified in the standard. For example, the bitstream should be such that the CPB does not overflow or underflow. In addition, at application level the bitstream should be such that it does not create unpleasant visual effects such as severe distortion in the quality of the pictures at the junction points. One also needs to take into account that commercials and main videos may be (and generally are) compressed at different locations and times and by different encoders with no knowledge of the parameters used when compressing the other stream. Therefore commercials and main videos may have different parameters such as number of reference pictures used, CPB delay used and the GOP structure used. A splice is called seamless when there are no artifacts introduced at the junction points. However, seamless splices are harder and hence more expensive to create. Let us go a layer deeper and see how all of this impacts upon the splicing point.

Consider the GOP of Figure 10.1 and the corresponding order of the pictures in the bitstream shown in Figure 10.2 and also in Figure 10.7 with reference picture dependency shown by arrows.

Let us assume that both the main video and the commercial use the same GOP structure. If we now cut the main stream just before picture 16 (I picture) and insert some other stream then pictures 14 and 15 (B pictures) loose reference picture 13 (P picture) and will not be decoded correctly. Therefore, there are two options, either re-encode pictures 14 and 15 with only picture 16 as a reference or just live with the fact that those two pictures will have distortion. In this case, the splicing is not seamless. As that distortion will be limited to only two pictures it may not be too noticeable. To minimize the visual annoyance at the seam, those two pictures may be replaced by gray pictures. If the I picture with picture number 16 is converted to an IDR picture then the picture order will not be as shown in Figure 10.7 as pictures 14 and 15, in that case, are not allowed to refer to the picture before the IDR picture (i.e. they can not refer to picture 13). In that situation,

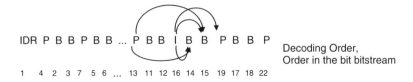

Figure 10.7   Picture order in the bitstream for GOP shown in Figures 10.1 and 10.2

IDR P B B P B B ... P   B B P P IDR P B B P ...

1   4 2 3 7 5 6 ... 13   11 12 14 15 16   19 17 18 22...

Figure 10.8 A possible picture order and picture type in the bitstream around IDR in the middle of the stream

many different possibilities exist. Either those pictures are compressed as P pictures using picture 13 as a reference with the GOP structure as shown in Figure 10.8 (those P pictures can also use IDR picture 16 for reference as AVC allows P pictures to use a picture in the future in display/capture order as a reference). Or, B pictures 14 and 15 use IDR and P picture 19 as references. These GOP structure decisions around the splice points can either be made by the encoder, expecting that the splicer will use these as splice points, or the splicer can modify the GOP structure around the junctions of the bitstreams.

Next, consider the case where the GOP structures of the main video and the commercial sequences are different. Let us assume that the GOP structure of the main video is as shown in Figure 10.7. In that structure, each P picture uses only one reference picture in the past and B pictures use one in the past and one in the future as reference pictures. Assume that the commercial bitstream is such that B pictures use two pictures in the future and one in the past as reference pictures, as shown in Figure 10.9. To keep the example simple, let the P pictures use only one picture in the past as a reference, as is done by the P pictures in the main program. Now the pictures in the bitstream appear as shown in Figure 10.10.

As an example, consider I picture number 16 in the commercial video sequence shown in Figure 10.9. The picture decoding order in the bitstream is shown in Figure 10.10. I picture 16 needs to be delayed and displayed after B picture 15 is decoded and displayed. In order to avoid gaps in the display, the entire sequence needs to be delayed accordingly before it can start to display. Clearly, this delay is longer than the corresponding delay in the main video with the GOP structure as shown in Figure 10.7. Therefore, during the transition from the main program to the commercial there will be no picture available for display for some time. A similar situation will arise if the commercial uses B_R as shown

I B B P B B P ... P B B I B B P

1 2 3 4 5 6 7 ... 13 14 15 16 17 18 19

Figure 10.9 GOP structure, in capture/display order, where B pictures use 2 future and 1 past pictures as references

... P I B B P B B P

... 13 16 11 12 19 14 15 22 ...

Figure 10.10 The picture order in the bitstream and decoding order for the GOP of Figure 10.9

in Figures 10.4 and 10.5. In this case one needs to either insert dummy gray picture(s) or repeat the previously displayed picture. While coming out of the commercial into the main program one will be transitioning into the bitstream with a smaller display delay. Therefore, there are now extra pictures to be displayed. One can either discard those pictures or delay all subsequent pictures. Hence, the complexity of the operation increases significantly when the GOP structures of the two sequences to be spliced together are not the same.

Another issue that needs to be taken into account in the splicing operation is the mismatch between the CPB delays of the main video and the commercial sequences. If the commercial has a longer CPB delay then additional dummy pictures need to be added. If commercial has a shorter delay then one or more pictures may need to be removed from the main video. Or, the streams need to be recompressed so as to have the same delay. During these processes, one has to also ensure that the CPB buffers do not overflow or underflow while decoding the spliced bitstreams. Considering the vastly greater number of possible variations which are available in the GOP structures in AVC than in MPEG-2, the commercial insertion process is significantly more complex for AVC streams than for MPEG-2. The industry wide practice of using the GOP structure of Figure 10.2 in MPEG-2 video compression based systems also makes the task simpler for those systems. Although the GOPs in Figures 10.3 or 10.4, are more commonly used, the digital TV community and application standards mentioned in section 10.1 have not converged on one single GOP structure for AVC in digital TV systems. Considering that various GOP structures are already widely used, the chances of that happening are small and commercial insertion devices need to be aware of the various possible GOP structures when inserting commercials.

10.4   Trick Modes

In addition to decoding and playing back a stored digital video, commonly used functions include fast forward, reverse and pause, etc. They are also called trick modes or special effects. There is no single standard that specifies how to perform these functions and various devices, such as DVD players and DVRs (Digital Video Recorders), take many different approaches.

When the compressed stream is being recorded by a DVR, an index table pointing to the start code within the stream can be created and saved. The table can be used for indexing locations of (I, P, and B) pictures within the stream in order to allow a decoder to further manipulate the stream so as to remove certain pictures without parsing the entire stream. However, if the compressed stream is scrambled, the picture start codes and coding types are also scrambled and are hard to separate from other picture header data. Therefore, in this case, the index table cannot be created without first descrambling the stream.

In AVC/H.264, access unit delimiter NAL units can be used to index the location of the pictures within a stream in order to support trick modes. To provide better indexing efficiency and greater security, the access unit delimiter NAL unit can be transferred without scrambling while all other AVC contents are scrambled. During the recording stage, the video streams do not need to be descrambled. Since the access unit delimiter NAL unit is not scrambled, the indexing engine can find the unit and build the index table without descrambling and processing the video content, i.e. recorded video content is still in its originally scrambled form.

10.4.1 Fast Forward

The simplest way to implement Fast Forward (FF) is to decode only the I pictures. The rate at which they are decoded and the duration for which each one is displayed depends on the fast forward rate. For example, if I pictures are 15 frames apart then to achieve about 8 ×, 4 × and 2 × speeds, each I picture has to be decoded and displayed for 2, 4 and 8 picture durations. To go higher than 8 ×, a decoder may skip one or more I pictures. Note that although I pictures are simpler to decode than P and B pictures owing to the lack of temporal dependencies, they consist of a large number of bits. Therefore, reading and decoding I pictures at video rate (30 frames per sec for a 525 line system) may surpass the bandwidth constraints in some designs. In those designs a trade off is done where more than one I picture may be skipped so as to allow the decoder more than one frame time to retrieve and decode the I pictures. However, only decoding I pictures in the FF mode provides slide show type display characteristics. This may be annoying at low FF speeds such as 2 × or 4 ×. Two approaches are taken in that situation. In one approach, only the reference pictures are decoded and non-reference B pictures are skipped. In the other approach, the full stream is decoded at faster than the real time. These provide better visual quality in FF mode at low FF speeds.

10.4.2 Reverse

Achieving the Reverse function is a harder task than FF. As the bitstreams are created in the decoding order with past references appearing before the current picture, to decode a picture that is not an I picture one has to go all the way back to the start of the GOP and decode all of the pictures up to the current picture in order to display it. This process becomes harder to implement and going from I to I picture becomes by far the simplest way to achieve the Reverse function.

10.4.3 Pause

Pause is easy to achieve as all it requires is to stop a decoder at a given place and display the current picture till the Pause function is removed.

10.5 Carriage of AVC/H.264 Over MPEG-2 Systems

The MPEG-2 standard consists of several parts in addition to Part 2 which specifies the digital video compression standard. One of those parts is MPEG-2 Systems [1], known formally as ISO/IEC 13818-1 (MPEG-2 Part 1). It provides the functions and capabilities critical for the operation of a digital TV system, including:

1. Packetization of the compressed video and audio streams (called elementary streams).
2. Synchronization of audio video streams in order to provide capabilities such as lip synchronization so that audio and visual movements are synchronized when presented after decoding.

3. Synchronization of clocks at the transmitting and receiving ends.
4. Multiplexing of several programs.
5. Tables to specify where to find various programs in a multiplex.
6. Hooks to support encryption and exchange of security keys.
7. Recovery of a packet in error prone conditions.
8. System target decoder model and system level buffering requirements at the receiving end, etc.

10.5.1 Packetization

Two layers of packetization are specified in the MPEG-2 Systems standard: Packetized Elementary Stream (PES) packets and Program Stream (PS) or Transport Stream (TS) packets.

The PES layer is the network agnostic packetization layer that provides basic functions, such as stream identification, audio video synchronization, copy right information, etc. PES packets are further packetized into Program Stream (PS) or Transport Stream (TS) packets. This layer of packetization includes the information necessary for a digital TV system to operate. In TS, this information includes timing information to synchronize the sending and receiving platforms, directory of the programs multiplexed together, encryption and scrambling information, etc. Programs that are multiplexed may have an independent time base (clocks used to generate the real time streams). PS is obtained by combining one or more PES packets with a common time base and wrapping them in PS headers. PS is more suitable for use in error free environment such as storage media (Hard Disk, CD, DVD, etc.). Generally, TS is used in what are called 'push mode' applications and PS is used for what are called 'pull mode' applications. Push mode is the name given to applications such as Broadcasting where the data is sent at the rate decided by the sending side and the receiver is in slave mode where its clock needs to be synchronized with the sender's clock. Pull mode is the name given to applications such as DVD playback where the decoder controls (based on its local clocks and timing) the rate at which the data is pulled. However, there is no fundamental technical limitation

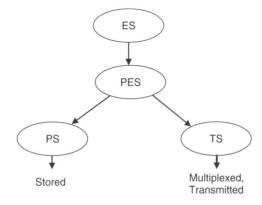

Figure 10.11   MPEG-2 systems based packetization

or reason for not using TS in pull mode applications and it sometimes becomes more convenient to store the streams in a TS format. Because of long packet sizes, PS tends to be a little more efficient.

10.5.1.1   Packetized Elementary Stream (PES)

A PES packet can be of fixed or variable length. It consists of a PES header followed by the pay load of ES.

As shown in Figure 10.12, a PES packet header consists of a Start Code Prefix, Stream ID, PES packet length and PES header elements (many of those elements are optional), which include Presentation Time Stamp (PTS) and Decoding Time Stamp (DTS). The Packet Start Code Prefix is a 24 bit code and the Stream ID is an 8 bit code. The Packet Start Code Prefix is the bit string 0x000001 (23 '0' bits followed by a '1' bit). The Stream ID specifies the type of the elementary stream. For example, 1100 xxxx ID specifies the MPEG-2 Video bitstream. The Packet Start Code Prefix plus the Stream ID constitutes a packet start code. It identifies the beginning of a PES packet. As the name indicates, PES packet length field signals the length of the PES packet in bytes. It is a 16 bit long field. A value of 0 is used to signal that PES packet length is not specified or is not bounded for TS carrying the video payload. PES Header data includes information such as PTS and DTS flags, PTS and DTS, copyright and PES scrambling information, etc. DTS is a 33 bit number and tells a decoder when to pull out the bits associated with a compressed picture from the input buffer in order to decode that picture. The value of DTS is specified in units of a 90 kHz period, which is the system clock frequency (27 MHz) divided by 300. PTS is also a 33 bit number in units of 90 kHz. It signals the decoder when to present that elementary stream, i.e. if it is a video stream then when to display it. This is used for Audio-Video synchronization. A Copyright flag indicates if the payload associated with that PES packet is protected by copyright or not. PES scrambling control signals if the payload is encrypted or not. When the scrambling is performed the PES packet header is not scrambled. Generally, PES is not used for interchanging the content among various devices. That is done via TS and PS.

10.5.1.2   Transport Stream (TS)

TS is constructed from PES packets. As shown in Figure 10.13, it consists of packets that are 188 bytes long with a 4 byte long header and 184 byte long payload that could contain

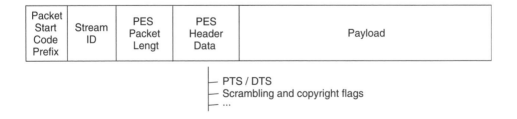

Figure 10.12   PES packet

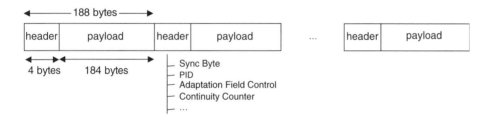

Figure 10.13 Transport stream

the adaptation field and PES header. The length of the packet was chosen so as to achieve a good compromise between the needs of providing low overhead, easy multiplexing of multiple streams and error-resiliency.

A transport stream may contain one or more programs, which may have an independent time base. A high level block diagram of a digital TV distribution system using multiple TS streams is shown in Figure 10.14.

As shown in Figure 10.13, the TS header includes information such as sync byte, packet identifier (PID), adaptation field control, continuity counter, etc. Sync byte is a fixed 8 bit field. Its value is '0100 0111' (0×47). PID is a 13 bit long number and packets associated with a particular compressed stream are assigned the same PID values. This allows the receiver (system decoder) to filter out the packets corresponding to an (video or audio) elementary stream and pass them on to the corresponding elementary stream decoder. Continuity counter increments by 1 from packet to packet to indicate the continuity of the packets. Loss of a packet will cause discontinuity in the counter. Intentional discontinuity in the continuity counter can be flagged in the adaptation field. Adaptation Field Control consists of 2 bits. They indicate whether an adaptation field after the header is present or

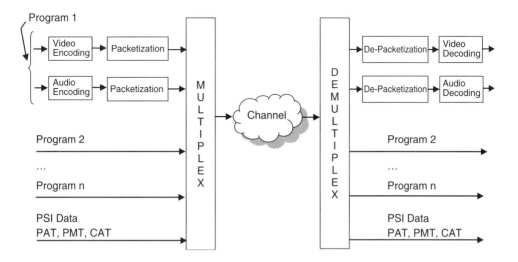

Figure 10.14 Digital TV distribution system

not and if present then whether is it followed by a pay load or not. An adaptation field, if present, includes information such as Program Clock Reference (PCR) which is used for the synchronization of the clocks at the sending and receiving ends (see section 10.5.3). The adaptation field also includes flags such as Random Access Indicator and Splice Point. The Random Access Indicator flag can be used to aid random access. For example, it can be set to '1' to indicate that the next PES packet in the video payload with current PID corresponding to the video stream contains the first byte of a video sequence header or contains the first byte of the audio frame with current PID corresponding to the audio stream. The Splice Point flag is used to indicate the presence of a Splice Countdown field. It is an 8 bit field that can be used to flag the location of the splice point. For example, a positive value informs the receiver of the number of remaining TS packets until a splice point is reached. The packet with zero countdown value contains the last byte of the audio frame or coded picture. Similarly, a negative value indicates the TS packet following the splice point.

MPEG-2 Systems also provide a theoretical hypothetical model for the system level decoding process. This serves as a guideline in order to verify the conformance of both encoder and decoder. A practical system decoder can be designed with the understanding that transport streams are created so that they do not violate this theoretical model. A high level block diagram of a system target decoder (STD) is shown in Figure 10.15. In STD, the TS packets corresponding to an ES (having the same PID number) are passed to the transport buffer TB. The transfer of the bytes from the STD input to TB is assumed to be instantaneous. Bytes entering the TB are removed at the rates specified in the standard.

There are TS packets, e.g. packets with PID values 0 or 1, which do not carry video and audio streams and contain system related information. These are transferred to TB_{sys}. PID number 0 is assigned for the stream carrying the Program Association Table (PAT). PAT provides for association of program number, the PID value of Program Map Table (PMT) and the PID value corresponding to the network information (physical network

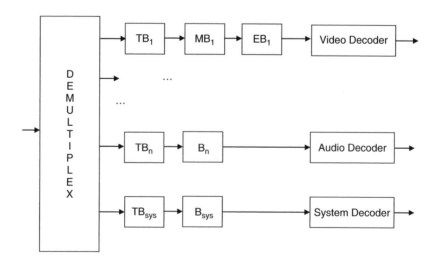

Figure 10.15   Transport stream system target decoder (STD)

parameters, transponder number, etc.). PMT contains complete information about all of the programs in the TS. It provides information which allows the receiver to associate various PID values to various programs in the multiplex, i.e. stream types, elementary stream PID values, PCR PID values, etc.

PID number 1 is assigned to the stream carrying the Conditional Access Table (CAT). These packets carry description related to conditional access and encryption. PAT, PMT, Network Information and CAT are called Program Specific Information (PSI).

The transport buffer's size is specified to be 512 bytes. Therefore, this puts a constraint on the network and the system on the packet rate – the number of packets per second that can be sent to the de-multiplexer corresponding to a particular ES. If packets are sent at a rate faster than that at which TB is emptied, then those packets will be lost. This allows one to design a real implementation with the maximum constraint on the input packet rates. The data is transferred from TB to the multiplexing buffer noted as MB for video ES, B for audio and B_{sys} for systems packets. For video ES an additional buffer EB is included and its size is equal to the CPB buffer size used by the encoder when creating that particular bitstream. At the time of decoding, bits corresponding to a compressed picture are removed instantaneously from MB and B. The time of decoding is specified by DTS (or PTS) in the stream. For B_{sys} the data is removed at the rate specified in the standard.

The sizes of the buffers in STD and the rates at which the data is transferred from one buffer to another are specified in the standard. The system and the delivery network are to be designed so that none of the buffers overflow. Underflow of the buffers is also not allowed except for specific modes such as low-delay and still video. These place a practical limitation on the required speed of the receiver hardware which can be designed by keeping in mind those constraints. In a real decoder the decoding and transfer of the data cannot be instantaneous. Therefore an additional margin, depending upon how far a particular design is from the theoretical model, has to be built-in when designing a real system considering that a system and a network are not allowed to break the STD.

10.5.1.3 Program Stream

PS is obtained by combining one or more streams of PES packets. PES packets are organized in 'packs'. As shown in Figure 10.16, a pack consists of a pack header followed by zero or more PES packets. The pack header starts with a 32-bit start code and also contains timing and bit rate information. The timing information includes the System Clock Reference (SCR). The header also includes the Program Mux Rate, the rate at which STD receives the PS during the pack which contains this information. This value

Figure 10.16 PS stream structure

can change from pack to pack. It is used primarily in environments that are error-free, such as digital storage media.

10.5.2 Audio Video Synchronization

As video and audio are sent as two different bitstreams, there is a need to insert markers in those streams to let a decoder know the time relationship between the two streams. Those markers are called PTS (Presentation Time Stamps). PTS are sent in the bitstream at regular rates and these time stamps tell the decoder when to display the video and present the corresponding audio. The MPEG standard specifies that the PTS must be sent at least once every 0.7 sec. The presentation times of the pictures between PTS can be estimated or calculated by the receiver based on the frame rate information known (or sent) to the receiver.

10.5.3 Transmitter and Receiver Clock Synchronization

As encoders and decoders are generally separated, they do not have a common clock reference. To minimize cost, receivers are not designed with highly accurate clocks. Therefore, there is a need to send the encoder clock reference in the bitstream so that a decoder can synchronize to the clocks at the sending end. Otherwise, buffers in the receiver will overflow or underflow depending upon whether its clock is slower or faster than that at the sending end. These clock references are called PCR (Program Clock Reference) stamps. Every so often an encoder captures the value of its clock counter and sends it as part of the bitstream. A receiver also keeps a counter at the receiving end. By comparing the values of its local clock counter and the received clock counter, a receiver can track whether its clock is running slower or faster than the encoder's clock. This allows a receiver to make appropriate adjustments to its clock speed and lock it to the encoder's clock. The MPEG-2 Systems standard requires that PCRs must be sent at least every 0.1 sec. When packets from multiple streams are multiplexed together, the multiplexing process adds jitter and that jitter is compensated for by measuring the amount by which the packets are delayed, which may not be the same for every packet, and re-adjusting the PCR values in order to compensate for that delay. The MPEG-2 Systems standard requires that inaccuracy in the received PCRs be within ±500 ns.

10.5.4 System Target Decoder and Timing Model

In a real time transmission system it is vitally important to have a mechanism to synchronize the sending and receiving systems and specify the model of the rate of video and audio bits entering the system. Note that, as is the case for Broadcasting applications, channels over which the streams flow can be one way only and may not have a common network clock that can be used by the sending and receiving ends. In addition those channels may consist of several different heterogeneous network types consisting of a mixture of optical communication, over the air communication and satellite based communication, etc. The MPEG-2 Systems provide these unique and complex capabilities by specifying a hypothetical System Target Decoder (STD) and a way to send a Program Clock Reference (PCR).

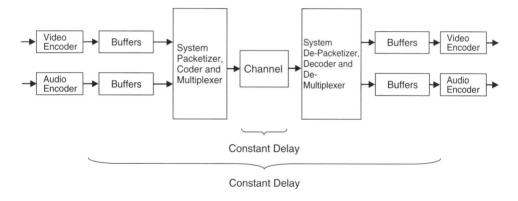

Figure 10.17 MPEG-2 systems timing model

Figure 10.17 shows the assumed timing model. Although small deviations are acceptable, the basic assumption behind the timing model is based on a constant delay between the points shown in Figure 10.17. If a network or a system does not have a constant delay then it needs to make an appropriate adjustment so that from the standpoint of MPEG-2 Systems it behaves like a constant delay system. Those adjustments include buffering of the packets at various stages in the system and correction of the time stamps as those packets pass through the parts of the system, such as multiplexers and switchers, which introduce jitters.

References

1. ISO/IEC JTC 1, "Generic coding of moving pictures and associated audio information – Part 1: Systems," ISO/IEC 13818-1 (MPEG-2), 1994 (and several subsequent amendments and corrigenda).
2. ISO/IEC JTC 1, "Generic coding of moving pictures and associated audio information – Part 2: Video," ISO/IEC 13818-1 (MPEG-2), 1994 (and several subsequent amendments and corrigenda).
3. ISO/IEC JTC 1, "Generic coding of moving pictures and associated audio information – Part 3: Audio," ISO/IEC 13818-1 (MPEG-2), 1994 (and several subsequent amendments and corrigenda).
4. ITU-T and ISO/IEC JTC 1, "Advanced video coding of generic audiovisual services," ITU-T Rec. H.264 and ISO/IEC 14496-10 (MPEG-4 Part 10), 2005 (and several subsequent amendments and corrigenda).
5. Digital Video Broadcasting (DVB) project, http://www.dvb.org.
6. Society of Cable Telecommunications Engineers (SCTE), http://www.scte.org/standards.
7. Advanced Television Systems Committee (ATSC), http://www.atsc.org.
8. 3rd Generation Partnership Project (3GPP), http://www.3gpp.org.
9. 3rd Generation Partnership Project 2, http://www.3gpp2.org.
10. Internet Streaming Media Alliance (ISMA), http://www.isma.tv.
11. DVD Forum, www.dvdforum.org.
12. Blu-ray Disc Association www.blu-raydisc.com.
13. Digital Living Network Alliance (DLNA), http://www.dlna.org.
14. A. Puri, X. Chen, and A. Luthra, "Video coding using the H.264/MPEG-4 AVC compression standard", Signal Processing: Image Communication, Vol. 19, pp. 793–849, June 2004.

11

Interactive Video Communications

11.1 Video Conferencing and Telephony

Video conferencing and telephony are typical examples of interactive video communications, as nowadays mobile phones and portable devices are very popular. The major difference between video conference and telephony with other multimedia applications, such as multimedia messaging (MMS) and multimedia streaming is that the video conferencing is delay sensitive while others are not.

In the past decade, point-to-point protocols (PPP) have been in common use for TCP/IP communications (the protocol used by the Internet) over a telephone line. Using the Internet, multimedia communication can be achieved without incurring any long distance charges. On the other hand, the usage of a cable modem along with DSL technology has enabled broadband Internet access, where the cable modem is used to deliver broadband Internet access taking advantage of the unused bandwidth of the cable television network. The bandwidth of the cable connection varies from 3 Mbits/s to 30 Mbits/s and the upstream speed ranges from 384 Kbits/s to 6 Mbits/s. The DSL modem, on the other hand, takes advantage of the unused frequencies in the telephone line and varies in speed from hundreds of kbits/s to few Mbits/s. With the development of the 3G and 4G, the constraint on bandwidth for a wireless system to carry video content has been lifted, in addition, more sophisticated access protocols such as HSPA, EDGE, etc. hit the market and give a strong impetus to the mobile telephony business.

11.1.1 IP and Broadband Video Telephony

Compared to the very low-bit rate (around 40 kbit/s) of video telephony over the PSTN (Public Switched Telephone Network), the cable modem and Direct Subscriber Line (DSL) connections offer Internet connections at much higher bit rates. Thus, IP video telephony in conjunction with DSL and cable modems can now offer video communications at a much higher quality than before. IP Video Telephony uses the H.323 standard which can be used over any packet data network, such as those using the Internet protocol (IP). Owing to the interest in video telephony over IP, many of the existing commercial implementations use the H.323 standard.

4G Wireless Video Communications   Haohong Wang, Lisimachos P. Kondi, Ajay Luthra and Song Ci
© 2009 John Wiley & Sons, Ltd

The Digital Subscriber Line (DSL) connection and cable Internet connections are both referred to as broadband since they use different channels to send the digital information simultaneously with the audio signal or the cable television signal. A DSL connection uses a dedicated line from the subscriber to the telephone company, while the cable Internet service is provided to a neighborhood by a single coaxial cable line. Hence, connection speed varies depending on how many people are using the service. Cable modems send the data signal over the cable television infrastructure. They are used to deliver broadband Internet access by taking advantage of the unused cable network bandwidth. DSL, on the other hand, uses a conventional twisted wire pair for data transmission. ADSL [1] uses two frequency bands known as upstream and downstream bands. The upstream band is used for communications from the end user to the telephone central office while the downstream band is used for communicating from the central office to the end user. ADSL provides dedicated local bandwidth in contrast to the cable modem which gives shared bandwidth. Hence, the upstream and downlink speed varies depending on the distance of the end user from the telephone office. Conventional ADSL has a downstream speed of approximately 8 Mbits/s and an upstream speed around 1 Mbits/s. Thus, acceptable quality video telephony is achievable with the advances in modem technology and audio and video compression.

11.1.2 Wireless Video Telephony

Video Telephony is offered in 3G networks in both the circuit-switch mobile core network and packet network. The former provides 64 kbits/s per circuit switched path, while the latter may provide greater bandwidth but the bandwidth is not guaranteed during the call as no dedicated circuit is reserved. 3GPP uses the 3G-324M protocol to support video telephony services. In January 2004, NTT DoCoMo (a Japanese operator) announced that its FOMA (freedom of mobile multimedia access) 3G video telephony service had passed the milestone of 2 million customers. Recently almost all mobile phones supporting UMTS networks can make videophone conversations with other UMTS users, and it was estimated that there were more than 130 million UMTS users in mid-2007. In the next section we provide more detail of the 3G-324M protocol that is widely used in videophone applications.

11.1.3 3G-324M Protocol

3G-324M [2] is the 3GPP umbrella protocol for video telephony in 3G mobile networks, which is based on the H.324 (a standard for low bit rate GSTN networks) specification for multimedia conferencing over circuit switched networks. 3G-324M is comprised of the following sub-protocols:

- H.245 for call control.
- H.223 for bitstreams to data packets multiplexer/demultiplexer.
- H.223 Annex A and B for error handling of low and medium BER detection, correction and concealment.
- Adaptation layers.

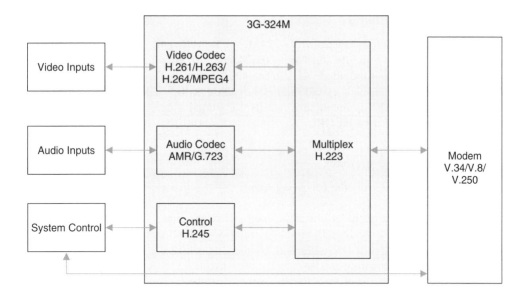

Figure 11.1   3G-324M basic structure

The basic structure of 3G-324M is shown in Figure 11.1, which consists of a multiplexer which mixes the various media types into a single bitstream (H.223), an audio compression algorithm (either a AMR or G.723 codec), a video compression algorithm (either a H.261, H.263, H.264 or MPEG4 codec) and a control protocol which performs automatic capability negotiation and logical channel control (H.245). The goal of this standard is to combine low multiplexer delay with high efficiency and the ability to handle bursty data traffic from a variable number of sources.

11.1.3.1   Multiplexing and Error Handling

3G-324M uses a multiplex standard, H.223, to mix the various streams of audio, video, data and the control channel together into a single bitstream for transmission over the modem. H.223 has a flexible mapping scheme suitable for a variety of media and for a variable frame length. In its mobile extension, it obtains greater synchronization and control of channel errors without losing its flexibility. H.223 consists of a lower multiplex layer and a set of adaptation layers. The lower multiplex layer mixes the different media streams, whereas the adaptation layers perform logical frame, sequence numbering, error detection and error correction by retransmission. Each adaptation layer is suitable for a different type of information channel. In H.223, there are 3 operation modes which are chosen according to the degree of error resiliency required in a 3G-324M system. In the first level, the multiplexing and QoS control are supported; in the second level, a 16-bit pseudorandom noise sequence is employed to improve the synchronization; in the third level, the payload length and FEC information are added in the header in order to improve error resilience capability.

11.1.3.2 Adaptation Layers

There are three adaptation layers in 3G-324M: AL1, AL2, and AL3. AL1 is intended primarily for data and control information transferring, in which no error detection and correction mechanism is provided. AL2 is intended primarily for digital audio transferring, which includes an 8 bit cyclic redundancy code (CRC). CRC is used to identify transmission errors. AL3 is intended primarily for digital video and includes provision for retransmission and a 16 bit CRC.

11.1.3.3 The Control Channel

The H.245 protocol controls the following items:

- Logical channel that opens or closes for media transmissions.
- Determines the master terminal at the beginning of a session.
- Exchanges the capabilities between both terminals, such as the mode of multiplexing, codec support, data sharing mode, etc.
- Operation mode that is sent from the receiver side to the transmitter side to convey the preference within its capability of the codec and the associated parameters.
- Call control commands and indications that check the status of the terminals and communications.

In addition, H.245 supports the numbered simple retransmission protocol (NSRP) and control channel segmentation and reassembly layer (CCSRL) sub-layer support in order to ensure reliable operation, therefore all terminals support both NSRP and SRP modes.

11.1.3.4 Audio and Video Channels

The 3G-324M specifications define the AMR codec as mandatory and G.723.1 as a recommended audio codec, it also declares the H.263 codec as mandatory and MPEG-4 a as recommended codec for video processing. The details of these video codecs have been discussed in Chapter 5.

11.1.3.5 Call Setup

There are seven phases in the call set up procedure, designated by letters A through G. In Phase A, an ordinary telephone connection is established. In Phase B, a regular analog telephone conversation can take place before the actual multimedia communication. When either user decides to start the multimedia communication, Phase C takes place. The two modems communicate with each other and digital communication is established. Then, in Phase D, the terminals communicate with each other using the H.245 control channel. Detailed terminal capabilities are exchanged and logical channels are opened. In Phase E, actual multimedia communication takes place. Phase F is entered when either user wishes to end the call. The logical channels are closed and an H.245 message is sent to the far-end terminal to specify the new mode (disconnect, back to voice mode, or another digital mode). Finally, in Phase G, the terminals actually enter the mode specified in the previous phase.

11.2   Region-of-Interest Video Communications

As mentioned in Chapter 6, ROI based video coding and communication has been very popular for wireless video telephony. As shown in Figure 11.2, the ROI based video communications system architecture provides users with greater flexibility and interactivity in specifying their desires and enables encoders to have greater efficiency in controlling the visual quality of coded video sequences. In this section, we demonstrate a few of the latest advances in ROI based bit allocation [3] and adaptive background skipping [4, 5] techniques.

11.2.1   ROI based Bit Allocation

In the literature, many ROI bit allocation algorithms [6–10] are based on a weighted version of the H.263+ TMN8 model [11], where a cost function is created and the distortion components in various regions in the function are punished differently by using a set of preset weights. As with most of the other video standards, as mentioned in Chapter 6, TMN8 uses a Q-domain rate control scheme, which models the rate and distortion with functions of quantization step size (QP). However, recent advances in rate control research and development have demonstrated that the ρ-domain rate control model [12] (ρ represents the number of non-zero AC coefficients in a macroblock in video coding) is more accurate and thus effectively reduces rate fluctuations. It is also observed that the ρ-domain rate control approach has already been used in industry trials [13–16]. To the best of our knowledge, so far there is no general optimized ρ-domain bit allocation model for ROI video coding, although [17] used the ρ-domain rate control model in their efforts to get an ad-hoc bit allocation solution. In this section, we introduce a ρ-domain optimized weighted bit allocation scheme for ROI video coding.

11.2.1.1   Quality Metric for ROI Video

Video quality measurement is still an open issue for ROI video coding. Most of the literature uses PSNRs on ROI and Non-ROI, respectively, as a measurement for evaluating regional visual quality, however, a quality measure for the whole image has not been addressed. In [18], a weighted mean squared error (MSE) metric was proposed in order to measure perceptual video quality. In this metric, the macroblocks are classified as activity macroblocks and static macroblocks, and different weights are

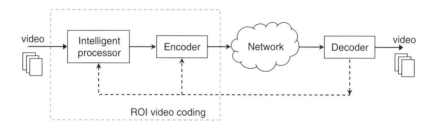

Figure 11.2   An example of ROI video coding and communications system

assigned to these macroblocks for calculating the weighted MSE for overall image. Although this measurement can be extended to use for ROI video, here we introduce a new quality measurement for ROI video coding which takes into account further aspects such as spatial/temporal visual quality.

In general, the evaluation of ROI video quality should consider at least three aspects: users' interest, video fidelity and perceptual quality of the reconstructed video data. The users' interest determines directly the classification of a video frame into ROI and Non-ROI parts and their associated perceptual importance factors. In video telephony applications, the speaker's face region is a typical ROI because a human being's facial expressions are very complicated and small variations can convey a large quantity of information. For the video fidelity factor, PSNR is a good measurement, which indicates the total amount of distortion of the reconstructed video frame compared to the original frame. In most cases, fidelity is the most important consideration for video coding, where any improvement might cause better subjective visual quality. However, it is not always the case, and that is why perceptual quality factors should also be taken into account. Perceptual quality considers both spatial errors, for example blocking and ringing artifacts, and temporal errors such as temporal flicker where the frame visual qualities change non-uniformly along the temporal axis.

Let us denote by D_R and D_{NR} the normalized per pixel distortion of the ROI and Non-ROI, and α the ROI perceptual important factor. If we assume that the relationship among the aspects mentioned above can be simplified into a linear function in video quality evaluation, then we can represent the overall distortion of the video sequence as:

$$
\begin{aligned}
D_{sequence} &= \alpha D_R + (1 - \alpha) D_{NR} \\
&= \frac{\alpha}{M} \left[\beta \sum_{i=1}^{M} D_{RF}(f_i, \tilde{f}_i) + \gamma \sum_{i=1}^{M} D_{RS}(\tilde{f}_i) + (1 - \beta - \gamma) D_{RT}(\tilde{f}_1, \ldots, \tilde{f}_M) \right] \\
&\quad + \frac{(1 - \alpha)}{M} \left[\beta \sum_{i=1}^{M} D_{NF}(f_i, \tilde{f}_i) \right. \\
&\quad\quad \left. + \gamma \sum_{i=1}^{M} D_{NS}(\tilde{f}_i) + (1 - \beta - \gamma) D_{NT}(\tilde{f}_1, \ldots, \tilde{f}_M) \right]
\end{aligned}
\tag{11.1}
$$

where f_i and $\tilde{f}_i$ are the ith original and reconstructed frames within the M frames in the video sequence, β and γ are weighting factors, D_R and D_{NR} are the total distortion for ROI and Non-ROI, D_{RF}, D_{RS} and D_{RT} are the normalized errors of ROI in fidelity, spatial perceptual quality and temporal perceptual quality, and D_{NF}, D_{NS} and D_{NT} are their counterparts for Non-ROI. It is clear that α, β and γ should be assigned real values between 0 and 1.

In low-bitrate video applications, such as wireless video telephony, blocking artifacts are the major concern of spatial perceptual quality. This kind of artifact is caused by the quantization where most of the high-frequency coefficients are removed (set to zero). The resulted effect is that the smoothed image blocks make the block boundaries quite pronounced. At the extreme low bit-rate cases, only DC coefficients will be coded which makes the decoded image piece-wise constant blocks. In this work, we define the D_{RS} (similar for D_{NS}) as the normalized blockiness distortion, that is:

$$D_{RS}(\tilde{f}) = \frac{\text{boundaries with discontinuities}}{\text{Number of boundaries}} \qquad (11.2)$$

where every boundary between blocks is checked to see if perceivable discontinuities exist. The discontinuity detection approach used in [19] is adopted, which checks the sum of the mean squared difference of the intensity slope across the block boundaries. The assumption of this approach is that the slopes on both sides of a block boundary are supposed to be identical and an abrupt change in slope is probably due to quantization.

In equation (11.1), the D_{RT} (or D_{NT}) is defined as an assigned score in the range of [0, 1] based on the variance of D_{RS} (or D_{NS}) for all the frames in the sequence. In this way, the terms on fidelity, spatial perceptual quality and temporal perceptual quality are normalized and can be bridged by weighting parameters α, β and γ to form a controllable video quality measurement. The selection of these weighting parameters is up to users based on their requirements and expectations. Again, this measurement is not a perfect metric, but it will be shown in the subsequent text that it helps the bit allocation process to favor subjective perception.

11.2.1.2 Bit Allocation Scheme for ROI Video

In video coding applications, a typical problem is to minimize $D_{sequence}$ with a given bit budget for the video sequence. The optimal solution for this complicated problem relies on an optimal frame-level rate control algorithm and an optimal macroblock-level bit allocation scheme. However, for real-time applications, such as wireless video telephony, where very limited information about future frames is available when coding the current frame, it is not practical or feasible to pursue an optimal frame-level rate control. Typically a popular greedy algorithm is resorted to which assumes that the complexity of the video content is distributed uniformly along the frames in the video sequence, and thus allocates a fraction of the available bits to each of the rest frames. For the same reason, taking care of $D_{NT}(\tilde{f}_1, \ldots, \tilde{f}_M)$ in the rate control is very difficult for these applications. Therefore, to find a practical solution and to simplify the problem we assume that good frame-level rate control is available and thus we narrow down the problem into a macroblock-level bit allocation problem. At the meantime, we propose a background skipping approach, which increases the chance of reducing the value of the term $D_{NT}(\tilde{f}_1, \ldots, \tilde{f}_M)$ because the skipped region will present the same perceptual quality as that of the previous frame and thus might reduce the fluctuation of the perceptual quality between consecutive frames. For measuring the image quality of a video frame, we use equation (11.1) by setting $\beta + \gamma = 1$.

Let us denote by R_{budget} the total bit budget for a given frame f and R the bit rate for coding the frame, then the problem can be represented by:

$$\text{Minimize } \alpha \left[\beta D_{RF}(f, \tilde{f}) + (1 - \beta) D_{RS}(\tilde{f}) \right]$$

$$+ (1 - \alpha) \left[\beta D_{NF}(f, \tilde{f}) + (1 - \beta) D_{NS}(\tilde{f}) \right]$$

$$\text{Such that } R \leq R_{budget} \qquad (11.3)$$

Clearly, this optimization problem can be solved by Lagrangian relaxation and dynamic programming in the same fashion as in [20]. However, the computational complexity is a great deal higher than a real-time system can bear. Therefore, a low-complexity near-optimal solution is preferred. We propose a two-stage bit allocation algorithm in ρ-domain to solve this problem. In the first stage, we are solving an optimization problem:

$$\text{Minimize } \alpha D_{RF}(f, \tilde{f}) + (1 - \alpha)D_{NF}(f, \tilde{f}), \text{ such that } R \leq R_{budget} \qquad (11.4)$$

After the optimal coding parameters for (11.4) is obtained, in the second stage we adjust the coding parameters iteratively to reduce the term $\alpha D_{RS}(\tilde{f}) + (1 - \alpha)D_{NS}(\tilde{f})$ until a local minimum is reached. Clearly, the result will be very close to the optimal solution when β is a relative large number. When $\beta = 1$, problems (11.3) and (11.4) are identical. In this section, we will focus on the first stage and solve problem (11.4).

11.2.1.3 Bit Allocation Models

In ROI video coding, let us denote by N the number of macroblocks in the frame, $\{\rho_i\}$, $\{\sigma_i\}$, $\{R_i\}$ and $\{D_i\}$ the set of ρ s, standard deviation, rates and distortion (sum of squared error) for the ith macroblocks. Thus, $R = \sum_{i=1}^{N} R_i$. We define a set of weights $\{w_i\}$ for each macroblock as:

$$w_i = \begin{cases} \dfrac{\alpha}{K} & \text{if it belongs to ROI} \\ \dfrac{1 - \alpha}{(N - K)} & \text{if it belongs to Non - ROI} \end{cases} \qquad (11.5)$$

where K is the number of macroblocks within the ROI. Therefore, the weighted distortion of the frame is:

$$D = \sum_{i=1}^{N} w_i D_i = [\alpha D_{RF}(f, \tilde{f}) + (1 - \alpha)D_{NF}(f, \tilde{f})]^*255^2*384 \qquad (11.6)$$

Hence the problem (11-4) can be rewritten as:

$$\text{Minimize D, such that } R \leq R_{budget} \qquad (11.7)$$

We propose to solve (11.7) by using a modeling-based bit allocation approach. As shown in [21], the distribution of the AC coefficients of a nature image can be best approximated by a Laplacian distribution $p(x) = \frac{\eta}{2}e^{-\eta|x|}$. Therefore in [11], the rate and distortion of the ith macroblock can be modeled in (11.8) and (11.9) as functions of ρ,

$$R_i = A\rho_i + B \qquad (11.8)$$

where A and B are constant modeling parameters, and A can be thought of as the average number of bits needed to encode non-zero coefficients and B can be thought of as the bits due to non-texture information.

$$D_i = 384\sigma_i^2 e^{-\theta\rho_i/384} \tag{11.9}$$

where θ is an unknown constant.

Here we optimize ρ_i instead of quantizers because that we assume that there is an accurate enough ρ-QP table available to generate a decent quantizer from any selected ρ_i. In general, (11.7) can be solved by using Lagrangian relaxation in which the constrained problem is converted into an unconstrained problem that:

$$\underset{\rho_i}{\text{Minimize}} \ J_\lambda = \lambda R + D = \sum_{i=1}^{N}(\lambda R_i + w_i D_i) = \sum_{i=1}^{N}[\lambda(A\rho_i + B) + 384w_i\sigma_i^2 e^{-\theta\rho_i/384}] \tag{11.10}$$

where λ^* is the solution that enables $\sum_{i=1}^{N} R_i = R_{budget}$. By setting partial derivatives to zero in (11.10), we obtain the following expression for the optimized ρ_i, that is:

$$\text{let} \ \frac{\partial J_\lambda}{\partial \rho_i} = \frac{\partial \sum_{i=1}^{N}[\lambda(A\rho_i + B) + 384w_i\sigma_i^2 e^{-\theta\rho_i/384}]}{\partial \rho_i} = 0 \tag{11.11}$$

which is

$$\lambda A - \theta w_i \sigma_i^2 e^{-\theta\rho_i/384} = 0 \tag{11.12}$$

so

$$e^{-\theta\rho_i/384} = \frac{\lambda A}{\theta w_i \sigma_i^2} \tag{11.13}$$

and

$$\rho_i = \frac{384}{\theta}[\ln(\theta w_i \sigma_i^2) - \ln(\lambda A)] \tag{11.14}$$

On the other hand, since

$$R_{budget} = \sum_{i=1}^{N} R_i = \frac{384A}{\theta}\sum_{i=1}^{N}[\ln(\theta w_i \sigma_i^2) - \ln(\lambda A)] + NB \tag{11.15}$$

so,

$$\ln(\lambda A) = \frac{1}{N}\sum_{i=1}^{N}\ln(\theta w_i \sigma_i^2) - \frac{\theta}{384NA}(R_{budget} - NB). \tag{11.16}$$

From (11.14) and (11.16), we obtain the following model:

$$\rho_i = \frac{384}{\theta}\left[\ln(\theta w_i \sigma_i^2) - \frac{1}{N}\sum_{i=1}^{N}\ln(\theta w_i \sigma_i^2) + \frac{\theta}{384NA}(R_{budget} - NB)\right]$$

$$= \frac{R_{budget} - NB}{NA} + \frac{384}{\theta}\left[\ln(\theta w_i \sigma_i^2) - \frac{\sum\limits_{i=1}^{N}\ln(\theta w_i \sigma_i^2)}{N}\right] \qquad (11.17)$$

As mentioned in [20], another model can be obtained if assume a uniform quantizer, then the distortion is modeled differently from equation (11.9), and thus the model can be derived as:

$$\rho_i = \frac{\sqrt{w_i \sigma_i}}{\sum\limits_{j=1}^{N}\sqrt{w_i \sigma_j}}\rho_{budget}. \qquad (11.18)$$

It is also indicated that both models have a good performance which is close to the optimal solution.

11.2.2   Content Adaptive Background Skipping

The concept of content-adaptive frame/object/macroblock skipping has attracted a great deal of attentions recently. The trade off between spatial and temporal quality was first studied in [21], where a perceptual rationale is employed: that the human visual system (HVS) is more sensitive to temporal changes when the frame contains high motion activities and otherwise is more sensitive to spatial details. The same logic is also used by [22–25] in determining the skip modes. In [22], a weighted function of motion and variance of the residue was used to evaluate the target bits for objects in bit allocation, which assigned more bits to objects with a more complicated texture (with a higher variance) or more activity (with a higher motion). The skipping decision of objects are based on an optimization process of a cost function which considers both coded distortion owing to quantization error and skipped distortion owing to skipped objects. This approach will be difficult for applying in real-time video systems which have tight time constraints and are not able to obtain future frames in advance. In [23], an adaptive macroblock skipping approach was proposed for ROI transcoding, where thresholds for motion and MAAD (mean of accumulated absolute difference) of the residue are used to skip those inactive Non-ROI macroblocks. In [24], the decision for frame skipping is dependent jointly on the temporal and spatial contents of the video, and on the fullness of the buffer by using empirical rules. In [25], considering the HVS model mentioned above, the decision for frame skipping is determined adaptively by motion, quantization parameter and buffer status. The motion is evaluated based on the sorted version of the most recent motion activities, and a dynamically adjusted threshold that is coupled with available resources, spatial quality, quantization parameters and motion activity. Utilizing the HVS model, by avoiding skipping frames during high-motion scenes, superior temporal quality is maintained. By skipping frames during low-motion scenes that are less temporally sensitive, coding bits can be saved for subsequent no-skipped frames, and spatial quality can be enhanced. Furthermore, in [25] overall temporal-spatial quality is enhanced when compared to the no-skipping and fixed-pattern solutions, given limited coding resources.

In this section, a low-complexity content adaptive background skipping scheme for ROI video coding is introduced. In this context, we use background and Non-ROI as exchangeable terms because Non-ROI in video telephony applications generally refers to background region. In this framework we consider background skipping jointly with frame-level and macroblock-level bit allocation. The skip mode is determined mainly by foreground shape deformation, foreground motion, background motion and accumulated skipped distortion owing to skipped background. A self-learning and classification approach based on the Bayesian model is proposed in order to estimate the number of skipped background (in the future frames) based on the context of motion and background texture complexity. In addition, a weighted rate control and bit allocation algorithm is proposed in order to allocate bits for the foreground and background regions.

In Figure 11.3, the system architecture of our ROI video coding system is shown, which follows a frame-by-frame processing sequence. The system adopts a ρ-domain frame-level rate control algorithm [12] and a weighted macroblock-level bit allocation algorithm. When a frame is fetched into the system, a greedy frame-level rate control module is called to assign a target ρ budget for the frame considering the remaining bits and the number of frames in the rate control window. The model is based on the assumption that the content complexity of the video frames in the rate control window is distributed uniformly and thus the bits should be allocated uniformly among the remaining frames. After that, the ROI of the frame is detected or tracked and the macroblocks in the frame are classified into ROI macroblocks and Non-ROI macroblocks. Then, motion estimation is conducted for all of the macroblocks in the current frame and the obtained motion information is used as a part of content cues in the following background skip mode decision. Once the decision, of whether or not to skip the current Non-ROI, is made, the ρ budget for current frame is adjusted, and then the macroblock-level bit allocation and the following DCT transformation, quantization and entropy coding are conducted in the same way as described in section 11.2.1.

11.2.2.1   Content-based Skip Mode Decision

Let us first define two filters $F(\{x_n\}, M, Th)$ and $G(\{x_n\}, M, Th)$, where $\{x_n\}$ is a set of real numbers in which x_n is the nth item, M an integer number and Th a threshed in the range of [0, 1], and

$$F(\{x_n\}, M, Th) = \begin{cases} 1 & x_n \text{ is greater than } Th *100\% \text{ of items in } x_{n-M}, \ldots, x_{n-1} \\ 0 & \text{otherwise} \end{cases}$$

(11.19)

and

$$G(\{x_n\} M, Th) = \begin{cases} 1 & \text{if } \dfrac{x_n - x_{n-M}}{x_{n-M}} \geq Th \\ 0 & \text{otherwise} \end{cases}$$

(11.20)

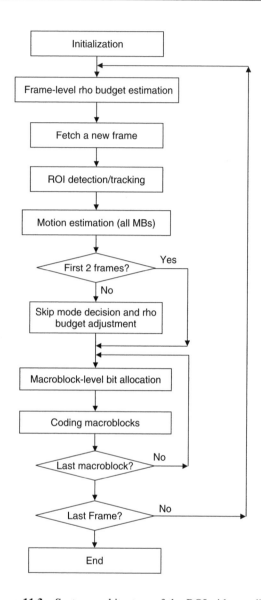

Figure 11.3   System architecture of the ROI video coding

Filter (11.19) detects within a local window (fixed length of M) if the current value x_n is in the top position (above more than $Th*100\%$ of items), and filter (11.20) detects if there is an increase from x_{n-M} to x_n by more than $Th*100\%$. These filters will be used in detecting the content status or status change, which indirectly affects the skip mode decision.

In [24, 25], the value of summed and averaged motion vectors in the frame (or recent frames) is used to represent the frame motion. The higher the motion the less skipping should be activated in order to protect possible content transition information. In ROI video coding, both foreground and background activities are considered. When a large amount of motion occurs in background regions, the frequency of background skipping should be reduced. On the other hand, when the foreground contains a large amount of activities, the skipping of background might be helpful so as to reallocate more bits to code the foreground. Let us denote by $\{\chi_n\}$ the amount of background activity, and $\{\zeta_n\}$ the amount of foreground activity for the frame sequences, then:

$$\chi_n = \sum_{i \in Non\text{-}ROI} (|MVx_i| + |MVy_i|) \tag{11.21}$$

where MVx_i and MVy_i are x and y component of the motion vector of ith macroblock in the nth frame, and:

$$\zeta_n = \mu_n \times \kappa_n \tag{11.22}$$

where $\{\mu_n\}$ is the ROI shape deformation factor and $\{\kappa_n\}$ is the ROI local movement factor, and

$$\mu_n = \frac{\textit{Number of pixels in nonoverlaped regions of ROIs of the (n - 1)th and nth frames}}{\textit{Number of pixels in ROI of the nth frame}}$$

$$\tag{11.23}$$

and

$$\kappa_n = \sum_{i \in ROI} (|MVx_i| + |MVy_i|) \tag{11.24}$$

Clearly, $\{\zeta_n\}$ can characterize the degree of the foreground activities because $\{\mu_n\}$ represents the degree of global activities such as object movement/rotation and shape deformation and $\{\kappa_n\}$ represents local activities such as change of facial expression. Two examples of these foreground activities are shown in Figure 11.4.

Let us denote by $\{\sigma^2_{B_n}\}$ the total energy of the background residue per frame for the frame sequence. Clearly it is also the distortion due to skipped background. So far, we can represent the skip mode decision:

$$S_n = F(\{\zeta_n\}, M_2, Th_{\zeta 1})G(\{\zeta_n\}, 1, Th_{\zeta 2}) + [1 - F(\{\zeta_n\}, M_2, Th_{\zeta 1})G(\{\zeta_n\}, 1, Th_{\zeta 2})]$$

$$\tag{11.25}$$

$$[1 - G(\{\sigma^2_{B_n}\}, p, Th_\sigma)][1 - F(\{\chi_n\}, M_1, Th_{\chi 1})][1 - G(\{\chi_n\}, 1, Th_{\chi 2})]$$

where Th_σ, M_1, $Th_{\chi 1}$, $Th_{\chi 2}$, M_2 and $Th_{\zeta 1}$ are thresholds and local window sizes defined by users, and $p-1$ the number of consecutive preceding frames of the current frame skipped background (in other words, the $(n - p)$th frame coded background but the $(n - p+1)$th, $(n - p+2)$th, ... and $(n-1)$th frames skipped background). When $S_n = 1$, the background of the current frame is skipped, otherwise, it is coded. Clearly from (11.25) it is observed that the system chooses to skip background when there is a sharp increase of the amount of foreground activity or the foreground contains large activity, otherwise, if background contains large motion or the accumulated distortion due to skipped background is rather high, then the background will be coded.

(a) Global activity (face movement)

(b) Local activity (change of facial expression)

Figure 11.4   Examples of frames with large activity in foreground

11.2.2.2 ρ Budget Adjustment

In Figure 11.3, the frame-level ρ budget estimation is based on an assumption that the whole frame is coded, however, in this system some backgrounds in the sequence will be skipped, therefore adjustment on ρ budget is necessary. Here we consider three types of strategies: 1) Greedy strategy, which simply reduces the ρ budget based on the texture complexity of ROI and Non-ROI when the skip mode is on, and does nothing if the background is coded; 2) *'Banker' strategy*, which reduces the ρ budget when the skip mode is on, but stores the savage of these ρ's for future frames. For a frame coding its background, it will obtain all the ρ's saved from the previous frames with background skipping; 3) *'Investor' strategy*, which estimates the future skipping events based on the statistics and patterns of the previous background skipping history, and then determines the ρ budget based on the estimation.

Let us denote by $\{\rho_n^{budget}\}$ the ρ budget obtained from the frame-level rate controller, $\{\rho_n^{adjusted}\}$ the adjusted ρ budget, and n the index of current frame. In the follows we describe more details of these strategies and compare them.

Greedy strategy
The $\rho_n^{adjusted}$ using this strategy can be calculated by

$$\rho_n^{adjusted} = \begin{cases} \rho_n^{budget} & if\, S_n = 0 \\ \dfrac{\sum\limits_{i \in ROI} \sqrt{w_i \sigma_i}}{\sum\limits_{i \in ROI} \sqrt{w_i \sigma_i} + \sum\limits_{i \in NON\text{-}ROI} \sqrt{w_i \sigma_i}} \rho_n^{budget} & otherwise \end{cases} \qquad (11.26)$$

where σ_i represents the standard deviation of the DCT coefficients of the ith macroblock in the current frame, and w_i is the associated weights for the macroblock in macroblock-level weighted bit allocation as defined in section 11.2.1. Equation (11.26) comes as an extension of equation (11.18).

'Banker' strategy

This strategy is a conservative approach similar to the traditional banking operation, where the customer can cash out the maximum of the total deposit of his account. In this case, the saving of ρ's in frames with background skipping seems to deposit the resource for the nearest future frame which codes its background. The calculation for adjusted ρ budget is obtained by:

$$
\rho_n^{adjusted} =
\begin{cases}
\dfrac{p\rho_{n-p+1}^{budget} - \displaystyle\sum_{i=1}^{p-1} \rho_{n-i}^{adjusted}}{\displaystyle\sum_{i \in ROI} \sqrt{w_i \sigma_i}} & \text{if } S_n = 0 \\[2em]
\dfrac{\displaystyle\sum_{i \in ROI} \sqrt{w_i \sigma_i}}{\displaystyle\sum_{i \in ROI} \sqrt{w_i \sigma_i} + \displaystyle\sum_{i \in NON\text{-}ROI} \sqrt{w_i \sigma_i}} \rho_n^{budget} & \text{otherwise}
\end{cases}
\tag{11.27}
$$

where $p-1$ is the number of consecutive preceding frames of the current frame with skipped background and the $(n - p)$th frame coded its background.

'Investor' strategy

A more aggressive approach is to predict future possible events and allocate resources based on the prediction. Here we assume that the future frames with skipped backgrounds have a similar complexity in foreground as the current frame, therefore, once we estimate that there will be q frames with skipped background following the current frame, we can calculate the adjusted ρ budget by:

$$
\rho_n^{adjusted} =
\begin{cases}
\dfrac{p\rho_{n-p+1}^{budget} - \displaystyle\sum_{i=1}^{p-1} \rho_{n-i}^{adjusted}}{\displaystyle\sum_{i \in ROI} \sqrt{w_i \sigma_i}} & \text{if } S_n = 0 \text{ and } n \le 50 \\[2em]
\dfrac{\displaystyle\sum_{i \in ROI} \sqrt{w_i \sigma_i} + \displaystyle\sum_{i \in NON\text{-}ROI} \sqrt{w_i \sigma_i}}{2(\displaystyle\sum_{i \in ROI} \sqrt{w_i \sigma_i} + \frac{1}{q+1}\displaystyle\sum_{i \in NON\text{-}ROI} \sqrt{w_i \sigma_i})} \rho_n^{budget} + \\[2em]
\dfrac{p\rho_{n-p+1}^{budget} - \displaystyle\sum_{i=1}^{p-1} \rho_{n-i}^{adjusted}}{2} & \text{if } S_n = 0 \text{ and } n > 50 \\[2em]
\dfrac{\displaystyle\sum_{i \in ROI} \sqrt{w_i \sigma_i}}{\displaystyle\sum_{i \in ROI} \sqrt{w_i \sigma_i} + \displaystyle\sum_{i \in NON\text{-}ROI} \sqrt{w_i \sigma_i}} \rho_n^{budget} & \text{otherwise}
\end{cases}
\tag{11.28}
$$

In equation (11.28), the 'investor' strategy acts exactly the same as the 'banker' strategy for the first 50 frames. In this period the statistics are collected for future q estimation. When $n > 50$ and $S_n = 0$, ρ is assigned an average value considering the previous saving and the predicted future saving due to background skipping.

We estimate q by using a Bayesian model and convert the problem into a multi-class classification problem, where the classes are represented by all possibilities of q (for example, classes 0, 1, 2, 3, 4, 5 if we limit q to be less than 6), and the feature vector used in making classification decision is $x_n = \chi_n, \zeta_n, \sigma_{B_n}^2$). By defining thresholds for χ_n, ζ_n and $\sigma_{B_n}^2$, we can map the space of $\{x_n\}$ into eight classes $\{y_n\}(y_n = 0, 1, \ldots, \text{ or } 7)$.

Therefore, for current frame, the best selection for q is the one maximizing the probability:

$$P(q|y_n) = \frac{P(y_n|q)P(q)}{P(y_n)}, \tag{11.29}$$

thus it is the q that maximizes $P(y_n|q)P(q)$. The probabilities of $P(y_n|q)$ and $P(q)$ can be obtained by a histogram technique based on the statistics of the previously processed frames. Let us denote by $H_q(y)$ the counts of frames with coded background that follows q frames with skipped background with feature vector y, then:

$$P(y_n|q) = \frac{H_q(y_n)}{\sum_y H_q(y)} \tag{11.30}$$

and $P(q)$ can be obtained by the similar approach. The diagram of the skip mode decision and ρ budget adjustment module with this strategy is shown in Figure 11.5.

In Figure 11.6, three bit allocation strategies are compared in coding the Carphone sequence. As mentioned in section 11.2.1, an ROI perceptual importance factor α is defined in order to bridge the distortion of ROI and Non-ROI so as to form a weighted

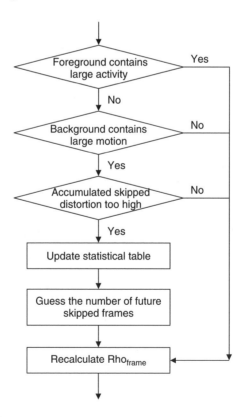

Figure 11.5   Diagram of the skip mode decision and rho budget adjustment module

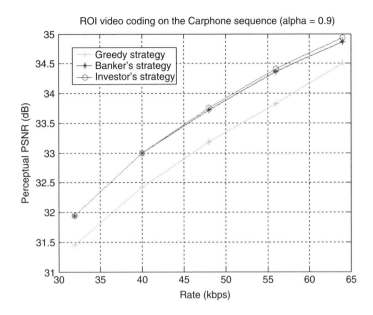

Figure 11.6   Comparison of three bit allocation strategies

distortion measurement for the frame. Therefore, the perceptual PSNR is defined as:

$$Perceptual\ PSNR = -10\log_{10}[\alpha D_R(f, \tilde{f}) + (1 - \alpha)D_{NR}(f, \tilde{f})] \qquad (11.31)$$

where f and $\tilde{f}$ are the original and reconstructed frames, and D_R and D_{NR} the normalized per pixel distortion of the ROI and Non-ROI. Clearly, both of the '*banker*' and '*investor*' *strategies* outperform the greedy strategy. The '*investor*' *strategy* slightly outperformed the '*banker*' *strategy* at higher bit rate end. Although it requires extra computational complexity for q estimation, this strategy might perform better for video sequences with repeated patterns or have self-similarity characteristics.

On the other hand, the 15 fps Carphone and other QCIF sequences at bit rates from 32 kbps to 64 kbps are tested in the H.263 Profile 3 simulations system. Four different rate control approaches are compared:

- Macroblock-level greedy algorithm [12] where the bits are allocated to the macroblocks in a uniformly distributed manner.
- Frame skipping algorithm that skips every other frame during encoding.
- Unit-based background skipping algorithm that groups every two frames into a unit and skips the background of the second frame within each unit.
- The proposed approach which content-adaptively determines the frames with skipped background, and uses the '*investor*' *strategy* for bit allocation.

As shown in Figure 11.7, the proposed approach outperformed all other approaches in the whole bit rate range and the gain is up to 2 dB. In Figure 11.8, the frame-level detail of

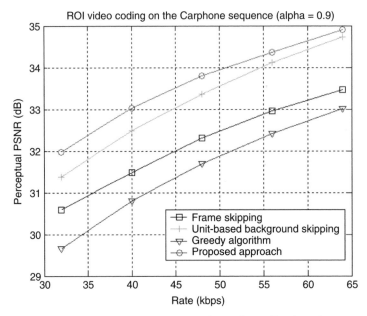

Figure 11.7   Comparison of various approaches in coding 'Carphone' sequence

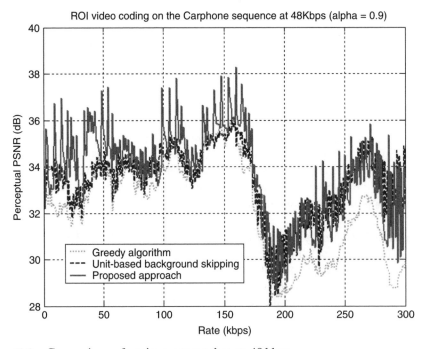

Figure 11.8   Comparison of various approaches at 48 kbps

(a) PPSNR = 32.05 dB
Greedy algorithm

(b) PPSNR = 33.99 dB
Unit-based background skipping

(c) PPSNR = 36.98 dB
Proposed approach

Figure 11.9 Comparison of reconstructed frames by various approaches at 48 kbps

(a) Original frame

(b) Reconstructed frame

Artifacts

Figure 11.10 Visual artifacts due to background skipping

these algorithms at the 48 kbps is demonstrated. Figure 11.9 shows the reconstructed 15th frame for the compared algorithms and the advantage of the proposed approach is almost 5 dB compared to the greedy algorithm and 3 dB compared to the unit-based background skipping approach.

We have to point out that background skipping sometimes might cause visual artifacts if more than enough number of backgrounds are skipped, for example, as shown in Figure 11.10(b), the coded foreground and the background copied from the previous frame are not aligned well thus causing artifacts at the collar. Clearly, this kind of artifact is very difficult to detect and be concealed because a certain degree of semantic information might be required in the processing. Further study on better background substitution or interpolation algorithms might be helpful in reducing such artifacts.

References

1. K. Maxwell, "Asymmetric digital subscriber line: Interim technology for the next forty years", *IEEE Commun. Mag.*, October, pp. 100–106, 1996.

2. Eli Orr, "Understanding the 3G-324M Spec", can be downloaded from the weblink: http://www.commsdesign.com/designcorner/OEG20030121S0009.

3. H. Wang, K. El-Maleh, "Joint adaptive background skipping and weighted bit allocation for wireless video telephony", in *Proc. International Conference on Wireless Networks, Communications, and Mobile Computing*, Maui, Hawaii, USA, June 2005.

4. H. Wang, K. El-Maleh, and Y. J. Liang, "Real-time region-of-interest video coding using content-adaptive background skipping with dynamic bit reallocation", in *Proc. IEEE International Conference on Acoustics, Speech, and Signal Processing*, Toulouse, France, May 2006.

5. Y. J. Liang, H. Wang, and K. El-Maleh, "Design and implementation of ROI video coding using content-adaptive background skipping", in *Proc. IEEE International Symposium on Circuits and Systems*, Kos, Greece, May 2006.

6. M. Chen, M. Chi, C. Hsu and J. Chen, "ROI video coding based on H.263+ with robust skin-color detection technique", *IEEE Trans. Consumer Electronics*, Vol. 49, No. 3, Aug. 2003. pp. 724–730.

7. C. Lin, Y. Chang and Y. Chen, "A low-complexity face-assisted coding scheme for low bit-rate video telephony", *IEICE Trans. Inf. & Syst.*, Vol. E86-D, No. 1, Jan. 2003. pp. 101–108.

8. S. Sengupta, S. K. Gupta, and J. M. Hannah, "Perceptually motivated bit-allocation for H.264 encoded video sequences", *ICIP'03*, Vol. III, pp. 797–800.

9. X. K. Yang, W. S. Lin, Z. K. Lu, X. Lin, S. Rahardja, E. P. Ong, and S. S. Yao, "Local visual perceptual clues and its use in videophone rate control", *ISCAS'2004*, Vol. III, pp. 805–808.

10. D. Tancharoen, H. Kortrakulkij, S. Khemachai, S. Aramvith, and S. Jitapunkul, "Automatic face color segmentation based rate control for low bit-rate video coding", in *Proc. 2003 International Symposium on Circuits and Systems (ISCAS'03)*, Vol. II, pp. 384–387.

11. J. Ribas-Corbera and S. Lei, "Rate control in DCT video coding for low-delay communications", *IEEE Trans. Circuits Systems for Video Technology*, Vol. 9, No. 1, pp. 172–185, Feb. 1999.

12. Z. He and S. K. Mitra, "A linear source model and a unified rate control algorithm for DCT video coding", *IEEE Trans. Circuits and System for Video Technology*, Vol. 12, No. 11, Nov. 2002. pp. 970–982.

13. H. Wang and N. Malayath, "Macroblock level bit allocation", US patent pending, May 2005.

14. H. Wang and K. El-Maleh, "Region-of-Interest coding in video telephony using Rho domain bit allocation", US patent pending, March 2005.

15. H. Wang and N. Malayath, "Two pass rate control techniques for video coding using a MINMAX approach", US patent pending, Sept. 2005.

16. H. Wang and N. Malayath, "Two pass rate control techniques for video coding using a rate-distortion characteristics", US patent pending, Sept. 2005.

17. T. Adiono, T. Isshiki, K. Ito, T. Ohtsuka, D. Li, C. Honsawek and H. Kunieda, "Face focus coding under H.263+ video coding standard", in *Proc. IEEE Asia-Pacific Conf. Circuits and Systems*, Dec. 2000, Tianjin, China, pp. 461–464.

18. C. Wong, O. Au, B. Meng, and H. Lam, "Perceptual rate control for low-delay video communications", *ICME'2003*. Vol. III, pp. 361–364.

19. S. Minami and A. Zakhor, "An optimization approach for removing blocking effects in transform coding", *IEEE Trans. Circuits Systems for Video Technology*, Vol. 5, No. 2, pp. 74–82, April 1995.

20. H. Wang, G. M. Schuster, A. K. Katsaggelos, "Rate-distortion optimal bit allocation scheme for object-based video coding", *IEEE Trans. Circuits and System for Video Technology*, July-September, 2005.

21. F. C. M. Martins, W. Ding, and E. Feig, "Joint control of spatial quantization and temporal sampling for very low bit rate video", in *Proc. ICASSP*, May 1996, pp. 2072–2075.

22. J. Lee, A. Vetro, Y. Wang, and Y. Ho, "Bit allocation for MPEG-4 video coding with spatio-temporal trade-offs", *IEEE Trans. Circuits and Systems for Video Technology*, Vol. 13, No. 6, June 2003, pp. 488–502.

23. C. Lin, Y. Chen, and M. Sun, "Dynamic region of interest transcoding for multipoint video conference", *IEEE Trans. Circuits and Systems for Video Technology*, Vol. 13, No. 10, Oct. 2003. pp. 982–992.

24. F. Pan, Z. P. Lin, X. Lin, S. Rahardja, W. Juwono, and F. Slamet, "Content adaptive frame skipping for low bit rate video coding", in *Proc. 2003 Joint Conference of the Fourth International Conference on Information, Communications and Signal Processing, and the Fourth Pacific Rim Conference on Multimedia*, Vol. 1, Dec. 2003, Singapore, pp. 230–234.

25. Y. J. Liang and K. El-Maleh, "Adaptive frame skipping for rate-controlled video coding", US patent pending, May 2005.

12

Wireless Video Streaming

12.1 Introduction

This chapter discusses the basic elements of wireless video streaming. Unlike download-and-play schemes, which require the entire video bitstream to be received by the client before playback can begin, video streaming allows a client to begin video playback without having to download the entire bitstream. Once video playback starts, it can continue without interruption until the end of the presentation. In order to enable playback without interruption even when the network bandwidth fluctuates, a client initially buffers the data it receives and begins playback after a delay of up to several seconds. This delay is fixed and does not depend on the length of presentation [1]. In order to achieve continuous playback, the interval between the time a video frame is transmitted by the server and the time it is displayed by the client should be the same for all frames. This means that there is a deadline for each frame when all packets that correspond to the frame must be available to the client for display. If some packets are missing at the deadline, they will be considered lost and error concealment will be employed in the decoding of the video frame. If packets that missed the deadline happen to arrive at the client later, they will simply be discarded. This concept of deadlines for the video packets is central to video streaming and will be discussed in detail later in this chapter.

With respect to the number of clients, video streaming may be classified as point-to-point, multicast and broadcast [2]. In point-to-point video streaming, there is one server and one client (unicast video streaming). Video conferencing is a special case of point-to-point video streaming, which requires low latency (low initial buffering delay). An important property in point-to-point communication is whether feedback exists between the client and server. If it exists, the server is able to adapt its processing based on the information it receives from the client regarding the quality of the channel.

Broadcast video streaming involves one server and multiple clients (one-to all communication). A classic example of this is terrestrial or satellite digital television broadcast. Owing to the large number of clients, feedback is usually not feasible, limiting the server's ability to adapt to changing channel conditions.

4G Wireless Video Communications   Haohong Wang, Lisimachos P. Kondi, Ajay Luthra and Song Ci
© 2009 John Wiley & Sons, Ltd

Multicast video streaming also involves one server and multiple clients, but not as many as with broadcast video streaming. Thus, it can be best characterized as one-to-many communication (as opposed to one-to-all). An example of multicast video streaming is IP-Multicast video streaming over the Internet. However, IP-Multicast is not widely available on today's Internet. Thus, multicast implementations at the application layer have been proposed.

As mentioned previously, there is typically an initial buffering delay before a client starts playback. In classic, non-interactive video streaming, this delay can be significant, up to several seconds. The viewer will have to tolerate the initial delay but, once the video starts playing, it will be delivered continuously at the correct frame rate. Again, this requires that the interval between the time a video frame is transmitted and the time it is displayed be constant for all frames. For interactive applications such as video conferencing, however, the delay (latency) cannot be large; otherwise it will be hard to have a conversation between two or more people. A typical maximum latency for interactive applications is 150 ms [2].

Depending on the application, video streaming may require real-time video encoding, or may only involve the transmission of an already encoded video bitstream. Applications that require real-time encoding are video conferencing (and all interactive applications), broadcast digital television and any streaming application where the input video sequence is live. However, the majority of video streaming applications utilize pre-encoded video. This removes any computational complexity constraints regarding video compression at the server and also allows for non-causal video compression techniques such as multi-pass encoding. However, real-time video encoding can adapt efficiently to changing channel conditions, whereas pre-encoded bitstreams offer limited flexibility. Scalable video coding may be used to encode these stored bitstreams so that transmission can adapt to changing bandwidth by adding or dropping enhancement layers.

12.2   Streaming System Architecture

The building blocks of a typical video streaming system are shown in Figure 12.1.

The video sequence to be transmitted is first source-encoded (compressed). Then, 'Application Layer QoS control' takes place [3], which adapts the bitstream according to network status and QoS requirements. Data are then transmitted using appropriate transport protocols. The inverse operations are performed at the client (receiver). We next discuss these operations in more detail.

12.2.1   Video Compression

As in all types of video communications, video compression has to be performed on the video sequence to be transmitted. Depending on the application, compression may be real-time or non-real-time. Real-time video compression is required for interactive applications, such as video conferencing, as well as in applications where the video sequence to be transmitted is live. In all other cases, non-real-time video compression may be used.

Since video transmission has to adapt to changing channel conditions, it is beneficial to use scalable video coding. Scalable coding provides an elegant solution for coping

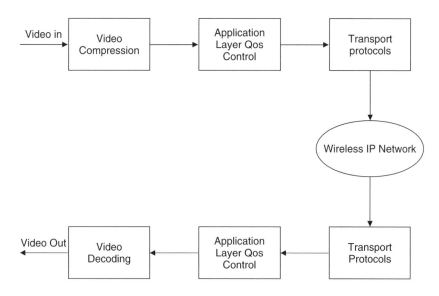

Figure 12.1 Video streaming system architecture

with bandwidth fluctuations, especially when compression is not done in real time and the bitstream is stored. In real-time compression, the target bit rate may be adjusted in non-scalable coding using rate control based on channel feedback. This is impossible to do in non-real-time compression, since the bistream has already been compressed at a specific target bit rate and stored. However, using scalable video coding, video transmission can adapt to changes in bandwidth by adding or dropping enhancement layers as appropriate, even if the bistream has already been compressed and stored.

Scalable coding is also very useful in the case of a server that serves multiple clients, which are connected to the network at different speeds (heterogeneous network). In that case, the base layer may be streamed to low-speed clients, while the base plus one or more enhancement layers may be streamed to faster clients.

Alternatives to scalable video coding for coping with changing channel bandwidths include *transcoding* and *multiple file switching* [2]. In transcoding, the video data are decompressed and recompressed again. The recompression may be done at a lower target bit rate in order to cope with bandwidth fluctuations. Other uses of transcoding are to achieve spatial downsampling, frame rate reduction or to obtain a video bitstream encoded using a different video compression standard (for example, MPEG-2 to H.264). Transcoding may address the problem of channel bandwidth fluctuations, but has two main drawbacks. The decoding and re-encoding operation typically results in some loss of video quality. Furthermore, the computational complexity of transcoding is significant, although there are ways to reduce it by reusing information selectively from the original bitstream (such as motion vectors and mode decisions) for the recompression.

In multiple file switching, more than one non-scalable bitstreams of the same video are encoded and stored. Each bitstream corresponds to a different target bit rate. In early video streaming implementations, the client had to choose from a set of bitstreams encoded at different bit rates based on its connection speed (e.g. dialup, DSL, T1, etc.) Once a choice was made, the same bitstream was used for the whole session. Later, multi-rate switching became available, which allows for switching between different bitstreams within the same session, should channel conditions dictate it. SI and SP frames in H.264 provide an efficient way of switching between video bitstreams. Multiple file switching does not have the drawbacks of high computational complexity and reduced video quality, like transcoding. However, it requires multiple bitstreams to be stored at the server, which leads to increased storage costs. Furthermore, in practice, only a small number of bitstreams are used, thus, the granularity of available bit rates is small. This limits the ability of the system to adapt to varying transmission rates.

Multiple Description Coding (MDC) may also be used in video streaming and be combined with path diversity. Thus, if multiple network paths exist between the server and client, each description may be transmitted over a different path. Thus, the client will receive video even if only one path is operational.

12.2.2 Application Layer QoS Control

Application layer QoS control aims to maintain good video quality in the presence of changing channel conditions. Several QoS control techniques have been proposed for video streaming. Some are specific to video while others are applicable to general networking problems.

Network congestion leads to bursty packet losses that are detrimental to video quality. Congestion control aims to reduce congestion by appropriately matching the bit rates of the video stream to the available network bandwidth. There are two main mechanisms for congestion control: rate control and rate shaping [3]. We briefly discuss both of them next.

12.2.2.1 Rate Control

Rate control in the context of networking refers to determining the appropriate transmission bit rate that will minimize network congestion. It should be emphasized that the term 'rate control' has a different meaning here than in the context of video compression. In video compression, 'rate control' refers to algorithms that adjust coding parameters in order to meet a target bit rate, while, in the context of networking, it only refers to determining the transmission bit rate. There are three main types of rate control: Source-based rate control, receiver-based rate control and hybrid rate control [3]. These techniques may be used in general networking problems and are not specific to video.

In source-based rate control, the transmission rate is adapted by the transmitter based on feedback information. Depending on how the current network bandwidth is determined, source-based rate control may be probe-based or model-based. With the probe-based approach, the transmitter probes for the available network bandwidth by adjusting the transmission rate so that the packet loss rate is maintained below a certain threshold P_{th} [4]. The sending rate may be adjusted using (a) additive increase and multiplicative decrease [4], or, (b), multiplicative increase and multiplicative decrease [5]. With the model-based approach, instead of probing for the network bandwidth, the transmitter uses the following equation (throughput model for a TCP connection) to estimate it [6].

$$\lambda = \frac{1.22 \times MTU}{RTT \times \sqrt{p}} \tag{12.1}$$

where λ is the throughput of a TCP connection, MTU (Maximum Transmission Unit) is the packet size used by the connection, RTT is the Round Trip Time for the connection, and p is the packet loss rate. The transmitter uses equation (12.1) to determine the video transmission rate. Using this model, video streaming data can compete fairly with TCP flows, thus decreasing the risk of congestion. Model-based rate control is also referred to as 'TCP-friendly' rate control.

In receiver-based rate control, the receivers (clients) adjust the receiving data rates by adding and dropping layers. Clearly, receiver-based rate control can only be applied to layered multicast. As in source-based rate control, receiver-based rate control may be probe-based or model-based. In probe-based receiver-based rate control, the receiver adds and drops multicast layers and monitors the resulting packet loss rate to probe the network bandwidth [7]. In model-based receiver-based rate control, the receiver uses equation (12.1) to estimate the network bandwidth.

In layered multicast, hybrid rate control may also be used, where the client adds or drops multicast layer and the server also adjust the transmission rate based on feedback by the clients [8].

12.2.2.2 Rate Shaping

Rate shaping refers to techniques which allow for the adjustment of the bit rate of a pre-compressed video bitstream in order to adapt to changing network conditions. Clearly, the most efficient and elegant way to do that is to use scalable video coding. Then, the server may add or drop enhancement layers according to the available network bandwidth. In the case of layered multicast and receiver-based rate control, the multicast layers are defined to coincide with scalable layers. Then, the receiver is able to add or drop layers in order to adjust the received bit rate.

There exist rate-distortion optimal techniques to determine a policy for transmitting and re-transmitting packets of a bitstream in the presence of feedback from the receiver. Such techniques will be discussed later in this chapter.

As mentioned earlier, multiple file switching may be used to adjust the transmission bit rate. Rate shaping may be performed (with limited success) even if only a single non-scalable pre-compressed bit stream is available. In [9], several 'rate filters' are proposed to adapt the bit rates of bitstreams. Most of these filters are also applicable to non-scalable bitstreams. We briefly describe these filters next.

- *Codec Filters.* Codec filters correspond to transcoding, as described earlier. The video bitstream is decompressed and then recompressed at a lower bit rate.
- *Frame-Dropping Filters.* As the name implies, frame-dropping filters drop frames in order to reduce the transmitted bit rate. B-frames are dropped first, since no frames depend on them, followed by P-frames and, finally, I-frames. Frame-dropping filters reduce the bit rate at the expense of a reduced frame rate.

- *Layer-Dropping Filters.* Layer-dropping filters drop enhancement layers from a scalable video bit stream, as described earlier.
- *Frequency Filters.* Frequency filters operate on the DCT coefficients of the compressed bitstream. Low-pass filters remove the high-pass coefficients from the bitstream. Color-reduction filters also perform low-pass filtering, but only on the chrominance components. Color-to-monochrome filters completely discard the chrominance information, thus converting the video to monochrome.
- *Requantization Filters.* Requantization filters perform 'inverse quantization' on the DCT coefficients and then requantize them using a larger quantization step size, thus reducing the bit rate at the expense of increased distortion. Frequency filters and requantization filters can be seen as special cases of transcoding.

12.2.2.3 Error Control

Error control is used at the application layer to ensure an acceptable QoS in the presence of adverse channel conditions. All the techniques that were described in Chapter 7 (error resilience, channel coding, error concealment) can be seen as forms of error control. If feedback is available from the receiver to the transmitter, retransmissions may also be employed. However, in video streaming, there is a strict deadline by which each packet needs to be received by the client in order to be useful. There is no point in attempting to retransmit a lost packet if, even it is received, it will be after the deadline. Thus, the retransmissions are *delay-constrained*. The problem of delay-constrained retransmission will be discussed in section 12.3.

12.2.3 Protocols

We next discuss the network protocols that are typically used in video streaming. We will assume that an Internet Protocol (IP) network is used. These protocols can be divided into three categories [3]:

- *Network-Layer Protocol.* IP is the network-layer protocol for video streaming and provides network addressing and other basic network services.
- *Transport Protocol.* The transport protocols provide end-to-end network transport functions. Some transport protocols used in video streaming are the User Datagram Protocol (UDP), Transmission Control Protocol (TCP), Real-time Transport Protocol (RTP) and Real-time Transport Control Protocol (RTCP). UDP and TCP are lower-layer transport protocols, while RTP and RTCP are upper-layer transport protocols, which are implemented on top of the lower-layer transport protocols.
- *Session Control Protocol.* Session control protocols are used to control data delivery during an established session. An example of a session control protocol is the Real Time Streaming Protocol (RTSP).

Figure 12.2 shows the relations between these categories of protocols. In a typical video streaming system, the compressed bitstream is packetized by the RTP layer, while the RTCP and RTSP layers provide control information. Then, the packetized stream is

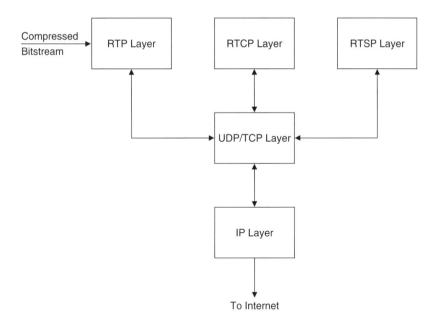

Figure 12.2   Protocols for video streaming

passed to the UDP/TCP layer and then to the IP layer. The inverse procedure is carried out at the receiver.

12.2.3.1   Transport Protocols

We next discuss the transport protocols in more detail. As mentioned earlier, these protocols include UDP, TCP, RTP and RTCP. UDP and TCP are lower-layer transport protocols while RTP and RTCP run on top of them. UDP and TCP provide multiplexing, flow control and error control. The main difference between UDP and TCP is that TCP guarantees that a packet will be delivered successfully via retransmissions, while UDP does not. However, the unlimited retransmissions that are employed by TCP make it impossible to meet the delay constraints associated with video streaming. Thus, UDP is typically used in video streaming. However, UDP does not guarantee packet delivery and an upper-level protocol such as RTP must be used to detect packet loss [3]. For more information on the delay constraints of video streaming, see section 12.3.

RTP is a protocol designed to provide end-to-end transport functions for real-time applications [10]. In addition to RTP, there is also RTCP, which provides QoS feedback to the participants of an RTP session. RTP provides the following functionalities:

- *Time-Stamping.* Time-stamping is used for the synchronization of different media types, for example, video and audio. RTP provides the time-stamping but the synchronization itself is done by the applications.

- *Sequence Numbering.* RTP assigns a number to each packet. Thus, if some packets are lost, this can be easily detected at the receiver. Also, if the packets arrive out of order, they can be put back in the correct order.
- *Payload Type Identification.* RTP puts payload type information in a packet header. Specific payload codes have been assigned for common payload types, such as MPEG-2, etc.
- *Source Identification.* RTP also puts information in the packet header regarding the source of the data so that the receiver may identify difference sources.

RTP is basically a data transfer protocol. Its companion protocol is RTCP, which is a control protocol. Participants of an RTP session typically send RTCP packets to report on the quality of the communication and other information. More specifically, RTCP provides the following functionalities [3]:

- *QoS Feedback.* QoS feedback is the primary function of RTCP. Applications receive feedback on the quality of data delivery. This feedback is useful for the senders, the receivers, as well as third-party monitors. QoS feedback is provided through reports from the senders and receivers. These reports contain information on the quality of reception, such as the fraction of lost RTP packets since the last report, the fraction of lost RTP packets since the beginning of reception, packet jitter and the delay since receiving the last sender's report.
- *Participant Identification.* RTCP transmits SDES (source description) packets that contain textual information about the session participants, such as name, address, phone number, email address, etc.
- *Control Packet Scaling.* In order to scale the RTCP control packet transmission to the number of participants, the total number of control packets is kept to 5% of the total session bandwidth. Furthermore, 25% of the control packets are allocated to sender reports and the other 75% of the control packets are allocated to receiver reports. At least one control packet is sent within five seconds at the sender or receiver in order to prevent control packet starvation.
- *Intermedia Synchronization.* RTCP sender reports contain timing information that can help in synchronization of different media types (for example, video and audio).
- *Minimal Session Control Information.* RTCP is also capable of transporting session information, such as the name of the participants.

In addition to RTP and RTCP, there is also RTSP, which is a session control protocol. The main function of RTCP is to provide VCR-like functionality, such as 'stop', 'rewind', 'fast-forward', etc.

12.2.4 Video/Audio Synchronization

In this chapter, we are concerned primarily with the streaming of video data. However, in most practical applications, video data are multiplexed and transmitted along with audio data. Thus, synchronization between video and audio is required in order to provide a pleasing user experience. If, for example, the audio is not synchronized with the lips of the speaker, this would be very annoying to the viewer. Or, in video streaming of a football (soccer) game, if the audio of the play-by-play announcer is not synchronized with the video, it is possible that the announcer will be heard shouting 'Goal!' before the ball can be seen passing the goal line.

A widely used method for synchronization is axes-based specification, or time-stamping [11]. In axes-based specification or time-stamping, timing information is inserted periodically in the media streams by the sender. These time-stamps provide a correspondence between the media streams and dictate how they should be presented together. This time-stamp information is used at the receiver to synchronize the media streams properly. As mentioned previously, the RTP protocol offers time-stamping functionality.

It should be noted that video/audio synchronization is not the only type of synchronization needed in streaming. For example, in distance learning, where slides are presented along with a narration audio stream, it is very important for the slides to be synchronized with the audio. If the narration does not correspond to the slide currently in display, the presentation will be very hard to understand.

12.3 Delay-Constrained Retransmission

The main characteristic of video streaming, which distinguishes it from off-line downloading is that the receiver begins playback before the entire bitstream is downloaded. Furthermore, once playback begins, it continues uninterrupted until the end of the presentation. The requirement for uninterrupted playback leads to delay constraints [2].

More specifically, there is a deadline by which each video frame has to be received and decoded. Let Δ be the time interval between displayed frames. Δ is the inverse of the frame rate. Thus, for a frame rate of 30 frames/s, Δ is equal to 33 ms, while for a frame rate of 10 frames/s, it is equal to 100 ms. Let us assume that the first frame in a video sequence (let's call it frame 0) that arrives at the transmitter at time $t = 0$ is displayed at the receiver at time $T > 0$. T includes the initial buffering delay that was mentioned in section 12.1. This buffering delay is designed to combat fluctuations in the channel bandwidth and also enable retransmissions. The initial buffering delay (and, subsequently, T) may be selected by the system designer (there is, of course, a minimum practical value of T, which depends on the encoding and decoding times and the minimum time required to transmit a video frame given the channel bandwidth and video coding target bit rate). A relatively small buffering delay is required for interactive applications, while a larger delay, of the order of several seconds, may be used for other applications.

Now, once the first frame has been displayed at the receiver, playback must continue at the original frame rate. Thus, the time interval between displayed frames must remain equal to Δ until the end of the presentation. This leads to the following deadlines for each frame of the video sequence:

- Frame 0 must be received and decoded by time T.
- Frame 1 must be received and decoded by time $T + \Delta$.
- Frame 2 must be received and decoded by time $T + 2\Delta$
- Etc.

In general, frame n must be received and decoded by time $T + n\Delta$.

This means that there is a strict deadline when all packets for a video frame need to be received. If one or more frame packets are not available by the deadline, error concealment will have to be used to estimate the missing information. If any packets arrive after their deadline, they will be discarded, since the video frame they belong to will have already been displayed.

For packet N that belongs to video frame n, we define the deadline:

$$T_d(N) = T + n\Delta \tag{12.2}$$

Thus, packet N needs to be delivered by time $T_d(N)$ in order to be useful.

Error control via channel coding (Forward Error Correction) can always be used in video streaming. If a feedback channel between the receiver and transmitter is available, retransmission of lost packets may also be employed. However, the above-mentioned delay constraints need to be taken into account. In general, retransmission may be used if the one-way trip time is short with respect to T.

We next discuss three non-rate-distortion-optimized techniques for delay-constrained retransmission for unicast video streaming: receiver-based, sender-based, and hybrid control [3].

12.3.1 Receiver-Based Control

In receiver based control, the receiver requests retransmissions from the transmitter. Its goal is to minimize the requests for retransmissions of packets that will not arrive before their corresponding deadlines. When the receiver detects the loss of packet N, it checks if:

$$T_c + RTT + D_s < T_d(N) \tag{12.3}$$

where T_c is the current time, RTT is the estimated round trip time, D_s is an appropriately chosen slack term, and $T_d(N)$ is the deadline for packet N. If inequality (12.3) is true, then the receiver requests the retransmission of packet N. The transmitter retransmits the packet upon reception of the receiver's request.

12.3.2 Sender-Based Control

In sender-based control, the transmitter makes the determination on whether to retransmit a packet. The main difference between the receiver-based and sender-based control is that, in the former, the receiver knows the deadline $T_d(N)$, while in the latter, the sender only has an estimate $T_d'(N)$. In sender-based control, once the sender receives information from the receiver that packet N is lost, it checks the inequality:

$$T_c + RTT/2 + D_s < T_d'(N) \tag{12.4}$$

If the inequality is true, then the sender retransmits packet N. It becomes clear that sender-based control also attempts to avoid retransmitting packets that are likely to miss their deadlines.

12.3.3 Hybrid Control

Hybrid control is a simple combination of sender-based and receiver-based control. Thus, the receiver may use inequality (12.3) before requesting retransmission and the sender may use inequality (12.4) to decide on whether to retransmit the packet.

12.3.4 Rate-Distortion Optimal Retransmission

It is clear from the above discussion that in video streaming using a non-scalable bitstream, the objective is to maximize the number of received packets through retransmissions while avoiding retransmissions that will likely prove useless due to delay constraints. Thus, each lost packet is retransmitted, if possible, subject to the delay constraints. However, in a scalable bitstream, there is a hierarchy. Thus, if a hierarchically more important packet is lost, it is pointless to try to retransmit a packet of lesser importance, since the latter will be useless, even if it is received successfully.

It should be pointed out that there are dependencies even in non-scalable bitstreams. For example, in a typical non-scalable compressed video sequence consisting of I and P frames (IPPPP...), decoding of the current frame requires successful reception of the previous frame. However, a P frame may still be decoded, with the use of error concealment, even if parts of the previous frame are not received.

In the general case of a video bitstream with dependencies, it is not clear how to design the best packet retransmission policy. For example, if several packets are lost, which ones should be transmitted first? There have been several research efforts in optimal streaming [1, 12–26]. Many of these works are not applicable strictly for video streaming and may also be applied to streaming of other types of media.

The framework in [1] assumes that the encoded data are packetized into *data units*. The server puts a data unit into a *packet* in order to transmit it. If the packet is lost, the corresponding data unit may be put into another packet and retransmitted. A packet may contain only a single data unit, whereas the same data unit can appear in multiple packets (through retransmissions).

As mentioned previously, there are dependencies in all scalable bitstreams. Such dependencies can be modeled as a *Directed Acyclic Graph (DAG)*. Figure 12.3 shows examples of dependency DAGs. Figure 12.3(a) shows the dependencies that exist in any embedded bitstream. Data unit 1 is the base layer while data units $2, 3, \ldots, L$ are the enhancement layers. In order for data unit 2 to be decodable, data unit 1 is required, etc. Figure 12.3(b) shows the dependency in a typical video sequence encoded using temporal scalability and I, P and B frames. A P frame depends on the previous I or P frame, while a B frame depends on two I or P frames. However, no frames depend on B frames. Figure 12.3(c) shows the dependencies for a typical MPEG Fine Granularity Scalability (FGS) video bitstream. Again, P frames depend on the previous I or P frame and also enhancement layers depend on the previous layer.

Thus, the dependence DAG specifies the dependencies between the data units in a video bitstream and is computed offline. Along with the DAG, the following information is stored for each data unit l: Its size B_l in bytes, its importance Δd_l (its differential distortion) and its timestamp $t_{DTS,l}$ (its deadline). The quantity Δd_l is the amount by

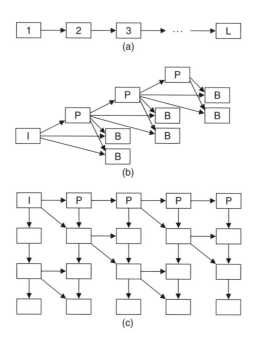

Figure 12.3 Directed acyclic dependence graphs. (a) Sequential dependencies typical of embedded codes. (b) Dependencies between IBBPBBPBBP video frames. (c) Typical dependencies for MPEG-4 progressive fine grain scalability mode. From [1], used with permission. Copyright IEEE, 2006

which the distortion of the received video will decrease if packet l is available to the decoder. t_{DTS}, l is the time by which data unit l must be received in order to be useful.

Chou [1] solves the video streaming problem in a rate-distortion framework. The rate R is defined as the expected cost of streaming the entire presentation (in our case, video sequence). The rate may refer to the total number of bytes transmitted, or, more generally, it may mean the total cost of transmitting the presentation. If the transmission of a data unit using a transmission option π has a cost per source byte $\rho(\pi)$, the cost of transmitting a data unit of size B bytes is $B\rho(\pi)$. Then, R is the expected value of the total cost, averaged over all possible realizations of the random channel for a given video sequence. The quantity D refers to the expected value of the total distortion of the video sequence, averaged over all possible realizations of the random channel for a given video sequence.

The objective of the video streaming algorithm is, given any presentation θ, to minimize the expected distortion $D = D_\theta(R)$ for an expected rate R. Each of the dots in Figure 12.4(a) denotes a possible (R, D) pair, which results from a specific retransmission policy. The dotted line shows the convex hull of all (R, D) pairs. The problem of minimizing the distortion subject to a given rate can be solved by minimizing the Lagrangian cost $D + \lambda R$ for some positive Lagrangian multiplier λ.

The problem of transmitting a single data unit is considered next. Chou [1] shows that the solution of the problem of transmitting a single data unit may be used as part of the solution to the main problem of minimizing the expected video distortion subject to a

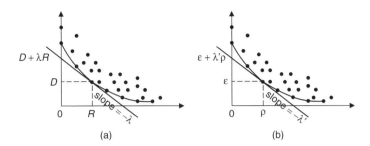

Figure 12.4 (a) Set of achievable distortion-rate pairs, its lower convex hull (dotted), and an achievable pair (R, D) minimizing the Lagrangian $D + \lambda R$. Each dot is the (R, D) performance of some algorithm. (b) Likewise, the set of achievable error-cost pairs, its lower convex hull, and an achievable pair (ρ, ε) minimizing the Lagrangian $\varepsilon + \lambda' \rho$. From [1], used with permission. Copyright IEEE, 2006

rate constraint. Let us assume that a transmission option (policy) π is used to transmit the data unit. The policy determines how the retransmissions are performed. Since we are talking about the transmission of a single data unit, it is possible to normalize its rate and distortion. Thus, the transmission of a data unit is associated with an expected cost ρ and an expected error ε. The cost ρ may be defined as the total number of transmissions of the data unit and ε may be defined as the probability of loss (unsuccessful reception) of the data unit (even after the retransmissions).

We assume that there are N transmission opportunities for data unit l before its deadline elapses. We denote these times as $t_{0,l}, t_{1,l}, \ldots, t_{N-1,l}$. Our problem is to determine which of these opportunities should actually be used to transmit the data unit. The objective is to minimize the error ε subject to a constraint on the cost ρ. This problem may again be solved using Lagrangian optimization, by minimizing the Lagrangian $\varepsilon + \lambda' \rho$ for some positive Lagrange multiplier λ'. Each of the dots in Figure 12.4(b) represents a (ρ, ε) point that can be achieved using a specific retransmission policy. The dotted line is the convex hull of these points.

Let us now concentrate on the scenario of sender-driven retransmission with feedback where the receiver sends an acknowledgement packet the instant that it receives the data packet. Obviously, once the sender receives an acknowledgment packet for a data unit, it will not attempt to transmit it again. The problem of transmitting a single data unit can be seen as a Markov decision process with finite horizon N. Such a process is represented by a trellis of length N. Any action that is taken at any state of the trellis influences the outgoing transition probabilities. A path through the trellis corresponds to a specific transmission policy for the data unit. Figure 12.5 shows an example of such a trellis. The trellis starts at time s_0, when the sender has to decide to either transmit the data unit, taking action $a_0 = 1$, or not to transmit it, taking action $a_0 = 0$. If the sender chooses to transmit the data unit, then, just before time s_1, it observes whether it has received a packet acknowledging successful reception of the data unit ($o_0 = 1$), or not ($o_0 = 1$). If the reception of the data unit has been acknowledged by time s_1, then the process enters a final state at time s_1. Otherwise, the sender decides again at time s_1 whether or not to transmit the data unit and observes before time s_2 whether it has received an

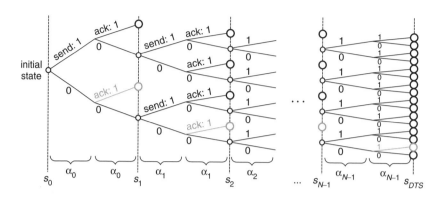

Figure 12.5 Trellis for a Markov decision process. Final states are indicated with double circles. From [1], used with permission. Copyright IEEE, 2006

acknowledgement. This process continues until an acknowledgement is received or the N transmission opportunities have been exhausted.

The trellis is used in the minimization of the Lagrangian cost $\varepsilon + \lambda'\rho$. This may be done using dynamic programming [14] or branch and bound algorithms [27].

We now concentrate on the original problem, which is the minimization of the distortion D subject to a constraint on the rate R for the presentation of the whole video sequence. This is equivalent to the minimization of the Lagrangian cost $D + \lambda R$. This problem is completely characterized by the dependence between the data units (the dependence DAG), the set of incremental distortions Δd_l, the packet sizes B_l, as well as $\varepsilon(\pi)$ and $\rho(\pi)$, the error and cost associated with a specific retransmission policy π for a packet. The dependence DAG, Δd_l and B_l depend on the source, source code and packetization, while $\varepsilon(\pi)$ and $\rho(\pi)$ depend on the transmission scenario and channel characteristics.

An iterative approach is proposed for minimizing the Lagrangian cost $D + \lambda R$, called the *Iterative Sensitivity Adjustment (ISA)* algorithm. This algorithm uses the solution of the problem of transmitting a single data unit (minimizing $\varepsilon + \lambda'\rho$) as part of the solution of minimizing $D + \lambda R$. More information can be found in [1].

In practice, a rate control algorithm is required for controlling the instantaneous rate of data packet transmissions dictated by the ISA algorithm. The rate control algorithm increases or decreases λ to adjust the number of data units selected for transmission at each transmission opportunity. Of particular interest is the case in which λ is adjusted so that exactly one data unit is selected for transmission at each transmission opportunity. Chou [1, 14] proposes an algorithm that selects λ in this case. The algorithm requires a series of approximations. The overall system proposed in [1] is called Rate-Distortion Optimized system (RaDiO).

12.4   Considerations for Wireless Video Streaming

The discussion on video streaming presented so far in this chapter is not specific to wireless transmission but may be also be applied to wired networks. However, wireless channels require additional considerations owing to the following problems [3]:

- *Bandwidth Fluctuations.* The available bandwidth of a wireless channel fluctuates with time. This may be due to a variety of reasons, including multipath fading, co-channel interference, noise disturbances and, mobile users, the changing distance between sender and receiver.
- *High Bit-Error Rate.* Wireless channels have a much higher bit error rate than wired channels due to fading and much higher noise levels.
- *Heterogeneity.* In a multicast scenario, different receivers may have different characteristics in terms of latency, visual quality, processing capabilities, power limitations and bandwidth limitations.

As mentioned previously, scalable video coding can deal effectively with the bandwidth fluctuations. Also, channel coding may be used instead of or in addition to retransmissions to cope with the higher bit error rates. Furthermore, the topics of error resilience and error concealment are more important in wireless video streaming due to the higher error rates. Scalable video coding used in conjunction with channel coding and unequal error protection is also very efficient for video transmission over wireless channels.

The ISA algorithm in [14] and [1] has been extended in [12] and [13] to wireless video streaming. In that work, forward error correction (channel coding) is used for error control in addition to retransmissions. An Incremental Redundancy (IR) scheme is used. In such a scheme, rate compatible channel codes are used. Thus, the bits for high-rate codes (less error protections) are subsets of the bits of low-rate codes (high error protection). The bits that correspond to the high-rate codes are transmitted first. If the bit errors cannot be corrected, more bits are transmitted. These bits are combined with the original received bits to create a lower-rate code, which may be able to correct the bit errors.

12.4.1 Cross-Layer Optimization and Physical Layer Consideration

The concept of *cross-layer optimization* and the related concept of *joint source-channel coding* are important in wireless video transmission. Shannon's *Principle of Separability* states that the design of source and channel coding can be separated without loss of optimality as long as the source coding produces a bit rate that can be carried by the channel (a rate that does not exceed the channel capacity). While being an important theoretical derivation, this principle relies on the crucial assumption that the source and channel codes can be of arbitrarily large lengths. In practical situations, due to limitations on the computational power and delay constraints, this assumption does not hold. Thus it is beneficial to consider the problems of source and channel coding jointly. Some representative works of joint source-channel coding for wireless video transmission include [28−32] and [33].

Cross-layer optimization can be seen as a generalization of joint source-channel coding. The *Open Systems Interconnection (OSI)* Reference Model specifies seven layers for communication systems [34]:

- Physical Layer.
- Data Link Layer.
- Network Layer.

- Transport Layer.
- Session Layer.
- Presentation Layer.
- Application Layer.

Traditionally, each of these layers has been considered separately. This makes system design easier. However, it has recently been shown that *cross-layer design and optimization* can be beneficial. Some recent work on cross-layer optimization for wireless video transmission includes [35–42] and [43]. Cross-layer design and optimization has been discussed in Chapter 9.

12.5 P2P Video Streaming

So far in this chapter, we have discussed video streaming from a server to one or more receivers. The topic of peer-to-peer (P2P) video streaming has gained interest in recent years, after the wide use and popularity of P2P file transfer. In P2P systems, there are no servers and all users are peers. Thus, the structure of a P2P system is decentralized. We will next discuss P2P video streaming systems briefly.

The advantages of P2P systems include their capability for self organization, bandwidth scalability and network path redundancy [44]. However, in P2P systems, peers typically join or leave the system rather frequently, making the system unstable. Furthermore, different peers may have different uplink and downlink bandwidths as well as different processing power. Thus, P2P systems lack any QoS guarantees and the problem of using them for streaming video data, which have strict delay requirements, is a challenging one.

P2P streaming systems rely on self-organization of the peers. There are two main network architectures used in P2P video streaming: *Tree-based overlays* and *mesh overlays*. In tree-based overlays, the peers are organized in a tree structure. The root of the tree is the source peer, while the leaves are the client peers. The intermediate peers in the tree push the video content from the source to the clients. Such architectures are easy to implement and maintain by the source. However, they suffer from high instability caused from peers joining or leaving the system. Also, each client is connected to the source over a single path. Thus, the available bandwidth is limited by the minimum upload bandwidth among the peers in the path.

In mesh overlays, the peers self-organize in a directed mesh. Thus, the data from the source peer are distributed among multiple paths. Each peer is connected to one or more parent peers and one or more child peers. Mesh overlays are more robust to peers entering and leaving the system than tree-based overlays. Furthermore, the existence of more than one path between the source and the client is very important.

The use of scalable video coding is appropriate in P2P video streaming in order to meet the constraints imposed by the bandwidth available at any given point in the network. Scalable video coding is also useful in cases where there is a large heterogeneity between the peers in terms of their access bandwidth and processing power [44].

Multiple Description Coding (MDC) may also be used in P2P video streaming. P2P systems typically offer multiple paths between a peer transmitting the video and a peer receiving it. MDC is a natural choice for this type of situation, since different descriptions may be transmitted via different paths. Thus, if a path becomes unavailable during the course of the presentation (perhaps due to a peer leaving the system), video reception will still be possible via the other paths.

A fundamental problem in P2P streaming systems is how to select the best subset of paths to use between the source and client, and also how to determine the optimal rate allocation between the selected paths. There are two main ways of dealing with this problem: *Receiver-driven streaming* and *distributed path computation*. In receiver-driven streaming, the client coordinates the streaming process. Content location information can be accessed by the receiver at super nodes/servers as in BitTorrent or PPLive. Alternatively, such information may be obtained from other peers using search algorithms adapted to decentralized systems, or the receiver peer may just probe the network connections toward candidate source nodes [44]. Then, the client makes an informed decision of source peers and network transmission paths based on the network connectivity information and streaming session characteristics it received [45].

In practice, it is impossible for a client to receive accurate information about the topology of the whole P2P streaming system, especially if it is very large. In distributed path computation, each intermediate node makes an individual routing decision for each upcoming packet, based only on local topology information [46]. However, distributed path computation may lead to suboptimal streaming strategies, since no peer has complete knowledge of the network status.

References

1. P. A. Chou and Z. Miao, "Rate-Distortion Optimized Streaming of Packetized Media," *IEEE Transactions on Multimedia*, Vol. 8, No. 2, April 2006, pp. 390–404.
2. J. G. Apostolopoulos, W.-T Tan and S. Wee, "Video Streaming: Concepts, Algorithms and Systems," Technical Report HPL-2002-260, HP Laboratories, 2002.
3. Y. Wang, J. Ostermann and Y.-Q Zhang, *Video Processing and Communications*, Prentice-Hall, 2002.
4. D. Wu, et al, "On End-to-End Architecture for Transporting MPEG-4 Video over the Internet," *IEEE Transactions on Circuits and Systems for Video Technology*, Vol. 10, No. 6, September 2000, pp. 923–941.
5. T. Turletti and C. Huitema, "Videoconferencing on the Internet," *IEEE/ACM Transactions on Networking*, Vol. 4, No. 3, June 1996, pp. 340–351.
6. S. Floyd and K. Fall, "Promoting the Use of End-to-End Congestion Control in the Internet," *IEEE/ACM Transactions on Networking*, Vol. 7, No. 4, August 1999, pp. 458–472.
7. S. McCanne, V. Jacobson and M. Vetterli, "Receiver-Driven Layered Multicast," in *Proc. ACM SIG-COMM'96*, August 1996, pp. 117–130.
8. S. Y. Cheung, M. Ammar and X. Li, "One the Use of Destination Set Grouping to Improve Fairness in Multicast Video Distribution," in Proc. *IEEE INFOCOM'96*, Vol. 2, March 1996, pp. 553–560.
9. N. Yeadon, F. Garcia, H. Hutchison and D. Shepherd, "Filters: QoS Support Mechanisms for Multipeer Communications," *IEEE Transactions on Selected Areas in Communications*, Vol. 14, No. 7, September 1996, pp. 1245–1262.
10. H. Schulzrinne, S. Casner, R. Frederick and V. Jacobson, "RTP: A Transport Protocol for Real-Time Applications," IETF, RFC 1889, January 1996.
11. G. Blakowski and R. Steinmetz, "A Media Synchronization Survey: Reference Model, Specification, and Case Studies," *IEEE Journal on Selected Areas in Communications*, Vol. 14, No. 1, January 1996, pp. 5–35.
12. J. Chakareski and P. A. Chou, "Application Layer Error Correction Coding for Rate-Distortion Optimized Streaming to Wireless Clients," in *Proc. International Conference on Acoustics, Speech and Signal Processing*, Vol. 3, Orlando, FL, May 2002, pp. 2513–2516.
13. J. Chakareski and P. A. Chou, "Application Layer Error Correction Coding for Rate-Distortion Optimized Streaming to Wireless Clients," *IEEE Transactions on Communications*, Vol. 52, No. 10, October 2004, pp. 1675–1687.

14. P. A. Chou and Z. Miao, "Rate-Distortion Optimized Streaming of Packetized Media," Microsoft Research, Redmond, WA, Tech. Rep. MSR-TR-2001-35, February 2001.

15. C. Luna, L. P. Kondi, A. K. Katsaggelos, "Maximizing User Utility in Video Streaming Applications", *IEEE Transactions on Circuits and Systems for Video Technology*, Vol. 13, No. 2, February 2003, pp. 141–148.

16. Z. Miao and A. Ortega, "Optimal Scheduling for Streaming of Scalable Media," in *Proc. Asilomar Conference on Signals, Systems and Computers*, Vol. 2, Pacific Grove, CA, November 2000, pp. 1357–1362.

17. Z. Miao, "Algorithms for Streaming, Caching and Storage of Digital Media," Ph.D. dissertation, University of Southern California, Los Angeles, May 2002.

18. Z. Miao and A. Ortega, "Expected Run-Time Distortion Based Scheduling for Delivery of Scalable Media," in *Proc. International Packet Video Workshop*, Pittsburgh, PA, April 2002.

19. M. Podolsky, S. McCanne and M. Vetterli, "Soft ARQ for Layered Streaming Media," University of California Computer Science Division, Berkeley, Tech. Rep. UCB/CSD-98-1024, November 1998.

20. M. Podolsky, S. McCanne and M. Vetterli, "Soft ARQ for Layered Streaming Media," *Journal of VLSI Signal Processing, Special Issue on Multimedia Signal Processing*, Vol. 27, No. 1-2, February 2001, pp. 81–97.

21. D. Quaglia and J. C. de Martin, "Delivery of MPEG Video Streams with Constant Perceptual Quality of Service," in *Proc. International Conference on Multimedia and Expo (ICME)*, Vol. 2, Lausanne, Switzerland, August 2002, pp. 85–88.

22. T. Stockhammer, H. Jenkac and G. Kuhn, "Streaming Video over Variable Bit-Rate Wireless Channels," *IEEE Transactions on Multimedia*, Vol. 6, No. 2, April 2004, pp. 268–277.

23. F. Zhai, R. Berry, T. N. Pappas and A. K. Katsaggelos, "A Rate-Distortion Optimized Error Control Scheme for Scalable Video Streaming over the Internet," in *Proc. International Conference on Multimedia and Expo*, Baltimore, MD, July 2003.

24. F. Zhai, C. E. Luna, Y. Eisenberg, T. N. Pappas, R. Berry and A. K. Katsaggelos, "A Novel Cost-Distortion Optimization Framework for Video Streaming over Differentiated Services Networks," in *Proc. International Conference on Multimedia and Expo (ICME)*, Barcelona, Spain, September 2003.

25. F. Zhai, C. E. Luna, Y. Eisenberg, T. N. Pappas, R. Berry and A. K. Katsaggelos, "Joint Source Coding and Packet Classification for Real-Time Video Transmission over Differentiated Services Networks," *IEEE Transactions on Multimedia*, Vol. 7, No. 4, August 2004, pp. 716–726.

26. J. Zhou and J. Li, "Scalable Audio Streaming over the Internet with Network-Aware Rate-Distortion Optimization," in *Proc. IEEE International Conference on Multimedia and Expo (ICME)*, Tokyo, Japan, August 2001.

27. M. Roeder, J. Cardinal and R. Hamzaoui, "On the Complexity of Rate-Distortion Optimal Streaming of Packetized Media," in *Proc. Data Compression Conference*, Snowbird, UT, March 2004.

28. M. Bystrom and J. W. Modestino, "Combined Source-Channel Coding for Transmission of Video over a Slow-Fading Rician Channel," in *Proc. IEEE International Conference on Image Processing*, 1998, pp. 147–151.

29. M. Bystrom and J. W. Modestino, "Combined Source-Channel Coding Schemes for Video Transmission over an Additive White Gaussian Noise Channel," *IEEE Journal on Selected Areas in Communications*, Vol. 18, June 2000, pp. 880–890.

30. G. Cheung and A. Zakhor, "Joint Source/Channel Coding for Scalable Video over Noisy Channels," in *Proc. IEEE International Conference on Image Processing*, Vol. 3, 1996, pp. 767–770.

31. I. Kozintsev and K. Ramchandran, "Multiresolution Joint Source-Channel Coding Using Embedded Constellations for Power-Constrained Time-Varying Channels," in *Proc. IEEE International Conference on Acoustics, Speech and Signal Processing*, 1996, pp. 2345–2348.

32. K. Ramchandran, A. Ortega, K. M. Uz and M. Vetterli, "Multiresolution Broadcast for Digital HDTV Using Joint Source-Channel Coding," *IEEE Journal on Selected Areas in Communications*, Vol. 11, January 1993, pp. 6–23.

33. M. Srinivasan and R. Chellappa, "Adaptive Source-Channel Subband Video Coding for Wireless Channels," *IEEE Journal on Selected Areas in Communications*, Vol. 16, December 1998, pp. 1830–1839.

34. A. S. Tanenbaum, *Computer Networks*, Prentice-Hall, 1996.

35. Y. Andreopoulos, N. Mastronarde and M. van der Schaar, "Cross-Layer Optimized Video Streaming over Wireless Multihop Mesh Networks," *IEEE Journal on Selected Areas in Communications*, Vol. 24, No. 11, November 2006, pp. 2104–2115.

36. S. K. Bandyopadhyay, G. Partasides, L. P. Kondi, "Cross-Layer Optimization for Video Transmission over Multirate GMC-CDMA Wireless Links", *IEEE Transactions on Image Processing*, Vol. 17, No. 6, June 2008, pp. 1020–1024.

37. Y. S. Chan and J. M. Modestino, "A Joint Source Coding-Power Control Approach for Video Transmission over CDMA Networks," *IEEE Journal on Selected Areas in Communications*, Vol. 21, No. 10, December 2003, pp. 1516–1525.

38. L. P. Kondi, D. Srinivasan, D. A. Pados and S. N. Batalama, "Layered Video Transmission over Multirate DS-CDMA Wireless Systems," *IEEE Transactions on Circuits and Systems for Video Technology*, Vol. 15, No. 12, December 2005, pp. 1629–1637.

39. E. S. Pynadath and L. P. Kondi, "Cross-Layer Optimization with Power Control in DS-CDMA Visual Sensor Networks," in *Proc. IEEE International Conference on Image Processing*, Atlanta, GA, 2006, pp. 25–28.

40. Y. Shen, P. C. Cosman and L. B. Milstein, "Error-Resilient Video Communications over DCMA Networks with a Bandwidth Constraint," *IEEE Transactions on Image Processing*, Vol. 15, No. 11, November 2006, pp. 3241–4352.

41. D. Srinivasan, L. P. Kondi, "Rate-Distortion Optimized Video Transmission over DS-CDMA Channels with Auxiliary Vector Filter Single-User Multirate Detection", *IEEE Transactions on Wireless Communication*, Vol. 6, No. 10, October 2007, pp. 3558–3566.

42. M. van der Schaar and D. S. Turaga, "Cross-Layer Packetization and Retransmission Strategies for Delay-Sensitive Wireless Multimedia Transmission," *IEEE Transactions on Multimedia*, Vol. 9, No. 1, January 2007, pp. 185–197.

43. Q. Zhao, P. C. Cosman and L. B. Milstein, "Tradeoffs of Source Coding, Channel Coding and Spreading in CDMA Systems," in *Proc. MILCOM*, Vol. 2, Los Angeles, CA, 2000, pp. 846–850.

44. D. Jurca, J. Chakareski, J.-P. Wagner and P. Frossard, "Enabling Adaptive Video Streaming in P2P Systems," *IEEE Communications Magazine*, June 2007, pp. 108–114.

45. A. C. Begen et al., "Multi-Path Selection for Multiple Description Video Streaming over Overlay Networks," *Signal Processing: Image Communication*, Vol. 20, 2005, pp. 39–60.

46. D. Jurca and P. Frossard, "Distributed Media Rate Allocation in Overlay Networks," in *Proc. IEEE International Conference on Multimedia and Expo*, Toronto, Canada, July 2006.

Index

1/4th Pixel Accuracy, 144
16x8, 140, 145, 146
1xEv-DO, 125
3D TV, 17
3D video, 17
3D-ESCOT, 90
3D-SPIHT, 89
3G-324M, 348
3GPP, 3, 98, 99
3GPP2, 125
4x4, 140
4x4 DCT, 149
802.16m, 98
8x16, 140, 146
8x4, 140
8x8 Transform, 151

AAA, 113, 126
Acquisition I Pictures, 330
Adaptive perceptual color-texture
    segmentation, 182
aGW, 102
Alternate Scan, 152
AMC, 23, 106
Application Layer QoS control, 370, 372
Arithmetic coding, 8, 71
ASK, 24
ASN, 113
ASP (Advanced Simple Profile), 136
Audio video synchronization, 344
AVC, 135, 138, 143, 325, 327
AVC/H.264, 61–64, 73, 78, 325
Average codeword length, 67

AWGN, 19, 25, 26, 28
Axes-based specification *See*
    Time-stamping, 377

B MB, 143
Background subtraction, 189
Baseline Profile, 160
BCH, 122
BCMCS, 132, 133
BER, 23
Blockwise correlating lapped orthogonal
    transforms, 232
Boundary-based segmentation, 181
BR Picture, 157
Branch and bound algorithms, 382
BSP, 129

CABAC, 138, 154
CAN, 126
Capture order, 327
Cathode ray tube, 64
CAVLC, 138, 152, 154
Change detection, 188
Channel change, 329
Channel coding, 223, 232, 374, 378, 383
Chroma, 60, 62–64
Chrominance, 60, 62, 64, 79, 86
CIR, 28
Closed GOP, 329
CNR, 37
Coded block patterns, 237
Coded Picture Buffer (CPB), 158, 159,
    164, 333

Codes
   BCH, 224
   block, 223, 232
   convolutional, 223, 232
   Rate Compatible Punctured
      Convolutional, 224
   Reed-Solomon, 224
Coding efficiency, 136
Coefficient splitting, 232
Color, 174
Color representation, 61
Compression, 59
Conditional coding *See* Context-based
    coding, 68
Conditional entropy, 65
Congestion control, 372
Constant Bit Rate (CBR), 332
Constrained baseline profile, 162
Constraint length, 233
Content analysis, 171
Content-adaptive frame/object/macroblock
    skipping, 356
Content-aware, 242
Content-based interactivity, 17
Context-based coding, 68
Context-based arithmetic encoding, 213
Control packet scaling, 376
Copyright flag, 340
Correlated predictors, 232
Correlating filter banks, 232
Correlating linear transforms, 232
COST, 187
Cost-distortion, 295
CQI, 121
Cross-layer, 241
Cross-layer design, 15, 241
Cross-layer optimization, 383, 384
CSN, 113

DAG *See* Directed Acyclic Graph, 379,
    382
Data hiding, 305
Data Partitioning, 237
DCT, 77–86, 93, 149
Deblocking, 155
Decoded Picture Buffer (DPB), 334

Decoding order, 336
Decoding Time Stamp (DTS), 340
Delay-constrained retransmission, 374,
    377, 378
DFT, 79, 80
DiffServ network, 297
Digital television broadcast, 370
Digital television broadcast:satellite, 369
Digital television broadcast:terrestrial, 369
Direct Mode, 146
Directed acyclic graph, 383
Discrete Cosine Transform *See* DCT, 77
Discrete Fourier Transform *See* DFT, 79
Discrete source, 64
Discrete Wavelet Transform *See* DWT, 86
Display order, 327
Distributed path computation, 385
Diversity gain, 35
DPCM, 73, 74, 76–79, 81
Drift, 89
DWT, 86–89, 95
Dynamic location management, 54
Dynamic programming, 249, 382

E-UTRAN, 101, 102
eBS, 125
Edge, 174
Edge detection, 174
Editable Structure, 331
eNB, 102
Encoding Order, 328
eNodeB, 106
Entropy, 65
Entropy rate, 66
EPC, 102
EPS, 102
Error concealment, 12, 223, 224,
    234, 291, 369, 374, 378, 379,
    383
   inter, 234
   intra, 234
Error control, 291, 374
Error propagation, 228, 229
Error resilience, 223, 224, 374, 383
Error resilient, 11
Error-Resilient Entropy Coding, 226

Exp-Golomb, 152
Extended Profile, 162

Fast forward, 338
FDD, 23
FEC, 29
    S,D
        102, 107
Filters
    codec, 373
    frame-dropping, 373
    frequency, 374
    layer-dropping, 374
    rate, 373
    requantization, 374
Fixed length coding, 224
Flexible marcoblock ordering, 238
FLSE, 125
Forward error correction *See* Channel
        coding, 223, 232, 378
Frequency-nonselective fading, 19, 20
Frequency-selective fading, 19
FSK, 24
Fully Open GOP, 329

GGSN, 110
Granularity, 372
Group of Pictures (GOP), 327

H.223, 348
H.245, 348
H.261, 10
H.262, 135
H.263, 10, 136
H.264, 135, 138, 325, 327, 328
H.264/AVC
    Error resilience features, 236
H.264/MPEG-4 AVC, 11
Hadamard transform, 151
Handoff re-routing, 54
HARQ, 37, 125
HDTV, 60–62
Heterogeneous network, 371
Heterogeneous networks, 229
Hierarchical GOP, 331
High 10 Profile, 163

High 4:2:2 Profile, 163
High 4:4:4 Predictive Profile, 163
High Definition Television *See* HDTV, 60
High profile, 162
HRD, 158
Huffman coding, 69
Hybrid Transform-DPCM Architecture, 77

ICI, 34
IDR, 157, 329, 330, 333, 336
Image segmentation, 179
IMT-Advanced, 97
Independent identically distributed, 64
Independent segment decoding, 223, 228
Information source, 64, 68
Information theory, 64, 73
Initial buffering delay, 369, 370, 381
Insertion of intra blocks or frames, 223,
        228
Inter frame, 81
Interlaced Video, 140
Intermedia synchronization, 376
Intra frame, 80
Intra only profile, 163
Intra placement, 236
IP, 370, 374, 375
iPhone, 1
ISI, 19, 28
ISO/IEC 13818-1, 338
Iterative sensitivity adjustment, 382
ITU, 97

Jakes model, 21
Joint entropy, 65
Joint source-channel coding, 310
JPEG-2000, 86–88, 94

K-means, 181

Lagrangian relaxation, 242
Latency, 370
Levels, 163
Lifting, 90–93, 95
Location management, 54
LOS, 20
lossy compression, 9

Low delay, 334
Low-level features, 174
LRU, 121
LTE, 98
Luma, 62, 64, 78
Luminance, 60, 62, 64, 86

Main profile, 162
Markov channel model, 21
Markovian movement, 54
Maximally smooth recovery, 234
MB, 140–142, 145
MB Pair, 156
MBAFF, 144
MCTF, 90
MIMO, 52, 106
MIMO beamforming, 35
Minimal session control information, 376
MMOG, 99
Mobile TV, 17
Mobile WiMAX, 4
Morphology, 178
Motion, 174
Motion compensated temporal filtering
    See MCTF, 90
Motion compensation, 8, 82, 84
Motion estimation, 138
Motion vector, 81–84
Motion-compensated temporal
    interpolation, 224, 235
Moving object tracking, 188
MPEG-1, 62, 63, 78, 95
MPEG-2, 62, 63, 73, 78, 135
MPEG-4, 11, 78, 86, 95
MPEG-2 Part 1, 338
MPEG-2 Systems, 338
MPEG-21, 219
MPEG-4 Part 2, 10, 135, 136
MPEG-7, 217
MQAM, 29
MU-MIMO, 123
Multiple description coding, 223, 230,
    372, 384
Multiple file switching, 371, 373
Multiple references, 143
Multiplex, 339

Mutual information, 65
MV Compression, 145

Nakagami-m, 22
NAS, 102, 107
Network abstraction layer, 14
Network-adaptive video encoding, 291
NGMN, 102
NodeB, 110
NTSC, 62, 79

Object-based video communications, 295
Object-based video, 209
Object-based video coding, 212
Object-based video representation, 211
Odd-even embedding, 307
ODWT, 88–90
OFDMA, 106, 122
Open GOP, 329
Open Systems Interconnection (OSI)
    Reference model, 383
OSI, 241
OTA, 125
Overcomplete Discrete Wavelet Transform
    See ODWT, 88
Overlays
    mesh, 384
    tree-based, 384

P MB, 142
P Picture, 161
P2P, 18
P2P wireless video streaming, 18
Packet Identifier (PID), 341
Packet scheduling, 293
Packet stat code prefix, 340
packetization, 247, 291, 338
Packetized Elementary Stream (PES),
    339, 340
PAFF, 144
PAL, 62
PAPR, 33
Parity bits, 232
Partially Open GOP, 330
Participant identification, 376
Pause, 337

Payload type identification, 376
PCRF, 126
PDCP, 107
PDSN, 126
PicAFF, 144
Picture segmentation, 236
PPS, 167
Prefix codes, 69
Presentation Time Stamp (PTS), 340
Principle of separability, 383
Program Clock Reference (PCR), 342
Program Stream (PS), 339, 343
Projection onto convex Sets, 234
PRU, 121
PSK, 24
PSNR, 137

QAM, 24
QoS, 14, 23, 241
QoS feedback, 376
Quality-driven, 243
Quantization, 9, 71–74, 76–78, 83–85,
    151
  overlapping, 232
Quantization bin, 72
Quantization levels, 72

Random access, 325
Rate control, 371–373, 382
Rate control:hybrid, 372, 373
Rate control:receiver-based, 372, 373
Rate control:source-based, 372
Rate shaping, 372, 373
Rate-distortion, 242
Rate-distortion optimal retransmission,
    379
Rate-distortion optimal source-coding
    scheme, 295
Rate-Distortion Optimized system, 382
Rate-Distortion Optimization, 142
Receiver shaping, 35
Receiver-driven streaming, 385
Recovery of coding modes and motion
    vectors, 224
Redundancy, 231
Redundant slices, 239

Reed-Solomon codes, 224
Reference picture selection, 237
Region growing, 181
Relay station, 116
Resynchronization markers, 223, 224
Retransmission, 291
Reverse, 337
Reversible Variable Length Coding, 223,
    226
RLC, 102
RLP, 129
RLSE, 126
RNC, 110
ROI, 265
ROI based video coding, 351
ROI bit allocation, 351
ROI video quality, 352
RRC, 102, 107
RRCM, 116
RSSI, 26
RTCP, 14, 374–376
RTP, 14, 374–377
RTSP, 374, 376
RTT, 373, 378

SAD, 142
SAE, 102
SAR, 129
SC-FDMA, 106
scalability
  SNR, 229
  spatial, 229
  temporal, 229
Scalable coding, 223, 229
Scalable video coding, 370, 371, 373,
    383, 384
Scalar lossless coding, 66
Scalar quantization, 72
Scanning, 60
SCH, 122
SD, 59
SDES, 376
SEI, 167
Sequence numbering, 376
Sequence parameter set, 238
SGSN, 110

Shape, 174
Shape feature, 176
Shift register, 233
SI Pictures, 157
SINR, 23
Smart phones, 1
SNR scalability, 86
Source identification, 380
SP Pictures, 158
Spatial direct mode, 146
Spatial interpolation, 224, 234
Spatial prediction, 147–148
Spatial scalability, 86
Spatial-temporal segmentation, 187
Spatially adaptive dominant, 182
SPIHT, 87–90, 95
Split-and-merge, 181
SPS, 167
SRNC, 126
Standard definition video See SD, 59
Stationary sources, 64
STBC, 37
STC, 36
Stream ID, 340
STTC, 37
SU-MIMO, 122
Sub-MB, 140
Subjective test, 168
System Target Decoder (STD), 342, 344

TCM, 29
TCP, 14, 244, 373–375
TCP-Friendly, 245
TDD, 23
Temporal direct mode, 146
Temporal scalability, 86
Texture, 174
Texture coding, 214
TFRC, 245
Threshold-based segmentation, 181
Time-stamping, 375, 377
Transcoding, 371–374
Transform coding, 9
Transmit precoding, 35
Transmitter and receiver clock
    synchronization, 344

Transport Stream (TS), 339
Trellis, 381, 382
Trick Modes, 337
Turbo codes, 224
type-I HARQ, 47
type-II HARQ, 47

UDP, 14, 245, 374, 375
UMB, 98, 130
UMTS, 99
Unequal error protection, 230, 294
Uniquely decodable, 66
Universal multimedia access, 16

Variable Bit Rate (VBR), 332
Variable length coding, 224
Vector quantization, 72
Video abstraction, 201
Video compression, 6, 370–372
Video compression:non-real-time, 370
Video compression:real-time, 370
Video conferencing, 347, 370
Video highlights, 201
Video object, 212
Video object segmentation, 185
Video redundancy coding, 232
video skimming, 201
Video streaming, 369–380, 382–384
Video streaming:broadcast, 369, 370
Video streaming:multicast, 370
Video streaming:P2P, 384
Video streaming:point-to-point, 369
Video streaming:unicast, 369
Video structure map, 201
Video summarization, 203
Video summary, 201
Video telephony, 347
Video transmission, 14
Video understanding, 200
Video/audio synchronization, 376
VoIP, 98

Weighted prediction, 144
WiMAX, 98, 112

Zig-zag scan, 152